The Management of Innovation and Technology

The Shaping of Technology and Institutions of the Market Economy

John Howells

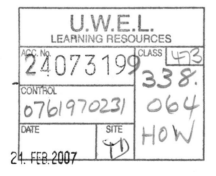

⑤SAGE Publications
London • Thousand Oaks • New Delhi

First published 2005

SAGE Publications
1 Oliver's Yard
55 City Road
London EC1Y 1SP

SAGE Publications Inc
2455 Teller Road
Thousand Oaks, California 91320

SAGE Publications India Pvt Ltd
B-42, Panchsheel Enclave
Post Box 4109
New Delhi 110 017

Library of Congress Control Number: 2004114266
A catalogue record for this book is available from the British Library

ISBN 0 7619 7023 1
ISBN 0 7619 7024 X (pbk)

Typeset by Selective Minds Infotech Pvt Ltd, Mohali, India
Printed and bound in Great Britain by TJ International Ltd, Padstow, Cornwall

The Management of Innovation and Technology

Contents

List of Figures and Tables

Acknowledgements

I would particularly like to thank James Fleck, Stephen Hill, Julia Engelhardt and Kristina Cornish for comments on all or part of the manuscript. As usual, I remain responsible for any remaining errors.

A previous version of the section in Chapter 1 on the technology complex was published with James Fleck under the title 'Technology, the Technology Complex and the Paradox of Technological Determinism' in the journal *Technology Analysis and Strategic Management*, 13: 523–31. A previous version of the section on the sailing ship effect in Chapter 4 was published under the title 'The Response of Old Technology Incumbents to Technological Competition – Does the Sailing Ship Effect Exist?' in the *Journal of Management Studies*, 39, 7: 887–907.

Grateful acknowledgement is made to the following sources for permission to reproduce material in this book.

Figures 1.1, 1.2 and 1.3 reproduced with permission of Taylor & Francis Ltd, www.tandf.co.uk/journals.

Figure 1.5 from Schmidt-Tiedemann's 'A new model of the innovation process', *Research Technology Management*, 25(2) (1982): 18–22; reprinted with permission.

Preface

In this book the 'management of technological innovation' is about the decisions and processes that generate, develop and shape technology.

It is argued at the beginning that a very broad definition of technology is appropriate for the study of innovation and this becomes a basis for organising the book. Thus there is a rough progression of topics from 'acts' of invention, to management choices within the R&D department, to patterns of long-term technological development and then the nature of technological competition between firms. There follows three chapters on the role of intellectual property, finance and technical knowledge. These cover the private firm's use and abuse of these institutions in technology development, but they also extend the idea of 'management of innovation' to cover the decision events that shape the form of such important institutions. The last chapter introduces some of the roles of the state in technology development.

The overall object of the book is to convey an understanding of technology as immediately shaped by the firm, but situated in 'society' – and situated in the particular form of society that is the market economy, understood as the working set of institutions and governance procedures that have evolved to sometimes limit and sometimes enable technology-shaping decisions by management and entrepreneurs.

1 Technological Innovation

Variety in the Meaning Attributed to Invention, Innovation and Technology and the Organisation of this Book

This chapter begins with a review of definitions of technology and innovation. This is then used to develop the device of the 'technology complex', a device that is exploited in the organisation of the rest of this book. This preference for the logical development of an argument from first principles does mean that discussion of the organisation of the book is postponed to half-way through the chapter. The second half of this chapter demonstrates the value of the technology complex as an intellectual tool by arguing for the comparative inadequacy of the more readily available concepts applicable to innovation and technological change.

Invention, Innovation and Technology

There appear to be almost as many variant meanings for the terms 'invention', 'innovation' and 'technology' as there are authors. Many use the terms 'invention' and 'innovation' interchangeably or with varying degrees of precision. At an extreme, Wiener prefers 'invention' to describe the whole process of bringing a novelty to the market (Wiener 1993). In contrast, Freeman prefers to restrict the meaning and increase the precision of 'innovation' by only applying it to the first commercial transaction involving a novel device (Freeman 1982: 7).

Some definitions are in order. In this book 'invention' will be restricted to describe the generation of the *idea* of an innovation. Innovation will describe some useful changes in technology, and technology refers to the organisation of people and artefacts for some goal. In this book then, technology is both the focus of analysis and yet is given a very broad and inclusive meaning. This usage is in contrast to many other authors and so deserves further explanation.

The term 'technology' is in a class of its own for variation in meaning and Figure 1.1 represents an effort to display some of this variety.

The 'spread' of definitions in Figure 1.1 has the striking quality that distinctly different elements appear in many of the definitions. The titles of the works from which the definitions are drawn show that the detail of the definition is linked to the discipline, or problem of study: industrial relations, organisational behaviour, operations management, and the problem of technology transfer. These are not 'wrong' definitions if one accepts the restricted focus of a subject discipline, problem or time frame and a general definition of technology should be able to incorporate such subdefinitions as special cases.

Collins Dictionary

1 The application of practical or mechanical sciences to industry or commerce.
2 The methods, theory, and practices governing such application.
3 The total knowledge and skills available to any human society.

Oxford English Dictionary

Science or industrial art; literally, the science of technique i.e. systematic knowledge of technique. Technique: the interaction of people/tools with machines/objects which defines a 'way of doing' a particular task.

Technology and Change – the New Heraclitus (Schön 1967)

Any tool or technique: any product or process, any physical equipment or method of doing or making by which human capability is extended.

The Trouble with Technology (Macdonald, et al. 1985)

Technology may be regarded as simply the 'way things are done'.

Technology Policy of Economic Development, IDRC, Ottawa (Vaitsos 1976)

Identifies three properties of technology:

1 The form in which technology is incorporated: machines/equipment/materials.
2 Necessary information covering patents and conditions under which technology can be used.
3 Cost of technology i.e. capital.

The Management of Technology Transfer, International Journal of Technology Management (Djeflat 1987)

Technology marketed as a complete entity: all technological components tied together and transferred as a whole: capital goods/materials/know how/qualified and specialised manpower.

The Business Enterprise in Modern Industrial Society (Child 1969)

The equipment used in the work flow of a business enterprise and the interrelationship of the operations to which the equipment is applied.

Competition and Control at Work (Hill 1981)

In the first place technology embraces all forms of productive technique, including hand work which may not involve the physical use of mechanical implements. Secondly, it embraces the physical organisation of production, the way in which the hardware of production has been laid out in a place of work. The term therefore implies the division of labour and work organisation which is built into or required for efficient operation by the productive technique.

The Sociology of Invention (Gilfillan 1935)

An invention is essentially a complex of most diverse elements – a design for a physical object, a process of working with it, the needed elements of science, if any, the constituent materials, a method for building it, the raw materials used in working it, such as fuel, accumulated capital such as factories and docks, with which it must be used, its crew with their skills, ideas and shortcomings, its financial backing and management, its purpose and use in conjunction with other sides of civilisation and its popular evaluation. Most of these parts in turn have their separately variable elements. A change in any one of the elements of the complex will alter, stimulate, depress or quite inhibit the whole.

Research and Technology as Economic Activities (Green and Morphet 1977)

> To sum up, the technology of a particular process or industry is the assemblage of all the craft, empirical and rational knowledge by which the techniques of that process or industry are understood and operated.

Operations Management (Schroeder 1989)

> That set of processes, tools, methods, procedures and equipment used to produce goods or services.

Figure 1.1 A range of definitions of 'technology' (from Fleck and Howells 2001: 524, reproduced with permission of Taylor & Francis Ltd)

Examination of the range of definitions in Figure 1.1 suggests that a suitable general and inclusive definition of technology becomes the 'knowledge of how to organise people and tools to achieve specific ends'. This is certainly general, but hardly useful, because the different elements of the subdefinitions have been lost. The technology complex[1] in Figure 1.2 has been suggested as a device that relates the general definition of technology to its subdefinitions (Fleck and Howells 2001).

The elements within the technology complex have been ordered to range from the physical and artefactual to the social and the cultural.[2] This captures the idea that there are multiple 'levels' within society at which people organise around artefacts to create working technologies. Any or all of these elements could be analysed in a working technology – a technology 'in use'. It is rather rare that a full range of elements are considered, but it will prove worthwhile to provide some examples of when it makes sense to extend the range of analysis over the range of the technology complex.

MATERIAL
ENERGY SOURCE
ARTEFACTS/HARDWARE
LAYOUT
PROCEDURES (PROGRAMS, SOFTWARE)
KNOWLEDGE/SKILLS/QUALIFIED PEOPLE
WORK ORGANISATION
MANAGEMENT TECHNIQUES
ORGANISATIONAL STRUCTURE
COST/CAPITAL
INDUSTRY STRUCTURE (SUPPLIERS, USERS, PROMOTERS)
LOCATION
SOCIAL RELATIONS
CULTURE

Figure 1.2 The technology complex (from Fleck and Howells 2001: 525, reproduced with permission of Taylor & Francis Ltd)

Use of the Technology Complex[3]

An example of how an apparently simple technology nevertheless includes a range of these elements is given by the Neolithic technology of stone-napping.

At the 'operations' level, skilled individuals use bones – tools – to shape the major raw material – flint – to produce stone tool artefacts. The stone tool artefacts then had a wider range of uses – preparing skins, weapons and wood. Their production, though involving high levels of skill, appears simple in organisational and material terms.

However, this 'simplicity' may be more the product of examining a simple context – here, routine production of the artefact. Other elements of the technology complex will be 'active' and apparent if non-routine changes in the production and use of the artefact are studied.

An excellent example of this is the account by the anthropologist Sharp of the effect of the introduction of steel axe-heads into Stone Age Aboriginal society (Sharp 1952), which reveals the complex interrelationship between artefacts, social structure and culture.

In this patriarchal society, male heads of households buttressed their social position through their control of the use of stone axes, primarily through the limitation of access to young males and women. The indiscriminate distribution of steel axe heads to young males and even women by western missionaries disrupted this balance of power within the tribe.

Steel axes wore away more slowly than stone axe heads and this physical property helped to disrupt the trading patterns that connected north and south Aboriginal tribes. The raw material for making axes existed in the south and it was progressively exchanged through a network of tribes for goods and materials from the tropical north. The annual gatherings when exchange took place had ritual, quasi-religious significance as well as economic exchange significance, but the arrival of steel axes removed the necessity to meet on the old basis and so undermined the cultural aspects of the annual ritual meetings. In these ways society and culture were disrupted by a change in the material of an important artefact.

When changes to stone axe technology were made the subject of enquiry it was clear that stone axe technology was not 'simple' in its social context. Within the society that generated and used this technology the artefact had complex significance that involved many elements of the technology complex for its description.

The technology complex warns us that what appears to be a simple technology may be simple only through the construction of the case analysis. 'Simple' here means describable through only a few of the elements from the technology complex.

Modern technologies are obviously more complex at the level of the artefact and organisation and they are sustained within a more complex society. As in the stone technology example, the study of their routine use is likely to yield relatively more simple descriptions than the study, for example, of the social process of their generation or implementation. An example of the latter is the account by Howells and Hine of the design and implementation of EFTPOS (Electronic Funds Transfer at the Point of Sale), an IT network, by the British retail banks (Howells and Hine 1993). This found that a complex set of choices of artefacts and social arrangements had to be defined by the banks. These choices ranged across the full range of complex technology elements, as shown in Figure 1.3. Decisions in these categories served to define the technology of the IT network that was eventually implemented.

Material/artefact

Variety in the designs of computer, software, communications and terminals offered imposed an expertise management problem on the banks. Isolated artefact design decisions had unexpected implications for other network component design and cost. This variety of offered designs imposed a learning process on the banks.

Topology/layout

The artefact components could be linked in different ways to represent preferences of the owning organisations. So transaction processing could take place in a jointly owned organisation, or the banks' IT departments. Terminals could be stand-alone or integrated into retailer equipment. Such decisions had implications for competition, ownership and control of the network and changed through the project.

Procedures/software

There were many ways of building security into the network. One possibility was to encrypt electronic messages, but then there were two rival proposals for encryption method that each had its own political and competitive implications.

Organisational structure and location of technical expertise – knowledge/skills

The banks had to decide on how to organise the development of the network. They first experimented with the option to contract out responsibility for network design, then a jointly owned hybrid organisation of technical consultants controlled by commercial bankers, then full scale development occurred through VISA and Switch network designs to which each bank made an individual decision to affiliate.

Cost/capital

The cost of IT networks is large and the returns depend on the speed with which paper-based systems can be closed down. There is a theoretic role here for sophisticated financial evaluation techniques, but the impetus for the technology derived from inter-bank competition rather than the immediate economics.

Industry structure

As an oligopoly, the banking industry had a range of more or less cooperative or competitive options available to it in the organisation of this large-scale project. The banks began by cooperating fully in their approach to EFTPOS, but cooperation broke down into two factions that would develop two rival designs of network.

Social/legal relations

In 1986 the UK government passed legislation that allowed Building Societies to compete with banks. This helped prompt the breakdown of the banks' cooperative approach to network design.

Culture

Past experience of government led many of the banks to expect the Bank of England to regulate the sector and to signal its approval or disapproval of bank strategies. On the EFTPOS project some banks continued to wait for Bank of England guidance, some interpreted Bank of England statements as signalling approval or disapproval; late in the project most banks agreed there was no longer the close supervision of earlier years.

Figure 1.3 EFTPOS technology as an example of the technology complex (from Fleck and Howells 2001, reproduced with permission of Taylor & Francis Ltd)

The Problem of the Attribution of Causality and the Design of a Study

These examples raise the general issue of how causation should be attributed in accounts of technology development. It is superficially attractive to express this problem in a general way as the issue of how technology and society may relate to each other. In particular the question 'does technology shape society?', sometimes described as the issue of technological determinism, has been intensively debated in the history and sociology of technology.[4] The technology complex would suggest that a good response to this question is 'what do you mean by technology?' For technology should be understood to necessarily include people, so that to ask how technology influences society is to ask how the people organised around a particular set of artefacts influence society. Expressed this way it is easier to see that whether and how technology changes society will depend on the activity of both those most directly attached to the technology, for example through their careers or ownership rights, and those who will be affected more indirectly, by the products or pollution that result from the development of the technology. The question 'how does technology affect society?' is only apparently simple, because it is short and is expressed as if there could be an answer with general validity. Any answer depends on the characteristics of both a specific technology and a specific society: there is no general answer.

As part of a collection of writings on this topic of technological determinism (Smith and Marx 1994), the historians Thomas Misa and Bruce Bimber argue that subject discipline and scale of account (Misa) predispose authors to attribute causation in the innovation process quite differently (Misa 1994; Bimber 1994).

A key difference between what Misa calls 'micro'- and 'macro'- scale studies is the degree to which the uncertainty of the decision process is captured. 'Micro-level' studies of managers' thought and behaviour, such as that of EFTPOS design and implementation, capture the messiness of real processes of change; the wrong ideas, the misconceptions, the failed bids for competitive change that formed part of the decision process. In contrast, a 'macro-level' study of technological change in retail banking over a long time span would find the series of technologies that were actually built – the uncertainty and the alternative paths of development are lost and long-run patterns of change would appear to have their own logic, related to the cheapening cost of computer power over time. Micro and macro studies capture different aspects of the process of innovation and in principle they should complement one another.

To caricature the possible problems of such accounts, unless micro-scale authors import values from outside their studies they may become unable to judge the value of the outcomes they describe – they tend towards 'relativism'. In contrast, Misa argues that it is those whose research is conducted at the macro level that tend towards overdetermined interpretations of their accounts, for example by attributing excessive rationality and purpose to the actions of actors (Misa 1994: 119). His example is the business historian Alfred Chandler's interpretation of the Carnegie Steel Company's vertical integration through the purchase of iron ore deposits. Chandler imputes a defensive motivation to vertical integration, a desire by Carnegie to defend his company against new entrants (Misa 1994: 12). Misa's micro-analysis of the decision to integrate vertically reveals that Carnegie and his partners were long reluctant to buy iron ore properties and did so only when convinced by a middle-rank manager, whose arguments were largely based on the serendipitous profits that he believed would accrue to the company from the purchase of iron ore *land* – land that had become temporarily cheap as a result of earlier company actions. As Misa comments:

> Vertical integration proved economically rational (lower costs for iron ore plus barriers to entry) for Carnegie Steel, but it was a two-decade-long process whose rationality appeared most forcefully to later historians and not to the actors themselves. (Misa 1994: 138)

With a broad concept for technology such as the technology complex, or Misa's essentially similar 'sociotechnical networks' (Misa 1994: 141), the effort to use and mix different types of account of technology development can be expected to generate many similar problems of interpretation.

Technology, its Uses and the Institutions of the Market Economy

The very term 'technology' implies usefulness of some kind, for someone, most obvious at the level of the artefact, an object by definition given form by humankind and whose creation necessarily implies some human purpose, some human use. Innovation in its widest sense is then some change in the way technology relates to its uses. The questions this formulation begs are of social context – useful how and to whom?

This book begins by following the convention that 'management studies' should be concerned with organisations operating within the market economy and in recent periods of time, a convention that follows the incessant demands for the subject to be immediately relevant to management practice. There is nothing wrong with such a demand in itself, but it does risk that we take the organisation and performance of the market economy for granted and forget that this is only one way of organising society and technology. One of the standard economics texts defines a market economy as one in which 'people specialise in productive activities and meet most of their material wants through exchanges voluntarily agreed upon by the contracting parties' (Lipsey 1989: 786). This clarifies the nature of the economic motivation for technology development for those directly dependent on the market economy for their financial resources – to improve, or to add to the stock of market tradable goods and services. Yet it ignores the institutional arrangements that are essential to a functioning market economy.

Institutions and state policy have always been central to the subject of political economy, and economic and business history have always found room to include the origin and development of such institutions as industrial investment banks and technical education. There is a tendency for the importance of such institutions to be rediscovered by subjects once narrowly conceived: the economics of innovation expanded its scope to include institutions and state policy in the 1990s, as symbolised by the publication of a collection of essays on *National Innovation Systems* (Nelson 1993); Whitley has similarly sought to expand the scope of organisational sociology in a series of edited essay collections on business systems culminating in his latest book entitled *Divergent Capitalisms* (Whitley 1999). As implied by these titles, institutional endowments and state policy style vary significantly between nations and the guiding question for much of this kind of work is whether these differences matter for economy and sometimes society. The use of the word 'system' with its implication of a thorough-going master design is unfortunate, but by the end of this book it will be obvious that if one wants to understand how and why firms innovate one simply cannot ignore such institutional arrangements.

It is a matter of common observation that even modern societies organised largely as market economies contain classes of technology that are socially useful but that are developed outside, or uneasily alongside, the market economy – for example, the military technologies or the technologies developed by the publicly funded institutions of science organised so that the generation of understanding is their most immediate motivation, rather than the direct generation of economic activity. Nevertheless, once technologies have been demonstrated as viable for military or scientific purposes, in principle they are available for development for market economic purposes. According to its inventor, the idea of the useful technology of the laser was motivated by a desire to create a device that would further scientific understanding (see Chapter 2). That is indeed how the inventor sought to apply his technology, but once a working device had been publicly demonstrated, private sector firms began to develop the technology for economic uses.

In other words, even in societies largely organised as market economies, while all technologies have their uses, they do not necessarily have an economic 'market' and they are not necessarily first generated to serve economic purposes. They may nevertheless come to have a significant economic impact. To discuss technologies and their 'uses' in principle allows a discussion of any technology, whatever its social and cultural context and whatever the human purpose motivating its creation and use. To discuss technologies and their relationship to the economic institution of the 'market' implies a restriction of the range of technologies to those that generate tradable goods and services and that serve the interests of the people and organisations participating in the market. It is primarily the second class of technologies that are of interest in this book, but as in the case of the laser, non-economic institutions will be of interest when they have a productive interface with the market economy.

It is one thing to be interested in the technologies developed within the market economy and developed for economic reasons, but quite another to be uncritically committed to economic theory as an aid to understanding. Economic theory may have little to offer the study of technology. Although this book does make use of ideas from the subject of economics, it is also the case that certain ideas within economic thought have been applied to the detriment of technological development. This is very apparent in the application of US antitrust and patent law, discussed at length in Chapters 4 and 5 on competition and intellectual property. It should be said that the revision of policy in these cases has also been driven by economists, but economists with particular theoretical commitments. The point here is that our interest in technological development in the market economy can lead quite logically into an interest in the construction and application of economic thought, as one of the many influences on the development of technology.

Socio-economic Perspectives on Technology and the Market Economy

The problem of how to include economics in the study of technology while not necessarily accepting the assumptions of economic theory has bothered the sociologists of technology. As the title of a collection of readings edited by MacKenzie and Wajcman, the 'social-shaping of technology' has been popularised as a term to describe some of the many ways in which technology can be influenced by society (MacKenzie and Wajcman 1999b). Like many others, MacKenzie and Wajcman understood that neoclassical economic theory, as the dominant economic theory of the relationship between technology and economics, could have little relevance when technology was changing, for then future prices, costs and competitive scenarios matter, and these cannot be known in full, as neoclassical economic theory requires (MacKenzie and Wajcman 1999a: 13). Nevertheless, MacKenzie and Wajcman recognise that economics matters to the social shaping of technology and so they take the position that the economic shaping of technology through price and cost considerations is a form of social shaping, and that such economic calculation depends on the organisation of the society in which it occurs. This is essentially the same position as the one developed here; however, none of the examples in MacKenzie and Wajcman's text concern the mutual 'shaping' between technology and market economy institutions that are of interest in this book.[5]

Of greater apparent relevance is the important stream of contemporary economic thought that has attempted to reintroduce the analysis of 'institutions' to economics-the-subject, the so-called 'New Institutional Economics' (Swedberg and Granovetter 2001: 15). However, Swedborg and Granovetter are critical of this stream of thought as part of their development of what we might term a 'counter programme' for the study of the 'sociology of economic life'.[6] According to Swedberg and Granovetter, central to all versions of the New Institutional Economics is

> the concept of efficiency. Institutions, it is argued, tend to emerge as efficient solutions to market failures ... another common feature is the assumption that while most of the economy can be analysed perfectly well with the help of rational choice (and with total disregard for the social structure), there do exist a few cases where you have to take norms and institutions into account. ... This way of arguing makes for a type of analysis that is only superficially social. Finally there is also the disturbing fact that when economists analyse institutions they feel no need to take into account how and why the actors, *in their own view* [author's emphasis], do as they do. ... These two limitations – inattention to social structure and to the beliefs of individuals – make the New Institutional Economics into a kind of analysis that can best be described as a science of what economists think about what is happening in the economy. (Swedberg and Granovetter 2001: 15)

This amounts to a general warning that this school of economists have a tendency to find what they presuppose is there. In Misa's analysis of Chandler's account of the rationality of vertical integration in the US steel industry we already have an example of the kind of bias to explanation that this might produce. However, one should be careful of accepting Swedberg and Granovetter's picture of the 'New Institutional Economics' as a clear 'school of thought'. The lesson I prefer to draw is that the interpretation of evidence is more fraught than one might have supposed, and mistakes are made. The remedy would be more care in analysis and awareness of the ways that prior theoretical commitments *may* lead interpretation adrift. However, there is certainly further justification here for other ways of studying economic life than solely through the lens of contemporary economic thought.

Implications for the Structure of this Book

This book adopts such another way. The technology complex and the issue of scale and attribution of causation will be used to organise the contents of this book. First, the book seeks to introduce the *variety* of reciprocal influences between social, organisational and artefactual forms that constitutes the design and creation of technology. Second, it seeks to move between micro- and macro-level accounts to capture the different forms of technological reality. Third, it introduces the shaping and reshaping of relevant institutional forms as part of the management of innovation.

The chapters are organised to flow from fundamental issues of technology development and growth to the relationship between institutions and the ability of the private sector to develop technologies. The institutional forms of intellectual property, finance and education that are the subject of the later chapters tend to be more stable through time compared with specific instances of innovation and so lean towards being taken for granted. They are nevertheless subject to their own processes of change, in which the private sector participates, but in collaboration with other actors such as the state. They are included here both because they influence the conditions for technology development in the firm and because they extend the coverage of elements of the technology complex.

In sum, this book seeks to emulate the success of the social shaping texts in offering a rich surrogate experience in the sheer variety of how social forms and human purpose influence technological design, but within the narrower range of the operation of the market economy. In response to the discussion above of the limitations of any one school of thought, there will be a preference for detailed contextual analysis of change processes. This makes it more likely that disjunctions between empirical material and theoretical commitments can be teased out of the sources. This also implies the choice of a relatively small number of texts and cases for their richness, detail and significance, rather than an attempt to comprehensively cover the management literature. This approach represents my definition of the management of innovation and technological change. It does not pretend to offer direct help for immediate practical problems, but if this text succeeds in its effort to make an original synthesis of material that explicates an extended understanding of the process of technological change, then because good understanding is the basis for good practice, it will have value.

A brief review of some contrasting, alternative approaches to innovation follows. This should reinforce the value of the broad concept of technological change outlined here; on this matter the problem is not one of a lack of research volume or of examples, but of fragmented and inadequate, often competing and overlapping, frameworks of understanding.

Some Contrasting Conceptual Approaches to the Representation of Innovation

The Structure of Management Education

The most obvious route to provide a broad understanding of innovation in management might be thought to be the MBA; so, for example, organisational behaviour might be thought to deal with the organisational design aspects of technology, strategic management to deal with strategies for technology development, and so on.

As forcefully argued by Rickards in his review of the typical MBA curriculum, there is no shared concern with innovation and change (Rickards 2000). The academic specialisations that comprise the MBA did not develop around the object of 'explaining' technology or technological decisions, but around their own professional agendas and method preferences. There is no single definition, or discussion of technology, shared between the subjects – at best there are partial and incompatible definitions of the sort collected in Figure 1.1. Taken as a whole, one is in danger of obtaining a fragmented and contradictory understanding of technological change and the role of management.

The distinct institutional history of management education compared with the engineering subjects (in the USA and Britain) is also likely to play a part in promoting a partial understanding of technological change and management. Artefacts and their design are the preserve and principal focus of the engineering disciplines.[7] Business and management education is provided in distinct institutions and so there is a strong tendency to define business and management subjects free from concern with particular classes of artefact; in other words, without roots in a technological context. The potential value that may come from building appropriate business and management material into a particular technological context is lost.

Diagrammatic Models of Innovation

An example of the effort to condense and simplify understanding of innovation is the effort to build a general model of the innovation process in diagrammatic form. A model is a simplified representation of a more complex reality and its object is to make that reality more readily understood in its essential features so that it can be used – to educate, or to instruct policy.

Forrest has reviewed many of the diagrammatic 'box and line' models of the innovation process and her review makes apparent the bewildering variety within even this modelling format (Forrest 1991). Some of this variety derives from the degree of complexity of models. Rather like the technology complex, the more complex models were the result of authors adding more elements of the innovation process to improve what we might call the representational coverage of the models – at the cost, of course, of the simplicity which is the object of this kind of modelling process (compare, for example, the models in Figures 1.4 and 1.5). Forrest observed that not even these complex models were complete, in that it remained possible to think of further elements that one should include in a model (Forrest 1991). Her 'minimum list' of such elements included:

A definite pre-analysis and pre-evaluation stage, definitive feedback loops, both internally within the firm and externally with the environment; the industry and life stage of the organisation within the industry; a recognition or the environmental variables – not only the marketing and technological, but the socio-cultural and political environmental variables and the internal environment (culture) of the firm; and the important dimensions of time and cost/resource commitment. (Forrest 1991: 450)

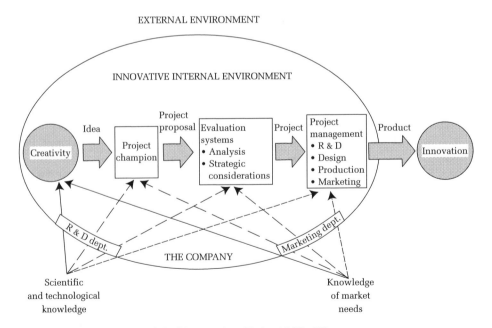

Figure 1.4 Twiss's 'Egg' model of innovation (Twiss 1992: 25)

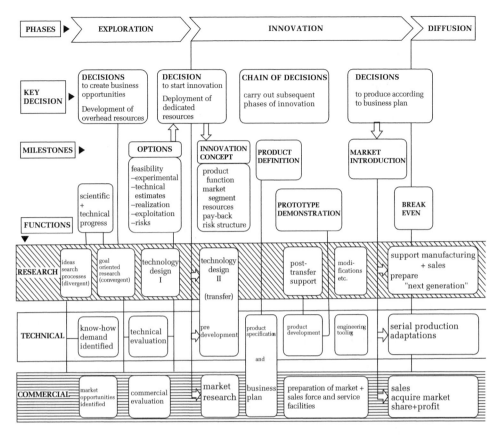

Figure 1.5 The Schmidt-Tiedemann concomitance model of innovation (Schmidt-Tiedemann 1982, reproduced with permission from *Research Technology Management*)

This begins to resemble the list of the technology complex, with the added complication that the modelling form is burdened with the object of representing a definite set of stages or functions *and their causal relation* to each other, usually by means of arrows. The more complex models modify any strong causal connections, for example by having arrows point in both directions to represent reciprocal or iterative influence of functions, or by many of the functions or stages being interlinked by arrows to show that innovation does not necessarily proceed in neat, ordered stages.

It may not be surprising that Forrest concludes by questioning whether this pursuit of a truly general innovation model is possible given the complexity and variety of the innovation process. Yet she does not dismiss outright the utility of this kind of modelling process – instead she urges management to create their own technology or industry-contingent innovation models. So the pursuit of a diagrammatic representation in a particular innovation context may be a useful aid to clarifying thought, but Forrest warns that one should be wary of any particular 'model' being taken too seriously and applied too rigidly as a guide to action out of the context in which it was created.

Technology or 'Manufacturing and Services'

The very *creation* of primary, secondary and tertiary categories of economic activity was made to demonstrate the trend with which we are all now familiar: the steady movement of employment from agricultural to manufacturing and now service activities[8] that has occurred in developed countries over many decades and centuries. Yet even an author like Bell who made these categories fundamental to his forecasts of a 'post-industrial society' comments that the category of services is a 'residual': that is, it is defined by *not* being agricultural or manufacturing activity (Bell 1973: 15). And here is the problem, for if the focus of analysis is technology, the artefact and associated social change, the category of manufacturing captures economic activity where the artefact is the end result of economic activity, but 'services' contains such diversity it is well nigh useless as a descriptive category; it contains personal services like hairdressing, but also software production, financial services, scientific research, health and government services.

The reality is that despite the general fall in manufacturing employment, manufacturing output continues to rise strongly in almost all industrial countries, with the exception of Britain: so between 1973 and 2000, manufacturing output increased by 114% in the USA and 14% in Britain (Rowthorn 2001). The widespread reporting of economic activity through the categories of manufacturing and services tends to promote the illusion that the decline in manufacturing employment means that the production and use of artefacts is of decreasing importance. Hence the counter-argument makes a periodic reappearance, evident in the titles of works such as 'The Myth of the Post-Industrial Economy' (Cohen and Zysman 1994) and *In Praise of Hard Industries* (Fingleton 1999). If 'industrial' means the ability to produce artefacts, then no rich, developed country is 'post-industrial' – they are more 'industrial' than ever.

It is worthwhile to explore how the artificial division between the statistical categories of manufacturing and services largely breaks down if we insist on thinking of economic activity in terms of technologies, artefacts, and their uses. This is perhaps most obvious with the imposed classification of software production as a 'service'. This is despite the fact that the output of software production is a 'made thing', an artefact, albeit with distinct properties, and also despite the fact that the organisation of software production has much in common with the organisation of production of other forms of artefact. This seemingly arbitrary act of classification allows the great recent growth in software production to add to the measured growth of service activity.

Some services are specialisations in the *provision of use* of highly complex artefacts, and their growth – and existence – are clearly dependent on improvements in the artefact; the airline industry is a good example of such a service. Other categories of service are 'tightly-linked' (in Cohen and Zysman's term (Cohen and Zysman 1994: 33)) to particular manufacturing industries; these include repair, maintenance and design and engineering consultancy services. In other words, 'tightly-linked' services are the characteristic activities organised around a specific kind of artefact that together give a technology its separate identity.

Such tightly linked activities have the property that they may be organised in-house, in which case they are classified by the statisticians as manufacturing, but if they are organised separately to the production of artefacts – in today's jargon, if they are outsourced – they are classified as services. One can imagine that if there were a net increase in outsourcing in the economy, it would increase the measured statistical trend towards more service activity – yet for such services, the 'organisational shell' from which, or within which, they are provided is less important than that they exist at all and that they continue to relate to a specific form of artefact.

Indeed, in order to make sense of recent organisational trends in complex product industries, some writers have dropped the terms 'manufacturing' and 'services' in favour of the 'business of systems integration' (for example, Prencipe et al. 2003). It is well known that companies such as the US car manufacturers and General Electric have moved 'downstream' into high value-added services such as the provision of finance to supplement their product incomes, but in the 1990s General Electric continued to expand the services offered by its financial services division, GE Capital, so that it could offer 'integrated solutions' to its customers' needs, including products, finance and maintenance (Davies 2003: 338). By 2002 GE Capital generated 49% of the firm's total revenue (Davies 2003: 338).

Ericsson offers a more radical case, where much of its product manufacturing has been outsourced to a favoured manufacturing contractor, Flextronics, while service offerings and business consulting activities have been combined into a new division, Ericsson Global Services, created in 2000 with the object of the provision of high-value integrated systems and services to all mobile phone operators (Davies 2003: 352). As some of Davies' cases show, while such firms as these are moving from manufacturing into integrated systems provision, other firms have moved from being 'pure' service providers into integrated systems provision: this is the case with WS Atkins, the former project management and technical consultancy provider (Davies 2003: 352).

What this reveals is that the appropriate bundle of activities for containment within the firm is very much in flux and very much a focus of management attention. Our focus of enquiry is management activity in a technological context and it will prove hardly possible not to refer to the specific 'services' that are tightly linked to artefacts. The custom of categorising economic activity by the end product of the firm so that firms are *either* manufacturing *or* service firms obscures more than it reveals.

Reading from the Artefact Alone – Common Sense and Some Limitations

It is an interesting exercise to consider what can be learnt from artefacts alone. An instructive and contrasting pair of thought experiments consists of trying to isolate significant innovation that on the one hand has no artefact component and on the other hand concerns the artefact alone.

So, for example, if the effort is made to think of a 'pure marketing' innovation, it is possible to imagine changes in marketing technique, but one soon realises that as marketing is a service to the major activity of the firm, where that activity involves changing artefacts, marketing activity is likely to be adapted to the nature of such changes. So the search for pure marketing innovation should seek examples where there is a minimum of artefact change in evidence. It is difficult to beat Foster's nice, laconic account of what innovation amounts to in cola drinks:

> We all remember the familiar 6.5 Coke bottle and its Pepsi equivalent. And we can also remember when the 6.5 ounce bottle was joined on grocery shelves and vending machines by the 12 ounce can. This advance – and some people considered it one – was so successful that it was followed with the 16 ounce glass bottle, and then the 26 ounce bottle, the 28 ounce bottle and finally the 32 ounce bottle. From there the competition got dirty. First one company went to the 48 ounce, then the other. Then both followed quickly with the 64 ounce bottle. Where would all this exciting competition lead? To the metric system when the 64 ounce bottle was replaced by the 2 litre bottle...now made of plastic, and then by the 3 litre bottle....What non-advance can we expect now? Five litre bottles with wheels? (Foster 1986: 254–5)

Such a business is driven by advertising and marketing as the means of manipulating consumer perceptions of symbolic and minor changes to the presentation of the product. These activities are interesting in their own right, but the scope for material change that might advantage the consumer is obviously largely absent. Indeed, the example sharpens another question: what can 'advantage' mean in the absence of some element of artefact change?

The complement of the 'imagine innovation without the artefact' exercise is to consider the meaning of a change in the artefact alone. But this is an absurdity because by definition change in an artefact is a result of human intention and activity. However, these activities are often complex with an elaborate division of labour and expertise; so this question can be changed to become a search for those individuals that experience changes in artefacts with a minimum of access to the human processes upon which these rely.

Common sense suggests that in our society it is in our role as consumers that we come closest to experiencing artefactual change without the allied knowledge of production and distribution. More than this, it is because of our individual social and cognitive limitations that we have necessarily restricted knowledge of the diversity and complexity of design and production of these same artefacts. There is a fundamental asymmetry between our degree of access to consumer and to producer technology.

And yet, as consumers, we have the 'knowledge of use' of many artefacts. We are familiar with the problems of acquiring this knowledge of use – these artefacts must be set up, used and maintained and they all fail in various ways. When they fail we tend to make judgements about the human design process that produced the artefact and a common struggle that occurs is over the attribution of 'fault' either to artefact design or to our manner of its use. In other words, we are all to some degree experienced in the *interpretation* of artefacts as indicators of the design and production process, *despite* our intrinsically limited ability to grasp the detail of all these production processes.

All this is basic and perhaps obvious, but it is worth restating here because it helps avoid the mystification of technology. The 'imaginative basis' for understanding the design process in its various contexts is widespread, yet usually limited by having access only to artefact forms. Of course experts in methods of production and design should make the better interpreters of artefacts – what is 'reverse engineering' but a highly and actively developed ability to read from artefact design the nature of the production and design process that created that artefact? And outside engineering we find the discipline of archaeology to be the study of human history through the 'science' of reading defunct artefacts (and other material remains).

Limits on the Interpretation of the Artefact – Skeuomorphs and Spandrels

From such disciplines' intensive relationship with artefact form and meaning we can derive useful and general 'limiting' concepts on what is probably our default assumption about any artefact form – that *every element of design* that we observe in an artefact is there by the designer's intention and for utilitarian reasons.

A skeuomorph is an element of design that has lost its original function but has nevertheless been retained in its artefact. The classic example in architecture is the square at the top of the much imitated Doric column (Adam 1989) and reproduced in Figure 1.6. Early Greek architecture was constructed from wood and this square had a stress-distributing function for wooden columns. The retention of the 'square' when marble and stone materials were used instead of wood was for reasons of style. Another example is the retention in

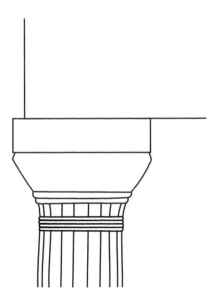

Figure 1.6 The square at the top of the Doric column is the skeuomorph

modern dust-pressed clay plates of the disc shape once necessarily obtained from wheel-spun pottery production. Many other shapes, such as squares, are in principle possible with the dust-pressing manufacturing process (Adam 1989). Adam suggests that the freezing of form in the plate may be partly aesthetic and partly prevalent usage; the spread of dishwashers designed around the standard form in crockery begins to limit the latitude in crockery design.

A quite distinct class of limitations on our ability to work out the design process from its resultant artefact is that what may appear to the artefact observer as an element of design may be the *byproduct of other design choices*. The difficulty of distinguishing between byproduct and intentional design elements, within the artefact alone and without production knowledge, is demonstrated in a controversy over the significance of the 'spandrel' as a byproduct of design in architecture.

A much discussed paper in evolutionary biology by Gould and Lewontin used the architectural 'spandrel' to argue that elements of biological form might also be byproducts of other design elements, rather than adaptations under natural selection, as conventionally assumed (Gould and Lewontin 1979). Their paper has been notoriously misused to undermine the public perception of evolution, but that controversy does not concern us here.[9] Gould and Lewontin's use of the spandrel in architectural form as a straightforward example does concern us and they comment that 'such architectural constraints abound, and we find them easy to understand because we do not impose our biological biases upon them' (Gould and Lewontin 1979: 581). On the contrary, they are not so easy to understand and for the same reason that the attribution of function may be complicated in biology: casual observation of the artefact or the biological form alone, without sufficient understanding of the process of construction or growth, is not sufficient to determine the function of all elements of form.

Gould and Lewontin's example of 'spandrels' is the Basilica of San Marco in Venice, which has a main dome supported by four arches. First it must be pointed out that the correct name for the roughly triangular segment of a sphere that tapers to points at the feet of its supporting arches is a pendentive, not a spandrel (Dennett 1996: 3). The difference between a spandrel and a pendentive is evident when Figures 1.7 and 1.8 are compared.

Now the historian of architectural technology, Robert Mark, has used arguments drawn from his understanding of then-contemporary construction knowledge to judge that the pendentives of San Marco do, after all, have a function.

In the earlier building (537 AD) of Justinian's Hagia Sophia in Constantinople (Figure 1.8), Mark found that pendentives had the function of stabilising the dome, which would otherwise be in danger of exploding outwards under *tensile* stress derived from its own weight[10] (Mark 1996; Mark and Billington 1989: 308). The pendentives performed this function when their outside surface was loaded with material to provide a stabilising weight – these weights are visible on the outside of the dome in Figure 1.8. Mark then argues that because construction knowledge in the prescientific era was developed by observation of the behaviour of existing buildings, the builders of the Basilica of San Marco in the late eleventh century would have been extremely unlikely to use any

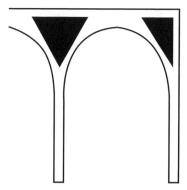

Figure 1.7 The roughly triangular area created when a viaduct is supported by arches is the spandrel

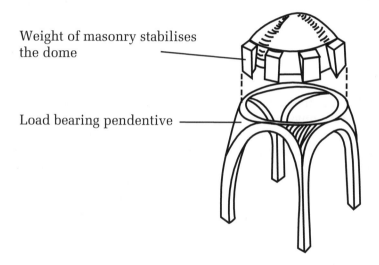

Weight of masonry stabilises the dome

Load bearing pendentive

Figure 1.8 Diagram of the construction of the Hagia Sophia. The pendentives are the curving, three-dimensional and tapering triangular areas between the arches and the rim of the dome. The tensile stress within the dome is stabilised by blocks of masonry that rest on the pendatives

other construction method than pendentives to stabilise its large, 30 m diameter dome (Mark 1996). In other words, in the Basilica of San Marco, pendentives were not a 'necessary architectural byproduct' (Gould and Lewontin 1979: 581) that resulted from the design *choice* of arches to support a dome, but they were a *necessary structural feature* of the proven system of support for large domed buildings: arches with loaded pendentives were the necessarily combined elements that together achieved the purpose of stabilising a dome. This leaves the Gould and Lewontin discussion of 'spandrels' not only wrong in its use of terms, but wrong in its conclusions on the significance of 'byproduct design' in both architecture and evolutionary biology.[11]

The significance of this is that even in architecture, whose product is that most visible of artefact forms, the building, mere inspection of the artefact is not a reliable means of working out the constraints on the design and construction process. In this example, the interpretation of architectural form was assumed to be simple; in reality, the problem of interpretation of architectural form proved as complex as some of the examples that could have been provided from evolutionary biology.

In conclusion, this pursuit of the feasibility of interpretations of the artefact tends to reinforce the conclusion that if innovation is most interesting when there is an element of artefactual change, the understanding of such change is necessarily accompanied by an understanding of change in other elements of the technology complex. The attainment of that understanding is often obscure and difficult to access, but in this book, in the search for understanding of patterns in technological change, we will be repeatedly drawn into investigations of the detail of the relevant technology, understood in its broad sense through the technology complex.

Conclusion

In this review of alternative conceptions of innovation, the problem revealed is not necessarily that innovation is inaccessible and neglected, but that many readily available conceptions and straightforward daily observation, give at best only a limited view of the innovation process. This is perhaps not surprising, but it reinforces the conclusion drawn in the first part of this chapter: that if general understanding is the object, then a broad conception of technological change is the appropriate one. It follows, then, that the rest of this book should be dedicated to an analysis of change in the elements of the broad definition of technology.

Notes

[1] Fleck developed the idea as a result of many years of teaching the management of technology at the University of Edinburgh. In 2001 he and I published a paper describing the technology complex and giving examples of its utility. Much of this section and the following section on the uses of the technology complex are from that paper (Fleck and Howells 2001).

[2] The list could be expanded through the addition of more subdefinitions of technology, for Fleck and Howells did not attempt a 'complete' review of disciplines with active and distinct uses of the term 'technology'. Such additions and the pursuit of an ideal 'complete review' would not change the essential pattern revealed here: that a given technology includes elements that range from the artefactual through various social forms.

[3] This section is almost entirely derived from Fleck and Howells (2001: 525–6).

[4] For example see Smith and Marx 1994; MacKenzie and Wajcman 1985.

[5] A branch of the sociology of technology based on the writing of Callon and Latour termed 'Actor Network Theory' (ANT) seems to be increasingly popular with prospective PhD students – as if it offered a coherent theory of technology. One of their virtues is that they are in favour of more 'thick description' of technological development, but they actively develop a jargon

all of their own that makes their writing often impenetrable (Callon 1991; Latour 1991). One example at the heart of their work is to allow 'non-humans' – by which they almost always mean artefacts – to be called 'actors' within certain stories. Callon gives the example of a story that might be told about radioactive fallout from Chernobyl, where he suggests that the Chernobyl reactor becomes an 'actor' in the story (Callon 1991: 142). This extension of the usage of 'actor' then becomes perfectly general – hence 'actor-networks' are about the chains of relationships between humans and non-humans – or rather humans and artefacts. The result of this generalisation of an extension in usage of the word 'actor' is that a decent English word that once usefully distinguished the human ability to act from the artefact's inability to act in like manner no longer can perform this task. This example of 'actor' is of course only the first of many other imprecise inventions and shifts in meaning given to a run of words such as 'actant', 'inscription', 'intermediary'. I remain sceptical of the value of this jargon. All these authors want to do is tell stories about technology and to tell the rest of us to tell stories about technology. If historians, economists and other sociologists can tell such stories without the burden of the ANT jargon, it is probably unnecessary. One should also keep in mind Andreski's book *Social Sciences as Sorcery*, paying special attention to the chapter 'The Smoke Screen of Jargon' (Andreski 1974). There is a growing critical literature on ANT – for example, Winner criticises ANT for its disdain for the many other writers and disciplines that have contributed to the serious study of technology in society and for its relativism – its abandonment of a general position on the value of the social and technological changes that it seeks to study (Winner, 1993).

[6] From the title to Granovetter and Swedberg's edited collection of essays (Granovetter and Swedberg 2001).

[7] In this institutional arrangement, engineers with MBAs would approximate to the 'few' with exposure across the range of relevant 'technology-shaping' subjects. See Chapters 7 and 8 for a more detailed discussion.

[8] Bell credits Colin Clark with creating the terms in his book *Conditions of Economic Progress* (Bell 1973: 14) and uses the criterion that a majority of a country's workforce be employed in the service sector as his first and most simple definition of that unfortunate term, the 'post-industrial' society.

[9] As Dennett has argued, most evolutionary biologists would welcome the point that biological design is not *exclusively* a product of adaptation to environment. The controversy arises because, in Dennett's words, 'Gould has persistently misrepresented the import of the Gould/Lewontin paper outside biology, and many have been taken in' (Dennett 1996: 3). The misrepresentation is the implication that one need not search for an adaptive explanation for complex biological attributes such as language (Dennett 1996). The significance of spandrels in biology appears to be the same as in artefacts – *elements* of biological design may result as byproducts of other processes, but *there must be other processes* – there must be conscious design in artefacts and there must be adaptation under selection in biology. Spandrels are merely evidence of constraints to these processes and serve only to warn us not to make over-hasty interpretations of the design process from the designed (or evolved) thing.

[10] Mark and Billington used computer modelling techniques to establish the stress patterns in the domes of the Roman Pantheon and the Hagia Sophia. These stresses allowed them to infer the function of the loaded pendentives (Mark and Billington 1989).

[11] This has been pointed out by Dennett in his article on 'The Scope of Natural Selection' (Dennett 1996: 2).

2 Invention, Science, R&D and Concepts of Use and Market

The examples in this chapter have primarily been chosen for the light they shed on the relationship between technology development and use. Through the examples of the laser and penicillin this chapter examines one type of inventive step, the type that generates great technology development potential. It continues with a discussion of how the creative inventive step necessarily involved in innovation relates to the established uses that make up existing markets. The second half revisits two classic accounts of radical technology development through the research and development department, again with emphasis on the role that concepts of prospective use play in technology development. In short, the chapter introduces some of the many ways technologies are made to vary in relation to their prospective uses.

Invention as the Creative Contribution to Innovation

The Cognitive Contribution in Invention

The inventive step is interesting in its own right and receives a great deal of scholarly attention when it generates spectacular new technologies. This special scholarly attention and the popularisation of spectacular examples of invention probably contribute to a widespread confusion that radical inventive steps are the most important part of the innovation process. The numerous acts of insight that are minor or that prove to lay a false innovation trail are quite reasonably less likely to attract academic or popular interest. Their relative underreporting is probably the reason for a further tendency to find the contribution of mental creativity, the 'eureka moment', or what Usher called the 'act of insight' (Usher 1972), the most glamorous aspect of the radical inventive step. However, invention, like innovation, will prove to be as much a social as a cognitive phenomenon.

The study of the radical inventive steps involved in the development of the laser and penicillin give us the chance to study not only the originating moment of highly novel technologies, but also the social context for the creative process of invention. It should be added at this point that it is within the discipline of psychology that we find the study of creativity is treated as a topic in its own right. However, a recent review of the contribution of psychology to the understanding of innovation, written for a management audience, argues that the difficulty of defining the degree of creative contribution in this tradition has led to widespread acceptance on the part of psychologists that the most productive definitions focus on the attributes of creative products (Ford 1996: 1114). In our terms, psychologists have begun to study creativity as cognitive change related to the artefact, in other words the inventive act as we defined it. Of course good scientific histories of invention provide an accessible means of studying both the social and cognitive aspects of invention.

The study of invention especially highlights the difference between the prospective or *intended use* that inspires and then guides the initial design of an artefact and the *actual use* people make of it. The two may be quite different, and this distinction in terms will prove particularly valuable for understanding the genesis of a science-based technology such as the laser.

Intended Use and the Invention of the Laser[1]

The laser immediately presents the problem of which of two inventive steps was the most significant. It might appear obvious that the greatest 'inventive significance' belongs to the first working device to exploit 'amplification through stimulated emission of radiation' (what we can call the 'laser effect', represented by the acronym ASER). However, the laser effect was first demonstrated using microwaves in 1954 in the device termed 'the maser' (Bromberg 1991: 66). In 1958 the effect was shown to be extendable in principle to optical wavelengths (Bromberg 1991: 73) and the first operating laser was demonstrated in 1960 (Bromberg 1991: 10). Although the physicist Richard Feynman commented on the laser in his *Lectures on Physics*, that it is 'just a maser working at optical frequencies' (Feynman et al. 1965, section 9-3), by Bromberg's and by Townes' accounts the extension of the effect to optical wavelengths was not entirely straightforward and the inventive effort behind the laser must be considered to have been distributed in time over several steps. Notwithstanding this caveat we shall focus on the first working device to demonstrate amplification through stimulated emission, a device that exploited the effect at microwave frequencies, hence the name 'maser'.

The inventor of the maser, Charles Townes, had trained as a physicist and was a pioneering developer of the scientific field of microwave spectroscopy (Bromberg 1991: 16). The Second World War diverted him from a pure physics career into the design of radar-guided bombing systems for Bell Laboratories. He therefore combined a professional physicist's interest in advancing spectroscopic understanding (science) with an understanding of the engineerin problems and approaching limits of the current technology for generating and receiving radar radiation. This unusual combination of physics and electrical engineering expertise would prove crucial to his invention of the maser.

Townes' 'intended use' for the maser is evident in the priority he gives to his basic science goals:

> I had myself been stubbornly pursuing shorter and shorter wavelengths. Because they interacted more strongly with atoms and molecules, I was confident they would lead us to even more rewarding spectroscopy. (Townes 1999: 54)

Some members of the US military were also interested in shorter wavelengths, but not for reasons of scientific advance. They understood that equipment which generated shorter wavelengths should provide, for example, lighter, more compact military equipment and shorter-range radar of greater information content (Bromberg 1991: 14). Townes' known interest in working to shorter wavelengths (albeit for reasons of scientific advance) led the Office of Naval Research in 1950 to ask Townes to form an advisory committee on millimetre wave generation with Townes as its chair 'to evaluate and stimulate work in the field of millimetre waves' (Townes 1999: 53).

This committee could not solve the problem of how to generate millimetre waves (Townes 1999: 55). The methods in existence used resonating cavities with dimensions similar to the wavelengths of the radiation they generated and at dimensions of a millimetre it was difficult to manufacture such cavities with useful power output and effective heat dissipation; below a millimetre it became effectively impossible.

> It was out of a sense of frustration over our lack of any substantial progress that the conceptual breakthrough came. (Townes 1999: 55)

The moment of insight came as Townes pondered the generation problem just before an all-day meeting of this committee in 1951. But to obtain some understanding of the 'focusing role' of this objective of millimetre-wavelength generation, some analysis of the scientific–technical content of Townes' inventive step is necessary.

Townes knew, as did other physicists, that molecules naturally resonate at the desired millimetre wavelengths. The focusing role of the intended use on this physical 'effect' is evident when Townes systematically set about thinking how molecular resonance might be exploited to generate millimetre-wavelength radiation (Townes 1999: 56). This systematic review enabled him to see significance in 'stimulated emission', a quantum physical effect with which physicists had also been long familiar, but that in *normal* physical circumstances could not be expected to be useful.[2] Townes' conceptual breakthrough (Townes 1999: 54) came when he realised that in *abnormal* physical circumstances,[3] stimulated emission *could* generate an amplified beam of radiation at the human scale.[4] Townes' engineering-derived knowledge of the design of cavities allowed him quickly to work out that a feasible device could be built based on a resonating cavity, into which would be pumped a source of excited molecules.[5]

In Townes' own words, his device 'employed only standard, known physics' (Townes 1999: 59) and this raises the question of why no other physicist invented the maser before him. Townes includes in his account a review of the relevant work of physicists that preceded his own and comments that 'ideas about stimulated emission were thus floating around, but they were not being pursued … no one had any idea that it could be useful' (Townes 1999: 62). Townes himself had apparently thought of demonstrating stimulated emission through amplification before (in 1948), but 'decided it was rather difficult to do and, because there was no reason to doubt its existence, I felt that nothing new would be proven by such a test' (Townes 1999: 57). In other words, at this time, the maser served no apparent *theoretical physics* purposes. When conceived by Townes, it would be as a useful tool: an instrument with a purpose, a potential technology for the investigation of microwave spectroscopy.

The Role of the Working Prototype

Only two years later in 1954 when the prototype maser had been built and could be seen to work was Townes' physics vindicated in the eyes of his colleagues, and only then were a range of organisations stimulated to develop the maser for a variety of applications (Bromberg 1991: 21) that included a precise clock (Bromberg 1991: 25), missile guidance systems (Bromberg 1991: 26) and in Bell Laboratories as a low-noise amplifier and communication device[6] (Bromberg 1991: 27). The long process of development of laser technology for a proliferating list of uses had begun.

The role of the working prototype in validating the idea of the maser is demonstrated by the status of Townes' idea before the construction of the device. Townes worked in the Columbia University Radiation Laboratory, funded by and dependent upon the military, and his colleagues had clear ideas about what was specifically useful to the military. According to Townes, they took the view that his work was *not* useful, because the physics was unsound, and that therefore it might endanger the laboratory's military funding. There were 'gentle suggestions that I should do something in tune with the Pentagon's interest in magnetrons' (Townes 1999: 52) and at one point the head and former head of the department attempted to persuade him to abandon his maser development work (Townes 1999: 65). Townes had tenure and could afford to ignore their intervention. In other words, by Townes' account, it was with a background of disapproval that he pursued the creation of a working device.

The Role of the Military

There is no doubt that military funding was very important in *facilitating* the invention of the maser; through the wartime work that had given physicists like Townes radar engineering expertise; through the radar equipment that had resulted and that could be put to spectroscopic uses; through their grants to the Columbia Radiation Laboratory where Townes was employed; and through their funding of the organisation of the advisory committee that acted as a prompt for his development of the maser idea.

With this variety of forms of funding intervention it is a temptation of hindsight to imagine that the military made *specific* demands for the new technologies that would 'meet their needs'. Townes is adamant that the military never specifically funded the development of the maser or laser and takes pains to correct what he clearly thinks is a deviant interpretation:

> Some science historians, looking back on those days, have concluded that we were being in some way orchestrated, managed, manipulated or manoeuvred by the military, as though the Navy already expected explicit uses for millimetre waves and even anticipated something like the maser and laser.... From our vantage point, the Navy didn't have any specific expectations at all about something like the maser or laser.... The military seemed quite uninterested in my maser work until some time after it was proven. (Townes 1999: 68)

The military were not funding research into radiation in a fit of altruism. They understood that such fundamental scientific research had proven useful during the war and they expected and hoped that it would prove useful again. However, they had no way of knowing when and how such research would prove useful: their method was to fund scientists and trust and hope that they would eventually generate useful knowledge. They had to refrain from attempting to select between different paths to the general end of developing improved means of generating and receiving radiation. In this way they acted as pioneers of the public funding of science.

The physical effects that the maser exploited may be obscure to the non-physicist; nevertheless it is clear that the cognitive act of insight was prompted by social context and prior experience and expertise. The laser also illustrates that an intended use need not relate to economic criteria, nor need its subsequent uses relate to the original intended use.

The invention of penicillin provides an interesting contrast to the laser. It can also be described using the terms of physical effect, intended and actual use, but intended use has a very different role in the penicillin story to its role in the invention of the laser. Thus the story of the invention of penicillin promises to extend our understanding of the process of invention.

The Penicillin Discovery Myth and the Unclear Significance of the 'Mould on the Plate'

The penicillin story has been popularised in the Anglo-Saxon world as an example of the good that science can do for society.[7] This popular and mythical story of the discovery of penicillin is usually represented as a classic of 'discovery through observation' and would typically run as follows: scientist Alexander Fleming observes a culture plate that contains both a mould and colonies of bacteria and observes that the bacteria have died in the area immediately around the mould. Fleming realises that the mould has secreted a substance that has killed the bacteria – and so he has discovered penicillin. The story is sometimes completed with the aphorism 'chance favours the prepared mind'. And it is implicit in this story that with this critical observation Fleming understood the significance of penicillin – in our terms, he at once connected the 'effect' demonstrated by the mould on the plate to the 'intended use' of penicillin-as-antibiotic.

Yet if he ever made this connection, he almost certainly dismissed it within a few months. Through its contrast with the popularised myth the 'true' story of penicillin both enriches our understanding of the inventive step and demonstrates how and why the innovation myth prospered.[8]

The Reinterpretation of Fleming's Discovery of Penicillin

The first great problem for the 'myth' of invention-upon-observation is that 11 years passed between Fleming's observation of the mould-on-the-plate in 1928 and the serious *development* of penicillin-as-antibiotic by Florey's research team at Oxford University. If Fleming made the correct link, what did he do with it for these 11 years?

Fleming wrote up his work on penicillin, but in the 27 research papers he published between 1930 and 1940, there are only two lone references to the potential therapeutic value of penicillin and these are really references to penicillin having a possibly significant *antiseptic* effect, not an antibiotic effect (MacFarlane 1984: 249). It is significant that in his 1931 paper entitled 'On the Indications for and the Value of the Intravenous use of Germicides' he does not mention penicillin at all (Hare 1970: 107). The absence of written or oral advocacy of penicillin as antibiotic is strong *prima facie* evidence that Fleming had not understood the potential of penicillin. If he had understood its significance, then his behaviour was extraordinary to the extreme.

MacFarlane used Fleming's publications, interviews with his colleagues and, most importantly, Fleming's original laboratory notes to reconstruct Fleming's thought at the time of his experimental investigation of penicillin (MacFarlane 1984). In their accounts of the invention, both MacFarlane and Fleming's former colleague Hare take pains to establish the state of bacterial expertise and Fleming's previous professional experience as influences on his behaviour at the time of the penicillin discovery.

Significant Available Professional Experience and Knowledge

In the 1920s the single existing effective systemic treatment for a bacterium was Salvarsan, an arsenical compound developed by Ehrlich that could cure syphilis. It worked as a poison that happened to be more poisonous to the delicate syphilis bacteria than to the cells of the body – but it had to be administered with care and with exactly the right dose to avoid human organ damage.

Fleming had more experience with the injection of Salvarsan than anyone else in Britain and MacFarlane comments that it was strange that Salvarsan appeared not to prepare Fleming or anyone else for the potential of penicillin (MacFarlane 1984: 251). Yet the probable reason why it did not is suggested by Hare, who comments that at the time 'even the most optimistic of bacteriologists' thought that any antibacterial infection compound would function essentially as antiseptics functioned (and as Salvarsan functioned) – as general poisons that destroyed bacterial cell proteins more effectively than the cells of the patient (Hare 1970: 142).

In addition to Salvarsan, Fleming (with the head of his laboratory, Wright) had spent years testing the new antiseptics that were constantly being proposed as effective treatments of surface infections. He had consistently shown that antiseptics applied to wounds worsened recovery rates because they killed the immune system's white cells faster than they killed infecting bacteria (MacFarlane 1984: 86). MacFarlane is even able to cite Fleming apparently drawing the general conclusion from this work that there was little chance of finding a chemical agent capable of destroying bacteria in the circulating blood (MacFarlane 1984: 109).

What no one anticipated was the radically different mode of action of penicillin and the other antibiotics. Not until 1957 was it understood that penicillin works by preventing synthesis of a polysaccharide that bacteria use to build their cell walls (MacFarlane 1984: 146). It therefore does not kill mature bacteria, but prevents the growth of new bacteria. In the body this allows the immune system to overwhelm the still living, mature bacteria, but in a test tube mature bacteria would persist in the presence of penicillin.

Fleming's Interpretation of His Penicillin Experiments

Fleming abandoned experimental investigation of penicillin only three months after the discovery of the mould-on-the-plate. According to MacFarlane the pattern of his penicillin investigation initially followed that of an antiseptic. Yet he extended his toxicity tests in a way that MacFarlane suggests shows that he did suspect penicillin might have had some systemic antibacterial activity.

Fleming had first demonstrated that, unlike most antiseptics, penicillin was non-toxic in the body, but his crucial 'extension' experiment was the injection of penicillin into a live rabbit to test its persistence in the body. He showed that within the short time of 30 minutes, penicillin was eliminated from the animal's body. He had also observed that penicillin took many hours to kill bacteria in culture and that this activity apparently diminished in blood serum experiments in glass containers *outside* the body. MacFarlane comments that these two results

> must have dashed any hopes that Fleming might have had for it as a systemic antibacterial agent. He did no further animal experiments and in consequence did not progress to the sort of protection tests that might well have encouraged him (and others) to greater efforts. ... Fleming had (or probably thought that he had) good reason to suppose that penicillin would be useless in the body. What would be the use of injecting something that takes over four hours to kill bacteria when it is itself destroyed in a few minutes? (MacFarlane 1984: 128)

The crucial 'animal protection' experiment that Fleming did *not* perform involved, as the Oxford team would later design it, the injection of penicillin into eight mice, four of which had been previously injected with a standard lethal dose of bacteria. The four mice given penicillin survived, the four without all died. This experiment would later do for penicillin what the first maser did for *its* technology: it showed that it *worked*, that penicillin was active and effective within the body against bacteria. The success of this experiment provided the motivation for a further scale-up of production to enable human trials.

For MacFarlane and Hare the best explanation for Fleming's failure to perform the animal protection experiment is that by early 1929 he had allowed himself to become convinced that penicillin was without therapeutic value as an antibiotic[9] (Hare 1970: 99). In our 'reconstruction' of Fleming's probable state of mind when he abandoned penicillin research, he had the experience of repeated negative results in his antiseptic work and he had Salvarsan as a 'model' of a working whole-body, antibacterial agent. This current state of disciplinary expertise and the *absence* of an understanding of, or hypothesis about, penicillin's true mode of action suggest that Fleming would have had a strong predisposition not to find the non-toxic penicillin a credible candidate as an antibiotic. His motivation for the experimental investigation of 'penicillin-as-antibiotic' was weak from the start and this coupled with his probable interpretation of his experimental results was enough to destroy his interest entirely (MacFarlane 1984: 128). Amid a welter of weakly antibacterial substances competing for attention, he simply did not believe it was credible that penicillin was something special – a true systemic antibiotic with a novel mode of action.

Fleming's Real Contribution to Penicillin 'as Innovation'

Fleming did find a use for penicillin. Fleming's laboratory was self-financing and drew much of its income from the production of vaccines. Fleming realised early on that penicillin's selective *inhibition* of common bacteria in culture could be used to grow pure bacterial cultures necessary for the preparation of some vaccines. This was Fleming's actual and enacted 'concept of use' for penicillin – he maintained penicillin in culture for *use as a laboratory reagent*.

If we ask what was his contribution to the 'innovation' of penicillin as antibiotic, then he was neither innovator (he never brought penicillin to market) nor 'inventor' (since he dismissed the major future use of penicillin). Nor was he the 'discoverer of an effect', since so many others had made similar observations before him – there was even a book published in 1928 that reviewed existing work on moulds' inhibition of bacterial growth, that he apparently never consulted or knew about (MacFarlane 1984: 136).

His contribution to the idea of 'penicillin-as-antibiotic' was therefore this maintenance and free distribution of a living culture of an extraordinarily powerful strain of mould. He never understood how rare or powerful it was (MacFarlane 1984: 264) and had he not found his bizarre use for penicillin and so kept it in culture, it would have been very unlikely to have been found again (MacFarlane 1984: 137). Without a sample of his mould, his 1928 paper would have been of little use to Florey and Chain at Oxford 11 years later.[10] So after all, the penicillin story remains an extraordinary illustration of the role of chance in invention.

Lessons from the Invention of the Laser and Penicillin

In both cases the creative mental step can be understood as the conception of a link between the effect and a use, but in neither story did the relevant physical effect immediately and unequivocally suggest its uses. In both cases professional expertise and experience mediated the inventor's estimation of the credibility of the link and so influenced the inventor's motivation to proceed with the laborious work of investigation. In both cases the conversion of scepticism to belief in the credibility of the 'idea' required an appropriate demonstration of feasibility; this was the essential feature of the experimental work on penicillin and the development of a working device for the laser. The categories of 'scientific experiment' and technological feasibility collapse into one another here.

In both cases the first real uses differed wildly from later uses, but served the vital functions of motivating the inventor to create something that worked (the maser), or to maintain something in use (penicillin). These 'initial' technologies became material and cognitive resources that the inventor and others could draw upon to develop the technology for other uses.

Because of subsequent development and evolution of uses, in both cases it requires some effort to disengage ourselves from our familiarity with some of the uses of the technology today, so that we do not prejudge the process of invention in the past. The impact of penicillin as an antibiotic has truly revolutionised medicine and it is our familiarity with this current important use and the trivialising popularisation of the penicillin myth that make it difficult to believe that another conception was possible and preferred when it was discovered. The laser differs in that it has been developed into a range of technologies with quite distinct uses; nevertheless, Townes' intention to use the laser as a spectroscopic instrument first probably appears to most of us as a relatively arcane use and striking for its professional scientific, rather than directly economic, origin.

The Role of Existing Patterns of Use in Invention – Reference Markets for Innovation

It is because of their subsequent importance as technologies that the invention of the laser and penicillin have become interesting and rightly attract much attention. It would nevertheless be a mistake to imagine that the creativity intrinsic to the inventive step was confined to such radical steps. The more normal context for the inventive step is a developing technology with established uses. When a new technology substitutes for an older technology, it again becomes interesting to analyse the inventive process, for the established uses of the old technology become a cognitive resource for the newer technology's development, even while the established capabilities of the old technology may remain largely irrelevant.

Some years ago I went so far as to invent the term *reference market* to describe the cognitive role of an existing pattern of use of an established technology, and distinguished this from the *innovation market concept*, the projected future market for the developing technology (Howells 1997). The innovation market concept is a mental construction of those qualities by which prospective users would value the new product and that managers use to guide the construction of new production technology. The reference market, on the other hand, is an existing market based on real, traded products; an existing pattern of production and use which is understood in most detail by those producing and consuming these products. The reference market is so called because it is the market conception to which the innovating firm managers refer most often in the process of constructing the innovation market concept; it is the major cognitive resource in building the innovatory market concept. The construction of the innovation market concept is a cognitive process of understanding the qualities that make up the reference market and *selecting* those that are valued, for inclusion in the innovation market concept.

The cost of generating new jargon then seemed worthwhile, because it allowed a discussion of how the old patterns of use set the context for the construction of the new, intended patterns of use. And perhaps it was worthwhile, for others have also found it necessary to invent a name to distinguish between how existing objects are used and how innovators conceive of future patterns of use; Hounshell and Smith appear to use the term 'target market' in essentially the same way as I use 'reference market' (Hounshell and Smith 1988).

The distinction between the market concepts accommodates the frequent observation that in innovation existing users' articulation of their needs is strongly influenced by the properties of the technologies that already exist and so cannot be entirely trusted as a basis for development. Some intelligent interpretation and amendment of understanding of the reference market may be necessary to obtain a viable innovation market concept. The original study furnishes an interesting example of how the reference market can be deficient as a source of ideas for the innovation market concept.

Reference Market for Bioprotein Innovation

Until the oil price rises of the 1970s there was a general fear of an impending protein shortage and this acted as a general stimulus for the development of bioprotein fermentation technologies. These involved the controlled fermentation by microbes of fossil-fuel-derived inputs such as methanol or North Sea gas to produce a high-protein biomass product.[11] The 'reference' or 'target' market for the fermentation technology was the existing animal feed market. This market is serviced by specialist 'feed compounders' that match varied proportions of inputs, largely soya and fish meal, to the various nutritional requirements of animals, whether goldfish, gerbil or veal calf. So the innovating firms guided their development process by reference to the compounders' understanding of how to combine protein sources to make a 'good' animal feed. So a good feed for veal calves is one that gives rapid weight and lean tissue gain without tainting the flavour of the flesh and this is understood to relate to the amino acid, mineral and vitamin content of the feed inputs. The compounders therefore valued feed inputs by these properties and their valuation procedures were the basis for the bioprotein innovators' anticipated valuation of their novel feed products. For example, economic value could be attributed to the 'property' that, in contrast to fish meal, high-protein novel feeds did not give animal flesh a fishy taste.

The role of the trace nutrient selenium in the novel feeds provides an example of a limit to the compounders' ability to articulate 'user needs'. In some of the feeding trials conducted by the chemical company ICI on its protein product, named Pruteen,[12] animals sickened when fed very high percentages of the novel feeds. ICI's R&D department then conducted experiments that showed that this was because Pruteen was selenium deficient. Once selenium was added to Pruteen-based feed, animals thrived on a high-Pruteen diet. The problem had arisen because Pruteen's production engineers had adjusted selenium inputs to the fermentation process to meet the nutritional requirements of the bacterium. Because this bacterium differed from multicellular life forms in having little need for selenium, the result was that compounded feeds containing very high percentages of Pruteen could then be selenium deficient compared with conventional feeds.

An understanding of the role of selenium in animal feed was an essential element of the innovation market concept for novel feeds, but one unavailable from the reference market articulated by existing 'users', because fish and soya meal contained more than sufficient selenium for all animal feeds; feed compounders had never had to specify minimum selenium content and so could not articulate such a requirement for the novel feeds.

The danger for the novel feed innovators had been that if government agencies produced adverse results from similar feed trials, they would not be motivated as the innovators were to understand their results. They would be more inclined to conclude that the novel feed products were toxic, or were potentially toxic. Even the suspicion of toxicity could be fatal to this kind of innovative effort.

The example shows that the reference market alone may be a deficient source of understanding of the market for innovation and that 'scientific' research rather than 'market' research can be the source of the future-market understanding on which successful innovation must be based. It is also a demonstration of how development involves inventive steps in its own right and cannot be thought of as merely the exploitation of the original inventive idea.

The User as Originator of Market Concept

If in the reference market case we have an example of the need for continued intelligent investigation of a strong guiding concept of existing use, then the point should be made that in general 'users' vary greatly in their ability to recognise and articulate their 'needs'. The 'work' that a technology developer must do on the market concept varies greatly depending on technology and circumstance. Within the innovation literature this is recognised in, for example, Souder's 'Customer Developer Condition' model and Souder gives examples of users presenting their requirements even in as detailed a form as a product specification (Souder 1987). In this case of 'sophisticated user' Souder argues that the developer firm's marketing department has little or no role, but that the less sophisticated the user, the more the marketing department of the developer must compensate by (in our terms) active development of reference and innovatory market concepts.

Von Hippel develops the idea that the relative contribution of users to innovation varies systematically by technology through his studies of innovation in scientific instruments. He studied the origin of not only the first, but also all subsequent major and minor improvements that occurred in a period of 20 years to four important types of scientific instrument (gas chromatograph, nuclear magnetic resonance spectrometer, ultraviolet absorption spectrophotometer and the transmission electron microscope). For 77% of his 111 innovations the user – typically a scientist working in a university – not only made the invention, but then researched and built the prototype to demonstrate its feasibility (von Hippel 1988: 13). In other words, what we found to be Townes' motivation in the invention of the laser is of common occurrence.

Von Hippel's examples of user-dominated innovation processes (others are the semiconductor and printed circuit board assembly processes) are another version of Souder's sophisticated user scenario, except that for scientific instruments another typical function of the developer, the creation of prototypes, the user now performs.

The picture that we now have is that stages of the innovation process may be variably located outside the firm that finally produces and markets the product. Like Souder, von Hippel points out that the work of R&D and marketing departments must be adapted to the typical source of innovation in these technologies (von Hippel 1988: 9) and that in general we must take care not to assume innovation takes place in a 'stereotypical' innovating firm possessing all the necessary functions to generate successful innovation.

This is fair enough, but there is an obvious response – that scientific instruments make a special case because *by definition* scientific 'users' have developer abilities at the laboratory-scale level – all scientists are trained in the manipulation of instruments for experimental purposes. This is fundamental to the practice of science whether in an R&D department or university laboratory. It makes sense that the evolution of science may involve the evolution of those instruments and techniques – the technology of scientific instruments. This is exactly what happened in the invention of the laser. Scientists and engineers are the one 'user group' that is in a position to engage effectively in its own product R&D and the scientific instrument 'industry' essentially complements their R&D capability with a mass manufacturing capability.

On the basis of his research von Hippel advises managers to assess their own industry for its characteristic location of innovation activities and to organise their innovative activities around this. One might argue that industries anyway evolve structures that account for the various characteristic strengths and weaknesses in user innovative abilities. An illustration in extreme contrast with scientific instruments is the development of nylon by the DuPont chemical company. DuPont not only produced novel textile fibres, but it worked closely with the fabricators of conventional textiles to create novel textile fabrication technology that would ensure that its novel materials could be turned into valuable textiles: for nylon, nearly every stage of the conventional, silk stocking fabrication process had to be changed to avoid the dangers of the finished product being wrinkled, discoloured or snagging easily. These intermediate users could not have done this without DuPont. It had to extend its organisational 'reach' into its users' fabrication technology so that it could secure the profits on its chemical innovations (Hounshell and Smith 1988: 264).

These examples certainly demonstrate that the ideas and various steps of development may be distributed between organisations in different ways, but more than this, the variation is a result of active accommodation to the innovative strengths and weaknesses of existing organisations.

Invention and Innovation as Socio-cognitive Processes

A striking feature of the stories of invention was the role of social context and prior expertise for the cognitive act of insight. A major point of contrast was that while these worked to prompt the act of insight that generated the laser, they also worked to weaken motivation to investigate penicillin-as-antibiotic. In neither case would it make sense to judge the cognitive part of the process as separate from social context and prior expertise. Of course, that is exactly what has happened in the generation of the myth of penicillin discovery by Fleming. Through simplification the context was dropped and the act of insight alone, falsely credited to Fleming, imagined to be the work of invention.

Because of the different outcomes in the laser and penicillin, it seems important to capture the sense that social context and expertise influence the act of insight that is more commonly understood as the inventive process. Several authors, including myself, have coined the term 'socio-cognitive' to capture the intertwined nature of social and cognitive context in invention and the development of technology (Howells 1995; Garud and Rappa 1994; Garud and Ahlstrom 1997).[13] Invention, then, can be understood as a socio-cognitive process. Examples from philosophies of science and organisation studies show that there is a degree of convergence among disciplines on this basic understanding: that the social and the cognitive interact, are not independent and the interaction can and should be studied.

Socio-cognitive Processes in the Philosophy of Science and Organisation Studies

There are many theories about science and the development of scientific knowledge, and as with the approaches to technology discussed in the last chapter, approaches to science have tended to be characterised by the academic discipline of the theoriser. It may be significant then that Thagard's recent work (1999) in the philosophy of science has also moved in the direction of a socio-cognitive understanding of science, so breaking with a tradition in that subject for a purely logical analysis of scientific method.

After a short review of the logical, psychological and sociological theories of science, as a philosopher of science, Thagard argues that cognitive and social explanations can be 'complementary rather than competitive, and can be combined to fit an Integrated Cognitive-Social Explanation Schema that incorporates both mental processes and social relations' (Thagard 1999: 19). For our purposes, the essential feature of this work is that it insists that science should be understood through the analysis of *interacting* cognitive and social practices and that it develops the argument most forcefully through the analysis of a principal, but complex case: the success of the theory that peptic ulcers are caused by a bacterium.[14]

The significant characteristics of the case can be outlined in social and cognitive terms. Gastroenterologists were the community of medical specialists that studied and treated ulcers. Marshall and Warren, two scientists from outside the gastroenterological scientific community and possessing expertise in bacterial treatment, thought of using relatively novel bacterial staining techniques to demonstrate that a bacterium, *Helicobacter pylori*, was present in the majority of gastric ulcer tissue samples. Acceptance by the gastroenterologist community of the bacterial theory of ulcer causation was delayed because of the novelty of the technique, the outsider status of the discovering scientists and the long-standing assumption on the part of this group that no bacteria could survive the extreme acidity of the stomach. Such was the strength of this belief that one of Marshall and Warren's early conference papers was rejected by the Australian Gastroenterological Society, and the paper they submitted to the *Lancet* was delayed by negative reviews of its plausibility (Thagard 1999: 88). However, these social and cognitive blocks served only to delay acceptance. Gastroenterologists first accepted the existence of the bacteria that the new staining techniques could reveal. It took a little longer for them to accept that *H. pylori* was the causative agent of ulcers, but controlled experiments with antibiotics showed a significant rate of cure and the experiment could be replicated effectively. The delaying effect of social and cognitive obstacles in this example was in the order of years, not decades.

In mainstream organisation research, March and Simon could write in the late 1950s that there was little concern with cognition; that is, with the adaptive, reasoning behaviour of human beings in organisations (March and Simon 1973: 233). Their book expressed the information and attention limits to human mental capacity through the term 'bounded rationality'. Cognition has received more attention since March and Simon, for example, the interaction between the cognitive and the social is recognised in Weick's work. Here individuals are understood to be theory-makers, or sense-makers on their own account, but what they understand to be useful knowledge and their ability to elaborate cognitive understanding are strongly influenced by how they are organised within the firm (Weick 1979). What this work generally ignores is the role of the artefact in the sense-making and negotiation over meaning and purpose that occur within the firm: artefacts, like organisations, are not 'facts' with entirely self-evident meaning; that has been the starting point of this book. Weick's work, drawing on the work of the psychologist Donald Campbell, can be applied to technology if the artefact is included as one of the elements of organising. So not only invention but also innovation should be understood as a socio-cognitive process, for the development of technology is a form of organising. The value of this approach in innovation is that it accommodates the micro political processes that occur, for example, when project conditions change and a new understanding of the viability and future of the project must be negotiated. Such questions may open and close throughout project lifetimes, as will be shown in the next chapter.

The way I have introduced 'socio-cognitive' is as a category, a new piece of jargon, and not a theory. For the study of innovation its value lies as much in the understanding it avoids as in the orientation towards understanding that it represents. So on the one hand it clearly breaks with the idea that invention could be no more than an isolated act of insight, from 'out of the blue'. On the other hand this term 'social' is loose and not deterministic, and this is deliberate. In our search for the creative element of innovation, we have found that the source of ideas and early idea-proving activities are not necessarily confined to one category of organisations. We found that there could be selective sourcing of ideas for a prospective new market in existing patterns of use – the reference market. Of course, just as the source of ideas can be variously sourced in different organisational forms, new ideas of use can become the basis for new organisations in development. For all these reasons it seems fair to increase the burden of jargon and to describe invention and innovation as socio-cognitive processes.

The term also has the virtue that it avoids the undue determinism that infects the most widespread alternative model of the relationship between technology and its uses, the characterisation of innovation as a product of either 'technology push' or 'market pull'. The push vs. pull metaphor survives in management studies despite its decline in the field of innovation policy dating from Mowery and Rosenberg's critical review of its abuse (Mowery and Rosenberg 1979). It will prove worthwhile to revisit this idea both because of its persistence and for the explanation of why it is not a good theory of the relationship between technology and its uses.

Popular 'Explanations' of Innovation – Market Pull, Technology Push Revisited

Some version of the push vs. pull 'explanation' of innovation can be found in texts as different as Simon's study of the success of small, innovative German companies, Fruin's field study of a Toshiba electronics plant and Kotler's textbook on marketing (Simon 1996; Fruin 1997; Kotler 2000). So in Simon's discussion of the 'driving forces' for innovation he uses survey evidence that sorts managers' choices of whether their companies are primarily market driven or technology driven (Simon 1996: 131). It is not clear whether the opposition between technology and market is assumed by his research method or represents the managers' own thinking about the sources of innovation, but the most significant feature of this research is that the technology and market are understood to be alternative sources of innovation.

Marketing texts are the other major site of survival of the idea that technology and market are in opposition, but first it makes sense to return to the problems with the original push vs. pull studies of the 1960s and 1970s.

Myers and Marquis' study of the technology-push or market-pull 'causes' of a collection of innovations provides an example of the kind of conclusion that these studies tended to generate.

> The primary factor in only 21% of the successful innovations was the recognition of a technical opportunity. Market factors were reported as the primary factor in 45% of the innovations and manufacturing factors in 30% indicating that three quarters of the innovations could be classed as responses to demand recognition. ... Recognition of demand is a more frequent factor in innovation than recognition of technical potential. (Myers and Marquis 1969: 60)

Even in this quote it is evident that it is the way the authors define 'market factors' that allows them to draw the simple and suggestive conclusion that recognition of demand is more important than technical potential – why should manufacturing factors be included as a subcategory of market factors? Myers and Marquis' other categories of 'demand' include: 'anticipated potential demand'; 'response to a competitive product'; 'attention drawn to a problem or inefficiency'. Reflection on these categories makes one uneasy with their classification as 'demand' factors: is 'response to a competitive product' not as much a response to the competitor's technological choice for that product as to that competitor's market choice? There is an evident degree of arbitrariness here.

Mowery and Rosenberg's critical review found that most of the innovation case surveys, like Myers and Marquis, drew the conclusion that demand/pull factors were more important than technical factors (Mowery and Rosenberg 1979). This tendency to privilege demand over technology as a cause of innovation mattered because it was used to support government policies of 'laissez-innover' – a non-interventionist policy of leaving innovation to the market. Mowery and Rosenberg drew the contrary conclusion that the evidence and methods of the studies did not allow one to privilege 'market pull' over 'technology push'.

They showed that the meaning of technology push and market pull had no inherent precision and varied between research projects, so that comparison of results was difficult. There was a tendency across studies to define the two categories such that most innovations were classified as 'market pull', but these studies did not recognise that their defined category of technology push tended to include more of the radical and important innovations than the 'market-pull' category.

According to Mowery and Rosenberg, much of the problem derived from the imprecise use of the economics term 'market demand', which suggests the 'market-pull' explanation for innovation, and in their view:

> Market demand must be clearly distinguished from the potentially limitless set of human needs. Demand expressed in and mediated through the marketplace, is a precise concept, denoting a systematic relationship between prices and quantities, one devolving from the constellation of consumer preferences and incomes. To be taken seriously demand-pull hypotheses must base themselves upon this precise concept and not the rather shapeless and elusive notion of "needs". (Mowery and Rosenberg 1979: 229)

This is a valuable critique of the 'push vs. pull' body of work, but it raises the question of whether it is really possible to use the concept of 'market demand' from economics to explain innovation.

There are problems with the precise concept of market demand; many of the assumptions required to make this concept precise have little to do with the real world and everything to do with creating a mathematically tractable concept (Ormerod 1994). These include assumptions about consumer preference functions, that supply and demand curves are (mathematically) continuous functions rather than discontinuous, that they can make incremental shifts and that they are knowable. Further, market demand is a static concept, that is it applies to a known good or service, and does not inform us of the mechanism by which shifts in supply or demand curves generate new products. Finally, it is this very concept of market demand that did inspire many of the push vs. pull studies, and their problems may be taken as evidence that it is difficult to operationalise the idea.

If the precise idea of market demand must be abandoned as a basis for explaining innovation, we appear, in Mowery and Rosenberg's terms, to be left with the 'rather shapeless and elusive notion of "needs"' and the continuing opposition between technology and market suggested by the popular push and pull metaphor. The trouble with the push vs. pull metaphor is that it is a *bad* metaphor. The worst feature is its statement of a false opposition between technology and market in the development of innovation. Technologies are designed for a purpose and so some idea of use is always implied, and when technologies are changed a change in use is also implied.

The push–pull formulation also confuses the role of agency in development; *all* technological innovation requires 'push' – human effort – for its development. It must be made to happen by motivated people, typically organised within the firm and this is obviously true even if this agency is distributed between actors and includes users as contributing developers. However, it was earlier argued that markets serve as patterns of use that are a source of ideas in the innovation process – they are not agents that can push, pull or demand specific innovation. To oppose developer push with market pull is therefore additionally misleading. Nevertheless, because certain innovations, such as barbed wire as discussed below, appear to strongly invite classification as 'market pull', a little more analysis will prove useful.[15]

Desperately Seeking 'Market Pull' – the Invention of Barbed Wire[16]

As cattle ranches spread west and into the great plains of the USA in the nineteenth century, a fencing crisis developed because there were few trees and traditional wooden fencing was prohibitively expensive in its role as an effective property demarcator. Farmers could not protect their crops from freely roaming herds of cattle, so

> fencing quickly became the farmer's primary concern. Between 1870 and 1880 newspapers in the region devoted more space to fencing matters than to political, military, or economic issues. . . . In 1871 the US Department of Agriculture calculated that the combined total cost of fences in the country equalled the national debt, and the annual repair bill for their maintenance exceeded the sum of all federal, state and local taxes. (Basalla 1988: 51)

Such was the scale of the problem that Hayter writes that westward expansion in the early 1870s was 'slowed down considerably' (Hayter 1939: 195).

There was then a pressing 'need' for a 'wooden fence substitute', but the duration of the problem shows that it could not immediately generate an adequate solution. For a while it was thought that the solution was a thorny hedge grown from osage orange. Osage orange is a bush with long tough thorns at right angles to their stems. It grew naturally in some southern US states and a small cultivation industry developed to export its seeds to the plains region. However, it took three to four years to grow an effective hedge and of course it could not be moved if property boundaries changed.

Bare wire fences were cheap and in use but were a poor deterrent to moving livestock. According to Basalla, it was the pattern of thorns on osage orange that inspired the 'thorny fence', as one inventor called his patented wire invention (Basalla 1988: 53). 'Barbed wire was not created by men who happened to twist and cut wire in a peculiar fashion. It originated in a deliberate attempt to copy an organic form that functioned effectively as a deterrent to livestock' (Basalla 1988: 55). In other words, the 'biofact' of osage orange inspired the artefact of barbed wire, or in our jargon, osage-orange-as-fence/hedge acted as the reference market for the invention of barbed wire.

If the case invites a 'market-pull' classification it is because there is so much evidence of the damage done to frontier cattle farming by the absence of an economic fence substitute. The trouble is that on this basis it can just as well be classified as 'technology pull'. The need for a fence-substitute was acute because the fully developed *technology* of cattle farming had been transferred into a new environment where just one, albeit critical, artefact component of the working technology was missing. The idea of what is being 'pulled' is also unclear. There is no 'pull' acting on the idea of barbed wire: if there is 'pull' it is conventional market demand for the artefact component that cannot be supplied. In contrast to the laser, the case is better understood as a defined problem. But although a defined problem invites puzzle solvers, like the famous case of Fermat's last theorem, that does not mean the problem is easier to solve. And it certainly does not imply that solutions to defined problems are better as a 'class' than proposed improvements to what already exists and only appears to be functioning well enough.

The case is a good demonstration of the problem of 'needs' in innovation. Genuine and general human needs such as shelter and food are relatively few, obvious, and tell us nothing about why and when specific innovation events occur. Acutely expressed 'needs' as in the barbed wire case often have a technological context and do not necessarily exist for people outside of that context.

We shall adopt the view that 'needs' have been proven to exist by innovation *success*: that is, if users are prepared to switch spending from current products to the innovation. When the firm develops its innovation it no doubt hopes that its intended use will prove to be a 'need' when the product reaches the market – but given the context-dependency of needs, it should not be fully confident.

In sum, the push vs. pull metaphor is objectionable because it is misused to falsely oppose technology to market as a cause of innovation, or to falsely oppose the development of technology as an alternative to development for a market. It is a misleading and widespread oversimplification that obscures understanding and the terms 'technology push' and 'market pull' are probably best avoided in all their many manifestations. Important in its own right is the manifestation of the opposition of technology and market in the marketing literature, for this is where a naïve student might look to find these issues discussed clearly.

In Practice the Split Lives On – the 'Marketing Concept' of the Marketing Literature and Technology

Many marketing texts begin with an effort to define the conceptual basis of the subject and this can drift into a statement that resurrects a version of the 'push vs. pull' metaphor. Early in Kotler's *Marketing Management* comes a statement of the 'marketing concept':

> the 'marketing concept' holds that the key to achieving its organisational goals consists of the company being more effective than competitors in creating, delivering and communicating customer value to its chosen markets. (Kotler 2000: 19)

Few would object to this, but then it is claimed that this is a 'business philosophy that challenges' both the 'production concept' that 'holds that consumers will prefer products that are widely available and inexpensive' (Kotler 2000: 17) and the 'product concept' that 'holds that consumers will favor those products that offer the most quality, performance, or innovative features' (Kotler 2000: 17).

Kotler has defined the product and production 'concepts' as if they are distinct from the marketing concept, and by that word 'challenge' it is implied that they are perhaps inferior. They are better understood as specific examples of the marketing concept itself, in some unspecified, but surely relevant, technological context. In other words, the distinctiveness of the defined concepts and the idea that they are 'competing' (Kotler 2000: 17) in some general sense as alternative concepts are misleading here. This is even implied by the examples used, which allow, for example, that the production 'concept' is a valid contemporary market orientation of Texas Instruments.[17]

Other marketing texts are less subtle than Kotler and claim there is an historic succession of business concepts that run in the USA from a production orientation until the 1940s, through a product orientation in the 1950s and 1960s, to the eventual marketing orientation of today (Lancaster and Massingham 2001: 4–7). Lancaster and Massingham use the Japanese-import-driven decline of the British motor cycle industry to 'quite simply' illustrate what they mean by a 'product orientation' (Lancaster and Massingham 2001: 5). They warn us that 'this orientation leads to a myopic view of the business with a concentration on product engineering rather than on the customer's real needs' (Lancaster and Massingham 2001: 5). Yet good product engineering does not necessarily exclude an understanding of the market, as is shown by this same example in the same textbook. The success of the Japanese derived in part from the 'product-engineering' replacement of motorcycle kick starting with electric starters. Even with the limited detail available (no source is given) the strong claims that success was derived through the choice of a market orientation rather than a product orientation can be seen to be false. In short, neither the claims of an historic succession of general orientation concepts that culminate in the marketing orientation, nor the distinctiveness of these concepts themselves, should be accepted. The real marketing issue in innovation is how to manage technology development for an articulated market concept, when the market concept is not fully articulated by users.

A parallel confusion dogs the presentation of the basic issue of 'needs' in the marketing literature. Marketing texts usefully acknowledge that users are not necessarily aware of their real 'needs', but then they tend to emphasise the value of marketing as the means of investigating and servicing the 'real needs' that are supposed to exist – indeed, this is the very definition of marketing as a business activity (Kotler 2000). Once again, the definition, but not the examples, is in isolation from any technological context. So in Kotler's text, Sony 'exemplifies a creative marketer because it has introduced many successful new products that customers never asked for or even thought possible' (Kotler 2000: 21). One can agree that Sony is a creative marketer, but hold an alternative interpretation of marketing to Kotler's, that the most creative and exciting marketing opportunities occur in those industries whose technologies are in a process of development; which is to say that the full set of stable technology–market matches have yet to be imagined and worked out in practice. In the view of this book, real marketing creativity occurs within the constraints and potential of a developing technology.

What these examples do show is that like any professional specialisation, marketing feels a need to justify itself by projecting an image of progression, development and universal applicability. Unfortunately the way the profession has chosen to do this is by the resurrection of the technology-push versus market-pull dichotomy and the associated claim of superiority for the market 'side' of the concept. It might therefore be foolish to expect this use of the push vs. pull metaphor to die out in the near future, but if it persists it will be at the cost of a marketing text that properly resolves conceptions of technology, market, need and use.

Return to Innovation as a Socio-cognitive Process

Innovation is fundamentally about how and why technologies are made to relate to various uses. In the market economy the common means of organisation of technology is that firms specialise in the creation and development of production technologies in order to trade and users are distanced from the organisation of production via the institution of the market. It is likely this organisation of technology that suggests the push vs. pull metaphor as a means of classifying innovation, but the story of the persistent use of push vs. pull instead serves to illustrate the danger of oversimplified and confused terms. When the false promise of such simplification is abandoned, one is forced back towards a craft model of knowledge in answer to the fundamental question of how and why technologies are made to relate to various uses. In other words, we need to accumulate experience in diverse examples.

The Management of Industrial R&D

Statistics of national and industrial sector R&D are available, but they cannot tell us about the experience of 'doing' R&D. If we want to understand the issues in the process of development – the sequence of events and decisions made by management that gradually shape a novel technological step – we can turn to the small body of business histories that cover research and development work.

Two of these will be considered here. First, Graham's account of RCA's VideoDisc project is one of those detailed historical studies that conveys the experience of radical technology project management in uncertain conditions.[18] The account here is organised around the perennial R&D management themes of the identity of competitors, the interpretation of market research and the role of project and company 'leadership'.

Second, Hounshell and Smith's unique history of the R&D function in the chemical company DuPont provides insights into the coevolution of corporate strategy and the R&D function. The account is organised chronologically over 80 years to highlight the interaction between competitive and regulatory environment and the understanding of internal capability and company strategy.

Leadership Lessons from Failed Project Development – RCA and the VideoDisc[19]

RCA successfully pioneered in succession radio, black and white television, and then colour television development to become the leading US consumer electronics innovator. Its dominance in television had been achieved by a strategy of simultaneously establishing a TV broadcasting standard and programme production for this standard – RCA owned the broadcaster NBC. The early availability of programme content ensured the dominance of RCA's choice of broadcasting standard and thereby the importance for rivals of gaining access to RCA's TV receiver standard. RCA then reaped profits from the lucrative market for TV sales through the licensing of its TV receiver standard to the manufacturers who might otherwise have become rival technology standard manufacturers in their own right. When this 'closed' or proprietary standard strategy worked, as it did for colour TV, it was immensely profitable, but it implied the coordination and control of major interacting technologies. RCA would bring this understanding of its past success to the VideoDisc project. It would tend to assume that it must continue with the closed standard strategy while it also mistook companies with similar 'systems-innovating' capabilities, such as the broadcaster CBS, as its most dangerous potential competitors.

Once a professional model of a videoplayer had been made available come the 1950s by the US company Ampex, the idea of a videoplayer for home use became an obvious one. The more difficult questions were *which* technology would make the better commercial product and *when* should this technology be advanced towards full-scale commercialisation? As befitted an industry leader, RCA Laboratories had three technologies as candidates for development: holographic, capacitance and magnetic tape technology. Each had its internal advocates, but external events played a pivotal role in forcing internal 'selection' of one technology over another – with the exotic outcome that each of the three enjoyed some period when it had been selected for commercialisation. How this could happen has significance for our idea of R&D leadership.

Competitor Perception and Project Selection

When CBS announced in 1967 that it was developing a photographic technology into a videoplayer product (Graham 1986: 102), RCA Laboratories management used the event to interest its senior management in their own candidate technologies for a videoplayer product. A process of product definition and selection was begun, but it was short-circuited when in May 1969 Robert Sarnoff, the chairman of RCA, learnt from watching a TV show that a Westinghouse colour camera (based on CBS technology) had been selected for the Apollo 11 space mission instead of RCA technology (Graham 1986: 114). The next morning he 'expressed his deep personal displeasure' and demanded to know what CBS would do next in a memorandum to his key appointment in new product planning, Chase Morsey (Graham 1986: 106). This pressure led Morsey to fear that CBS's planned public demonstration of its videoplayer technology for the autumn would be seen by Sarnoff as yet another successful challenge by CBS. To pre-empt this danger he brought forward a planned public presentation of the RCA videoplayer project from December, which had been the internally agreed earliest possible date for demonstration of the laboratories' preferred capacitance technology, to September, before the CBS presentation.

The laboratories had seen the holographic technology as a 'distant second-generation approach' (Graham 1986: 110) until this time, but it was closer to a demonstrable working prototype system than the capacitance technology and it was an ideal candidate to demonstrate RCA's 'continued technological leadership' (Graham 1986: 115) – this had now become the purpose of the public demonstration, and the audience was as much RCA's own chairman as the industry and public who would attend.

The public demonstration of this technology at the press conference naturally led observers to believe that this was RCA's *preferred* technology for a commercial videoplayer. Worse, the marketing staff had taken the laboratories' estimates for earliest product introduction and 'manipulated the data into a plausible, though highly optimistic, plan' (Graham 1986: 231). The marketing group went on to forget the dubious origin of the technology selection decision and began to implement the business development timetable that they had prepared as part of the demonstration (Graham 1986: 119).

> In the end, when the compressed goals proved to be unrealistic, the R&D organisation was charged with misrepresentation and Robert Sarnoff, never personally involved in the preliminary discussions, lost faith in the process. (Graham 1986: 231)

The credibility of the holographic approach was further undermined as the CBS project was abandoned and as it became clear that most other potential competitors had chosen magnetic tape technology (Graham 1986: 134).

In 1970 there came another announcement of an apparently credible VideoDisc threat, this time by the joint venture 'Teldec' using electromechanical technology – a version of the established 'needle-in-groove' record technology. The effect this time was to strengthen RCA management support for the laboratories' capacitance technology and to provide the development team with a set of benchmarks against which they could judge their own work. More researchers were added and a project manager appointed, reflecting the more serious 'weighting' the capacitance technology had gained (Graham 1986: 135).

Then in 1973 Philips demonstrated a radical, optical, laser-based VideoDisc system at an industry show that convinced RCA managers that RCA had an inferior product in features and performance (Graham 1986: 161). RCA's top management commitment to the VideoDisc project 'evaporated' (Graham 1986: 162). The VideoDisc development team responded with a crash prototype development programme to restore senior management belief in their technology. They duly demonstrated a capacitance prototype by the end of 1973 that beat the Philips technology, but at the cost of six months' development time. The example further confirms the vacillating nature of senior management's attitude towards the R&D department (Graham 1986: 162).

The problem with competitor announcements was accurate evaluation of their credibility and degree of threat. The tendency was to attach greater credibility to technologies supported by established powerful companies like CBS, mediated by internal research to check the claims of rivals – and, as we have seen, perceived internal political imperatives. More than 10 different videoplayer technologies were developed for consumer use in the 1960s and 1970s (Graham 1986: 22) and this long history of aborted efforts clouded judgement of what would turn out to be the real threat – from Sony and Matsushita.

Market Research and the Missed Japanese Competitive Threat

So Graham writes that when Sony's Betamax system was launched for $1300 in 1975, it was first seen by RCA as another opportunity to learn about the market by observing consumer response. Then when Betamax established itself in a high-priced niche, opinion was 'evenly divided at RCA between those who saw Betamax as direct competition for VideoDisc and those who maintained that the two products were so different that they complemented each other' (Graham 1986: 181). The idea of two distinct markets was supported by both the initial very high Betamax price and the belief that it was not credible that magnetic tape technology could be developed so as to achieve low prices. Although RCA ran extensive market research from the 1960s onwards and this always confirmed that consumers would be interested in a videoplayer product, in retrospect it misled RCA managers, not for reasons of 'quality' or 'fault' in the conduct of this form of research, but because the very understanding of what the market would be depended in part on technological forecasts.

In 1976 RCA market research on the performance of Betamax and the proposed VideoDisc remained 'inconclusive as to the relative appeal of the two different product concepts – recording, versus only playback – but it left very little doubt as to the importance of price' (Graham 1986: 190). The market forecast depended in turn on an economic forecast for the developing rival technologies. And the economic forecast in turn incorporated an industry-wide technological understanding that magnetic tape technology could not be developed to support a mass market product, but would remain high priced, leaving the mass market to some kind of VideoDisc technology. This expectation continued into the 1980s, as demonstrated by GE, IBM and even JVC (the successful developer of the VHS standard of magnetic tape video recorder) all having VideoDisc products in development by the early 1980s (Graham 1986: 212).

In sum, market research failed to predict the huge scale of the eventual market for videoplayers and it failed to predict the 'revealed preference' of consumers for video rental. The scale of the market was partly explicable by the sale of pornographic videos, which accounted for nearly half of pre-recorded videotape sales (Graham 1986: 214), while Graham suggests that the very proliferation of competing technologies may have shifted consumers' preferences towards rental (Graham 1986: 214).

R&D Leadership Lessons from the VideoDisc Project

Technological transformation had been fundamental to RCA's success and its founder, David Sarnoff, had created the corporate R&D laboratory as the means of continuing this strategy for success.

However, his successors began a programme of diversification into unrelated businesses (for example, Hertz car rentals) that absorbed their time and distanced them from the R&D laboratory. Graham comments that 'the most destructive effects of diversification on R&D must be assigned to a failure of leadership, the unwillingness or inability of top management to define a new mission for R&D when major change takes place' (Graham 1986: 232).

The pursuit of diversification led to a neglect of the R&D department – yet it continued to exist, product and symbol of RCA's past strategy of technology-led growth. In these conditions the laboratories were never secure in their future except through what they could promise to deliver, or as Graham expresses the situation:

> When the survival of the Laboratories depended on its clear identification with a proprietary, revenue-producing "blockbuster" project, the Laboratories could not be depended upon for reliable judgements about competing technologies. (Graham 1986: 225)

One of the results of this situation for the VideoDisc project was that:

> When there was too little money to fund exploratory work on high-resolution recording methods for its own sake, for instance, the electron-beam recording technique became the tail that wagged the VideoDisc dog. (Graham 1986: 225)

The laboratories had become committed to the capacitance technology with electron beam mastering for its prospective corporate strategic properties, not because it was the best technology for a VideoDisc product. Despite the implied lack of investigation into alternative mastering techniques in Graham's account, electron beam mastering had been defended by the laboratories just *because* it was difficult – the prospective advantage being that precisely because it was so difficult to perfect, it promised to make RCA's VideoDisc technology extremely difficult to copy and hence highly profitable for RCA to license *if* successful. This thinking also reflected RCA's history as a pioneer and leader of technological systems.

When senior management for the third time turned to the laboratories for a winning technology to compete with the developing threat of the Japanese magnetic tape products, they were prepared, for the second time, to back the development of capacitance technology. But it would then be found, very late in the day, that it was impossible to perfect capacitance technology as a mass manufacturing technology. It would eventually be abandoned leading to another protracted senior management crisis of confidence in the R&D laboratory's development ability. This time several years were lost before the former marketing manager now running the company realised that RCA's very future depended on it having a viable VideoDisc product – the patents and profits on colour TV were rapidly expiring. The fourth time around RCA chose, for reasons of developmental speed, a conventional electromechanical mastering technology as the basis for its VideoDisc player that was introduced commercially in 1981. This would prove several years too late. The huge costs of commercial launch were nevertheless incurred, contributing directly to the company's later dismemberment and the relegation of the corporate R&D laboratory to a mere contract research operation.

Close contact of some senior manager with the developing project has been noted as a feature of successful innovation by many studies, for example the comparative survey research of Project SAPPHO (Rothwell 1977). Graham's principal conclusion is also about the necessity of committed leadership and the integration of R&D into its overall strategy and the 'only way to do it is for top management and R&D management to engage in a constant process of mutual education' (Graham 1986: 230). This means that top management must take the lead in forming the relationship and be interested in *making* the R&D organisation educate them about the issues it faces.

Managing the R&D Function in DuPont

The story of the expansion of the DuPont research function from a handful of research personnel employed in 1902, to over 6000 supported by a budget of over a billion dollars by 1980 (Hounshell and Smith 1988: 9) is also the story of the growth of this one-time 'high-technology' company. Hounshell and Smith's unique[20] history of 80 years of R&D management in the DuPont chemical company allows us to explore the degree to which R&D can be 'directed' as part of a conscious company strategy.

Before the Second World War

The practice of DuPont research before the Second World War and in a period of high growth for the company was anything but a focused search for original breakthroughs. By 1911 DuPont's success and near monopoly in its original explosives business had generated the threat of antitrust action, and at the same time the US military were expanding their own explosives manufacturing interests. DuPont's response was to adopt an aggressive policy of 'research-led diversification' by its 'Development Department'. Research was used to assess companies as potential acquisitions and if a company was acquired, research was organised to improve the operations of these businesses. If there was little that was spectacular about it, DuPont found that this kind of research could be costed and evaluated and therefore controlled – and that it paid handsome returns.

The company grew so quickly that it encountered problems of administration which led to four restructurings (Hounshell and Smith 1988: 13) but by the end of 1921 it had resolved into its essentially modern structure: a central coordinating office with autonomous operating divisions, with research also organised into decentralised research divisions located in the operating departments and supported by a small central department. DuPont, with General Motors, is famous as a pioneer of this 'modern' organisational form of the multidivisional structure. The historical account leaves us in no doubt that this organisational innovation was an evolved response to the problem of size, which itself resulted from success in the exploitation of research – but research deployed in the improvement of the operations of acquired businesses. One might guess that this organisational structure would come under renewed pressure for change and perhaps break up, if research gradually ceased to yield the high returns that supported that structure.

DuPont was intent on growth and there were two major departures from this research model before the Second World War – neither involved original R&D breakthroughs. In response to the creation of the giant German chemical and dyestuffs company IG Farben, DuPont in 1929 signed an extraordinary technology transfer agreement with Britain's ICI. This involved the free exchange of patents, staff, confidential research reports, secret process knowledge and a geographical division of markets to limit competition between the two companies. The US Justice Department eventually forced the cancellation of the agreement in 1948 (Hounshell and Smith 1988: 204) but it succeeded in greatly extending the research capability and effectiveness of the two companies when compared with the German colossus (Hounshell and Smith 1988: 196).

The second departure was also an attempt at technology transfer, but the much greater ambition was to become a full-range dyestuffs manufacturer by using DuPont research to copy German dyestuffs technology. The sheer chutzpah of this attempt to build an entire research-dependent business from nothing and in the face of the dominant German industry remains impressive. Like the technology transfer agreement with ICI, it was stimulated by unusual circumstances, in this case the British blockade of German dye exports during the First World War and the absence of any substantial US dye manufacturers that DuPont could acquire. There was also a more timeless reason. Such was the success of DuPont's research-led 'acquisition-improvement' strategy that

> prideful DuPont research personnel had convinced executives that the problems would not be insurmountable. For a long time after the dyestuffs venture, executives were more skeptical of the opinions of research men. (Hounshell and Smith 1988: 77)

The results showed the limitations of a strategy of 'research as a tool of technology transfer'. Massive injections of capital were necessary ($40 million before any profit returned; Hounshell and Smith 1988: 96), much of this to organise dyestuff-specific R&D departments. DuPont research could not recreate the production know-how that German companies had built over decades and success required a formal know-how transfer agreement with a British dye company and the illegal seduction (through salary offers) of German chemists to work for DuPont.

The scale of the effort was only possible because of DuPont's wartime super profits and when the company stopped losing money on dyestuffs in 1923, it was because by 1921 it had successfully lobbied Congress to impose a tariff on returning German dye imports (Hounshell and Smith 1988: 95).

It is very important for any judgement of the later research strategy involving 'academic' research that it is understood that DuPont's rise to a polar position in the chemical industry was attained through other means; primarily through a research-driven acquisition and technology transfer strategy. Yet even before the Second World War we see DuPont taking risks with major departures in research strategy and so learning the limitations to the corporate 'uses' of the research function.

'Innovation-led Growth' after the Second World War

Wartime conditions had promoted the diffusion of much of DuPont's proprietary technology and the company understood that competition would therefore increase in existing markets with a return to peace. It was also apparent that acquisition targets were becoming exhausted, but this former route to growth was anyway understood to be closed because of the renewal of antitrust action against the company.

If the 'pillars' of the pre-war growth strategy had fallen away, the company had acquired a model of what it wanted – the stunning commercial success of nylon, 'a paradigmatic invention for DuPont in the post war era' (Hounshell and Smith 1988: 317). The whole point of the post-war reorganisations of the research function was to produce more fundamental breakthroughs 'like nylon', and so recreate DuPont's proprietary technology advantage. How then had nylon been discovered within the older research structure?

The discovery of nylon was a result of a very limited experiment in the use of academic scientists. When DuPont adopted its divisionalised structure most central research department staff were allocated to the divisions and its staff fell from 300 to 21 (Hounshell and Smith 1988: 109). Central research was therefore marginal and might have been abolished (Hounshell and Smith 1988: 120) had not an exceptional research manager, Charles Stine, been appointed who sought and found a means of making central research contribute to the company's growth.[21]

It was Stine that gained reluctant Executive Committee approval for a small programme of fundamental research that *nevertheless related to the industrial divisions*. Stine selected the areas of chemistry he believed offered most prospect for returns and hired 25 of 'the best' academic chemists to run his programmes (Hounshell and Smith 1988: 229). The great breakthroughs of neoprene and nylon would come from the group run by W. H. Carothers working on the 'theory' of polymerisation that Stine had identified as potentially useful to the fibres department (Hounshell and Smith 1988: 135).

When the company needed a new strategy for post-war growth, nylon provided it with a model of the means, as well as the desired ends of the new strategy for achieving 'control' of scientific breakthroughs – an increase in the employment of academic scientists to do 'fundamental research'. When, in the 1950s, evidence began to mount that the company's competitive position was continuing to erode, it launched its 'New Ventures' programme and

> raised the ante one more time and assigned their research divisions to develop products of an unprecedented degree of technical complexity for special high-priced uses. DuPont continued to see its technological capability as its major advantage over the competition. (Hounshell and Smith 1988: 500)

This heroic phase of DuPont's research strategy represented an attempt to escalate the R&D arms race with its rivals to such a degree that they could neither copy the strategy, nor close the proprietary technology gap. What was the result?

The Experience of Maturity

While research costs stayed constant into the 1960s at about three years and $700 000 per project, development and initial commercialisation losses rose from 4% of earnings in 1955 to 25% in 1967 (Hounshell and Smith 1988: 534). An internal analysis listing the source of DuPont's earnings in 1972 found that nylon continued to be DuPont's biggest earner (Hounshell and Smith 1988: 577).

However, this was 'failure' only relative to DuPont's ambitious yardsticks. 'Venture worth' of commercialised products was calculated by charging pre-commercial expenses against profits in the first 10 years, discounted at some set interest rate. Although only three of the products introduced between 1952 and 1967 had positive net worth at an interest rate of 10% (Hounshell and Smith 1988: 533), that does not mean they were not 'successful' and 'useful' products over longer timescales and lower 'imposed' interest rates. So Kevlar, discovered in 1964, took a record-breaking 15 years of technology development before commercialisation and had cost $500 million by 1982 (Hounshell and Smith 1988: 431). Whereas nylon had been cost-justified on the high-value substitution market for silk stockings alone, Kevlar was developed 'without assurance that any single market would sustain the venture' (Hounshell and Smith 1988: 431), prompting *Fortune* magazine to comment that the fibre was a 'miracle in search of a market' (Hounshell and Smith 1988: 431). Rather, it was a miracle that found many markets; as a 'portfolio' of niche markets was developed over time the product became a long-run success in defiance of standard ROI valuations; gloves, bullet-proof vests, brake pads and fibre optic cables are some of the diverse applications today.

Another blow to the 'innovation-led growth' effort was that research-active competitors proved able to follow similar research routes to DuPont, either simultaneously or in imitation, and this destroyed the carefully constructed proprietary position on which returns to high development costs were predicated. High-density or 'linear' polyethylene and polypropylene were discovered as a result of the use of radical new catalysts – but these were discovered almost simultaneously by DuPont and Karl Ziegler, a Max-Planck Institute director in Germany (Hounshell and Smith 1988: 493). Ziegler's patents were registered slightly earlier and he licensed 'generously', so prompting many new manufacturing entrants – even a shipping line seeking to 'diversify' entered the market.

The result was that supply grew even faster than the rapidly growing demand and nobody could take a profit – DuPont had registered a $20 million cumulative loss by 1975 from 15 years of production (Hounshell and Smith 1988: 495). In the case of polypropylene five companies claimed patent priority on the 1954 catalysts and it took 30 years to settle the ensuing patent war, or 'interference' proceedings (Hounshell and Smith 1988: 496). While the patent rights were at issue, it was clearly not in anyone's interest to develop the production technology.

One could continue with examples, but the essential point is that even when DuPont had a good product – and these were very good products – it could not be sure of reaping what it considered an adequate return. These were the problems of approaching technological 'maturity', but there was no signpost to signal when 'maturity' had arrived. That was a matter of judging both that new product opportunities were 'relatively' exhausted and that an equality of technological developmental capability had been achieved between DuPont and its rivals. In other words, expectations of future proprietary control and financial returns to R&D would be lowered. Together with other development experiences at this time, Kevlar, high-density polyethylene and polypropylene provided cumulative blows that gradually exhausted the DuPont Executive Committee's commitment to the 'innovation-led growth' strategy.

Increased Management Control as a Response to Relative Decline

By 1969 the Executive Committee 'considered the entire effort a failure' (Hounshell and Smith 1988: 504). Only now and *as a response* to the decline in returns did the Executive Committee abandon its reactive role of vetting project suggestions filtering up from the R&D departments, and take responsibility for the direction of research into new fields.[22] Fundamental research was cut everywhere and absolute R&D levels held constant so that expenditure per dollar of sales fell from 6.6% to 3.3% by the end of the 1970s (Hounshell and Smith 1988: 573).

This was no comment on the *quality* of the fundamental research – DuPont scientists would eventually win Nobel Prizes for some of the work performed at the central laboratory and by the 1960s this laboratory had become

> one of the outstanding industrial basic science laboratories in the US ... once the drift had begun it proved nearly impossible to contain or control. DuPont's top management tolerated and even encouraged this type of research because, as one research administrator put it, they had an almost 'mystical belief' that a 'new nylon' would be discovered. (Hounshell and Smith 1988: 376)

The historical sequence of events is significant – it would be an easy mistake to think that the existence of a formal control system implied an *ability* to control, in the important sense of being able to preselect the winning over the losing projects for development. Budgetary constraints reduced the finance available to play the research game, but they did not mean that project selection could be managed better. The significant phrase in the above quote on fundamental research is 'nearly impossible to contain or control'. Reduction of this kind of research was managed by requiring the central laboratory to move to be 50% funded by the industrial departments, as it had been in the early 1950s, rather than the 10% typical of the early 1970s (Hounshell and Smith 1988: 583). The problem of how to deselect fundamental research projects was therefore devolved to the laboratory itself.

Despite the restrictions on absolute R&D spending, DuPont remained the chemical industry leader because other chemical firms also reduced their R&D growth. In other words, the industry managed a collective 'de-escalation' of the competitive R&D arms race in response to the growing evidence of changing returns on R&D expenditure.

The story of DuPont provides a strong warning not to take measures of current R&D activity in any company or industry as evidence of some calculated, knowable and optimum level. The period of escalated expenditure represented a strategic choice based on an understanding of the post-war competitive context and a 'hope' that the example of nylon could be reproduced. Although this case is certainly exceptional because DuPont was an industry leader, a very rich company and had an unusual degree of discretion to raise R&D expenditure, the general warning is that *actual* levels of R&D expenditure are likely to contain a strategic, variable component related to future expectations of returns – and those future expectations will be based on an understanding of the past and current company and industry circumstances in a non-predictable manner.

If Hounshell and Smith's work finishes in the late 1980s with the advent of maturity in the established businesses and the beginning of an effort to move into the biological sciences, the newspapers provide some clues to later developments. Most relevant to an understanding of the evolution of maturity was DuPont's announcement in 2002 of the divestment of its fibre and textile division. It seems reasonable to take this date as symbolic of the end of the struggle to maintain innovative gains and innovation-derived profits for this former jewel within the corporation – a struggle that can be assumed to have continued for much of the preceding decade and a half (BASF FAZ Institut 2002).

Rise of the Organisational Innovation of the R&D Department

Comparisons between DuPont and Research-for-invention in the German Dye Firms

Elements of DuPont's experience of the returns to its R&D organisation and expenditure can be couched in terms of the relationship between science and technology. The discovery and development of nylon did derive from academic science, but this polymer science had been selected and brought inside DuPont for its promise in terms of useful new technology. Other science projects so selected did not deliver. DuPont's experiment in the support of academic science can be seen as the attempt to systematise the relationship between the development of scientific understanding and technological development opportunities in this industrial field. Its end can be seen as a measure of the erratic and unsystematic nature of the relationship. The management problem should not be seen as wholly one of the internal organisation of research-for-invention, but as part of the larger problem of how to relate industrial research activities to the publicly funded and separate institutions of the world of science. DuPont was only able to make the expensive incursion into academic science because of its leading position in the technology. Had DuPont not experimented with polymer science, nylon would surely have been invented one day by researchers elsewhere, perhaps in the universities.

Historical studies of the pioneering R&D departments of the German dye manufacturing firms in the second half of the nineteenth century confirm the essential pattern of development of industrial research shown by the example of DuPont. A brief comparison shows that first, like DuPont, firms such as Bayer only entered the uncertain field of research when they had already achieved a degree of market control and were financially secure (Meyer-Thurow 1982). Second, research-for-development, not research-for-invention, was an early priority in these firms as it was in DuPont. Chemists were first employed as either works chemists, to supervise production departments, or 'laboratory' chemists: the latter were most often lowly chemical analysts performing tests on materials entering and leaving stages of the production process (Homburg 1992). Once the research institutions had matured in form after 1890, 'a maximum of 20% of all employed chemists were working in the centralized research institution [IG Farben's main laboratories] with the other 80% in other laboratories, technical departments, or in the large sales agencies' (Marsch 1994: 30). This distribution of research activity persisted when these firms merged to form IG Farben, which by 1926–7 had 1000 research scientists working in 50 laboratories including 10 main laboratories inherited from the merging firms (Marsch 1994).

A difference between the German firms and DuPont lies in the nature of the first invention search tasks that these firms encountered. As described above, DuPont allowed a limited experiment with the import of selected academic research work into a central laboratory: the enormous commercial returns on the resultant discoveries of nylon and neoprene encouraged the company to expand academic research for invention. In contrast, the German dye firms needed to conduct a systematic search for the few commercially valuable azo-dye compounds amongst the hundred million possible products of the diazo coupling reaction (Meyer-Thurow 1982: 378). It was the nature of the work inherent in this search task that first brought forth the response of the centralised department for research-for-invention. Once the range of research organisations had been established and had proved their economic worth, the German dye companies extended inventive research into what appeared to be promising new areas such as pharmaceuticals, nitrogen, fertilisers and synthetic fibres. IG Farben never embarked on anything like the academic science experiment of DuPont in the 1950s: the search work of IG Farben's main laboratories remained orientated around their separate product areas. They were also physically located near the technical development laboratories and production sites to which they related (Marsch 1994: 56). However, according to an IG Farben director, between 5% and 10% of main laboratory research was what we might call 'basic' or 'fundamental' research unrelated to products and this element of their work is closer in content to DuPont's academic science experiment (Marsch 1994: 43).

So in both DuPont and IG Farben research-for-development was closely linked both physically and organisationally with production. In both, research-for-invention was organisationally separated and insulated from the immediate demands of production, but to different degrees: IG Farben used both product–area constraints and a greater degree of decentralisation as a means of control-ling research-for-invention. It would be unwise to attempt to make a strong judgement of the relative success of the different strategies of control of research-for-invention given the great differences between the firms – IG Farben was the gigantic, established firm that engaged in all forms of chemical research many decades before DuPont. All that can be said here is that difficult as it is to manage this form of research to yield useful discoveries, there are nevertheless disciplines that can be imposed upon it.

If we have a tendency to privilege the importance of research-for-invention over what appear to be the more mundane production technology development activities, it is probably because we know of such spectacular examples of invention as the laser and penicillin and because we understand that in the dye business, the development activities could not have continued for long without success in invention. In response it should be said, first, that in contrast to the idea of creative heroism attached to the penicillin and laser inventions, the exploitation of the diazo coupling reaction was tedious 'scientific mass work' that demanded no creative ability from the scientist and that has its modern parallel in the tedious gene sequencing work that was necessary to complete a scientific description of the human genome. Second, that without organised development capability these firms would have been unable to exploit the commercial value of their inventions. Marsch makes the pertinent comment on IG Farben that 'No clear borderline was drawn between research and develop-ment, neither by the management nor by the scientists themselves. Development was seen as part of the scientific process' (Marsch 1994: 46).

A similar argument applies to the precedence relationship between such broad and overlapping terms as 'science' and 'technology': one should be wary of claims of any specific and general relationship between them. In particular, one should be wary about privileging the status, or granting *general* precedence, of one over the other in the innovation process. To return to another problem with our 'great invention' stories such as that of the laser, penicillin or nylon, it is perhaps because the inventors were scientists that the stories can be taken to suggest the general priority of science over technology, at least for invention.[23] 'Science' as a search for understanding is less likely to be privileged over technology in these stories if proper attention is paid to the 'technologies' of science itself, to the role of scientific instruments and experimental technique: the scientist is also a kind of technologist, most especially when engaged in invention. The priority relationship between science and technology may also be perceived to operate in reverse; working technologies have often inspired scientific research for the sake of understanding. For example, it was not under-stood why laser light was as coherent as it proved to be, and 'coherence theory' was developed years after working lasers had been established (Bromberg 1991).

A scientific field such as metallurgy, now 'materials science', grew out of the study of the properties of alloys and metals that were discovered through trial and error in practice (Mowery and Rosenberg 1989: 32). So industrial research conducted with the object of gaining understanding can generate significant practical opportunities and practical development can generate significant new scientific opportunities. However, as Rosenberg has pointed out, the USA's National Science Foundation (NSF) collects industrial research data under the categories 'basic' and 'applied', with 'basic' research defined as research conducted without commercial objectives and 'applied' research defined as research intended to meet a 'recognised and specific need' (Rosenberg, 1990: 170).

Given what we know about the variable and unplanned outcomes of research with these original management motives, Rosenberg finds their use as a means of defining distinct 'types' of research to 'less than useful' (Rosenberg, 1990: 171).

The Economy-wide Rise of In-house Industrial R&D

The history of DuPont's R&D facilities is an important example of a general trend that begins at the end of the nineteenth century for manufacturing companies in developed countries to found R&D departments. They spread rapidly through US industry in the early twentieth century and typified the high-growth industries of chemicals, electrical and aircraft manufacturing, later of electronics and pharmaceuticals (Edgerton 1996: 34). However, by absolute volume of expenditure, R&D was always more highly concentrated than production, being dominated by a few large leading firms in each sector (Edgerton 1996: 34). Most, perhaps two-thirds of the 'work' of these departments, was development (Mowery and Rosenberg 1989: 57).

From the beginning, industrial research was not only organised in-house, but also provided by free-standing laboratories on a contractual basis. However, in the USA the percentage of scientific professionals employed 'out-of-house' in this way fell from 15% in 1921 to 7% in 1946 (Mowery and Rosenberg 1989: 83). Mowery and Rosenberg reviewed the relationship of the two types of research organisation in the USA in the pre-war period and concluded that:

> Rather than functioning as substitute, the independent and in-house research laboratories were complements during this period and performed different research tasks. ... The foundation of an in-house laboratory resulted in ... a substantial expansion in the range of research possibilities and projects open to the firm. The growth of industrial research within US manufacturing reflected the shortcomings of market institutions as mechanisms for the conduct and distribution of research and development. (Mowery and Rosenberg 1989: 91)

As DuPont found from experience, contract R&D was even more difficult to control than in-house R&D. Independent laboratory owners could not always be trusted in their evaluations of research prospects – they sought to benefit themselves. In contrast, the in-house organisation of R&D activity encourages the generation and solution of problems likely to be useful to the firm. It aids the coordination of the project development that will necessarily involve most of the other functions of the firm. And the secure employment conditions that in-house organisation makes possible help secure the loyalty of the scientific professionals against temptations to defect and transfer their ideas to rivals, or to become independent developers themselves (Mowery and Rosenberg 1989: 108). For all these reasons in-house organisation of R&D remains an important way of organising innovative activity today.

A recent study of trends in the organisational location of R&D has been framed as a response to the occasional claims in the 1990s that R&D would show a tendency to become more decentralised in the 'post-bureaucratic firm'[24] (Hill et al. 2000). The authors were unable to find a clear trend towards decentralisation of R&D within their sample of mechanical engineering and food and drink firms. Instead there was a bewildering pattern of change: some firms had historically had centralised control over R&D, others decentralised control to operating divisions; some firms were busy decentralising R&D, others were centralising control over R&D. Two observations from the historical studies are relevant here. First, one is reminded of the many changes in emphasis between central and decentralised R&D organisation in DuPont as the company's research strategy evolved; second and related to the first, that in general a shift in research strategy would imply some shift in emphasis between centralised and decentralised R&D organisation because R&D consists of many types of research and some types, such as research for development, are suited to a decentralised organisational location. The degree of internal centralisation of R&D is probably not so significant in itself, but something that changes through time with varying emphasis on particular R&D projects. These authors preferred quite rightly to stress that where R&D was thought to be of strategic importance to the firm the management issue was not centralisation versus decentralisation in itself: it was how to achieve good integration between R&D and other functions – something that remains difficult to achieve (see Chapter 7).

The Interpretation of Aggregate R&D Statistics

Certain forms of aggregate data on R&D expenditure receive a high public profile through their compilation and diffusion by government. In Britain there is an annual 'R&D Scoreboard' compiled from the R&D expenditure reported in public companies' annual reports (DTI 'Future and Innovation Unit' and Company Reporting Ltd 2001).

These claim to support several generalisations. So 'sustained high R&D intensity' (R&D per unit of sales revenue) is positively correlated with company performance (measured for example by sales growth or productivity) (DTI 'Future and Innovation Unit' and Company Reporting Ltd 2001: 12). Much of this correlation can be associated with which business sector a firm occupies and the consequently varying innovation opportunities (the highest R&D intensity sectors are pharmaceuticals, health, IT hardware, software and IT services, and aerospace (DTI 'Future and Innovation Unit' and Company Reporting Ltd 2001: 4)).

Some stark international differences are also apparent, although their significance is less so. For example, the scoreboard analysis states that the 'UK pharmaceuticals sector is above best world levels in intensity but the UK average R&D intensity, excluding pharmaceuticals and oils, is significantly less than 50% of the US, Japanese or international averages' (DTI 'Future and Innovation Unit' and Company Reporting Ltd 2001: 3). This does raise the question of why the differences exist and the temptation for the scoreboard authors to offer policy advice proves irresistible:

> It is crucial for companies to benchmark their R&D against best international practice in their sector and to understand the ways in which their R&D investment will affect future business performance. (DTI 'Future and Innovation Unit' and Company Reporting Ltd 2001: 1)

At one level this appears unexceptionable – who could disagree with a maxim that urges greater awareness of best practice? Yet it is also an interpretation of the data that suggests managerial ignorance as the problem; if only low R&D intensity British firms realised what others were doing, and if only they understood the benefits of greater R&D investment, they would spend more. It is implicit in the above comment that more spending on R&D is a 'good thing'.

But perhaps low R&D intensity firms are already aware of their best practice competitors and are seeking to close an economic performance gap, or at least to maintain their trailing position, through technology transfer strategies that require less intensive R&D expenditure – rather as DuPont's rivals were forced to do for many decades. Or perhaps the national institutional environments are very different and help or hinder in different ways in different countries. For example, if one were to begin to investigate the British relative success in pharmaceuticals, it would surely be relevant to consider the effect on private activity of, first, Britain's possession of the Wellcome Trust, the 'world's largest medical research charity'[25] and second, the British higher education system's tendency to excel in the production of pure scientists.

The point here is that such R&D statistics reveal gross patterns, but in themselves cannot provide explanations. The imported explanation lurking behind the quote above is the assumption that 'more industrial R&D would produce better performance', but it is also possible that high R&D intensity is the result of past performance success.

The historian David Edgerton has produced an invaluable historical review of the various forms of the argument that insufficient science and technology explain the relative British industrial decline from the late nineteenth century to the present (Edgerton and Horrocks 1994; Edgerton 1996). His work forms a more secure basis for drawing conclusions about the meaning of relative international expenditures on R&D.

At the national level there appears to be no correlation between rates of civil R&D expenditure and rates of economic growth. For example, in 1963, Japan spent 1% of GNP on R&D for 8.3% economic growth, Britain 1.2% of GNP on R&D for 2.5% growth, Germany 0.9% GNP on R&D for 4.1% growth (Edgerton 1996: 57). There is, however, a strong correlation between GDP per capita and rates of civil R&D expenditure and rates of patenting.[26] So the pattern is that high rates of economic growth with low R&D intensity are associated with relatively poor countries (in terms of GDP per capita) that are rapidly catching up with the relatively rich. The classic recent example is Japan, whose post-war history is one of organised technology transfer into the country and associated high rates of economic growth, slowing as the output per capita gap with the leading rich countries closed. Japanese R&D expenditure accelerated as the output gap disappeared and as the country entered a decade of relative economic stagnation in the 1990s. High national civil R&D expenditure is therefore associated with being rich, the exhaustion of major technology transfer possibilities from other countries, and therefore the need for a more vigorous search for genuinely novel development opportunities through R&D.

Britain provides an apparent exception to this pattern. Post-war[27] private British industrial spending on absolute amounts of industrial R&D was second only to the USA, even if British government civil R&D is excluded to remove the effect of government-funded prestige civil R&D projects such as Concorde (Edgerton 1993: 41). British industrial R&D stagnated from the mid-1960s and research intensity actually fell in the 1970s, so that despite some recovery in the 1980s, research intensity at the end of that decade had only returned to the level of 30 years earlier:

> There has been a catastrophic relative decline in British industrial R&D over the last twenty or so years. (Edgerton 1993: 42)

This relative decline occurred as German, Japanese and later French and Italian industry surpassed British industry levels of output per head with lower levels of R&D spending. The post-war higher British national industrial R&D spending did not translate into higher economic growth and Edgerton suggests that

> it could be argued that in the 1970s British industry scaled down the R&D input to a level where it could be translated into new products and processes, given levels of investment and profitability. (Edgerton 1993: 49)

Industrial R&D is an expensive overhead and 'more' is not always 'better'. It is only one input into economic growth, and one that is relatively more important with increasing levels of relative economic wealth. The British example is consistent with this analysis.

Our reaction should neither be that Britain has now 'insufficient national R&D spending' nor alarm at the relative decline in British R&D expenditure. Such a decline is to be expected in a country that no longer leads the tables of output per capita. It would be more pertinent to examine how British industries have adapted better to manage technology and its transfer from best practice overseas companies.

Concluding Comments

The main topic of this chapter was the way technology may be conceived and changed in relation to prospective uses and markets. When innovation is the issue, it makes no sense to divorce the market concept from technology development. It is rather how the market concept is articulated and related to technology development that matters. With this in mind it was argued that the distinction between a reference market and an innovation market concept has some utility in modelling the creative relationship between existing technologies and markets and prospective technologies and their uses. The creative or inventive step obviously varies in its significance case by case, but even in the examples of the laser and penicillin, characterised by a breakthrough step, there were many other, subsequent creative contributions to development.

The popular technology-push versus market-pull metaphor is downright misleading as a means of understanding this process and if a shorthand term must be used, invention and innovation are better understood as socio-cognitive processes.

RCA as an exemplar of poor innovation leadership practice showed that as a significant organisational and economic venture, development needs senior managers committed to understand and interrogate the choices of the R&D function on behalf of the firm's long-term interests.

The experience of DuPont was that one could not scale up organised academic-oriented R&D and expect a proportional return in breakthrough innovations. In other words, companies cannot expect to 'manage' major opportunities into existence, although they can excel at exploiting the ones they find.

At a macro level, the interpretation of aggregate, national, civil R&D expenditure is that it rises with wealth measured relative to other countries and as a means of further increasing that wealth. Technology transfer from leading countries is a more important issue for poorer countries that wish to increase their wealth to match the richer countries.

In the next chapter other macro-level patterns in technology development come under scrutiny.

Notes

[1] I have taken Bromberg's history of the laser and the inventor Charles Townes' own account as my two major sources for the laser sections (Bromberg 1991; Townes 1999).

[2] Stimulated emission occurs when radiation of some wavelength is incident upon an atom, molecule or electron in an excited energy level of the same energy as the incident radiation. The atom, molecule or electron is 'stimulated' to drop into a lower energy state and at once emit a radiation wave packet (photon) of the same energy (and wavelength) as the incoming photon. So there would be amplification of the incoming radiation. This could not normally be expected to be useful because in a material in *thermal equilibrium*, there would be a characteristic distribution of energy states, ranging from low (absorptive) to high (capable of stimulated emission). Incoming radiation was as likely to be absorbed as to stimulate emission and what radiation was produced by stimulated emission was also likely to be reabsorbed by the many low-energy levels.

[3] Abnormal physical circumstances here mean that the population state of the energy levels of matter were not in 'thermal equilibrium' and instead an unstable 'inverted population' of these energy levels existed.

[4] A full description of stimulated emission, thermal equilibrium, population inversion and energy levels can be found in physics undergraduate texts. There is of course more to the 'laser effect' than in this simple description; Hecht and Zajac write of stimulated emission that 'A remarkable feature of this process is that the emitted photon is in phase with, has the polarization of, and

propagates in the same direction as, the stimulating wave' (Hecht and Zajac 1974: 481). In other words, stimulated emission radiation is 'coherent'. Bromberg comments that it was not understood *why* laser light was so many orders of magnitude more coherent than other forms of light at the time of the laser's invention, and that 'Coherence theory, even when it became elucidated in the course of the 1960s, held low priority for many laser workers. It was not needed to make lasers work, and it was hard to master' (Bromberg 1991: 109).

[5] This account is of course much reduced – see Townes on the steps and circumstance of the invention of the maser, especially pages 55–68.

[6] The maser was the basis for sensitive receivers that for the first time allowed ground to geosynchronous orbit communication – they therefore helped to make working satellites possible (Bromberg 1991: 56).

[7] Whereas it is difficult to find an American or British person who has not heard some version of the penicillin story, it is difficult to find a German or Dane who has heard the story – at least, if my students are any guide.

[8] Many accounts of the discovery of penicillin have been written, but two stand out for their painstaking method of reconstructing the inventive process. Ronald Hare, once a colleague of Fleming, reproduced the physical conditions that would generate the famous mouldy plate and Gwyn MacFarlane reconstructed Fleming's thought primarily through the available written evidence.

[9] MacFarlane finds that the penicillin concentrate that Fleming's two assistants prepared was certainly strong and stable enough to allow Fleming to perform the animal protection experiments with success had he been motivated to try them (MacFarlane 1984: 175). This matters because after the successful development of penicillin by Florey's team, Fleming 'took it upon himself to complicate the matter very badly ... he used such expressions as "even skilled chemists had failed in their attempts to concentrate, stabilise and purify penicillin (Hare 1970: 102). This was not the case."' Fleming had two assistants, Ridley and Craddock, who worked on concentrating the 'mould juice' and MacFarlane comments that their methods were essentially the same as those independently developed by the Oxford group and two other independent researchers on penicillin. All these groups worked with Fleming's mould and so the difference between them was whether they could conceive of the animal protection experiment and then be motivated to perform it.

[10] Eleven years after Fleming's discovery, Florey's team should have had the advantage over Fleming that by then the sulphanomides, the first great group of antibiotics, had been discovered by Domagk working for Bayer in 1935 (Hare 1970: 148). The discovery had been made because Domagk had used live animal experiments and he reported the then-surprising finding that the dye Prontosil was lethal to bacteria in the body, *but harmless to them in culture*. Although the precise mode of action would be worked out later, it was now clear that one could *not* extrapolate from antiseptics and experiments outside the body to draw conclusions about the antibiotic properties of drugs. However, Clark's investigation of Florey and Chain's motivation for beginning their 1939 investigation into penicillin suggests Florey saw it as a way of attracting long-term private sector funding (Clark 1985: 38) and Chain was motivated by an interesting biochemical problem – not by its prospective use as a drug.

[11] See Howells (1994) for a full analysis.

[12] The source is Howells (1994) which is based on anonymised interviews with managers of bioprotein projects.

[13] Raghu Garud and myself seem to have independently coined the term for similar reasons in the field of management studies. Some years ago I searched for the term in the Proquest database and found six apparently independent uses of the term in different subjects.

[14] A full quarter of this book is dedicated to a thick description of the *Helicobacter* story, organised by the categories of discovery, acceptance, experiments and instruments and social interactions.

[15] A colleague uses barbed wire as a particularly clear demonstration of the concept of 'market pull'.

[16] I have relied upon Hayter and Basalla's derivative account for the detail of the barbed wire case.

[17] Although even here Texas Instruments' main product is not stated, again as if the market orientation floats free of the technological context.

[18] The book itself was a major project taking almost 10 years. There were more than 40 interviews, access was gained to all relevant company documents and Graham's interpretation is based on a factual narrative agreed with the RCA participants (Graham 1986: xiii).

[19] The single source for this account is Graham's work (Graham 1986).

[20] The editors write in their preface that 'heretofore no historian has written a full length, intensive study of the origins and evolution of the R&D establishment in one of the nation's leading, high-technology firms' (Hounshell and Smith 1988).

[21] Hounshell and Smith comment on the 'decision' of whether to site research in a central or devolved department that 'These issues had changed little between 1904 and 1921. They have not changed much since then either' (Hounshell and Smith 1988: 595). The tension that generated many changes was that between a need to make research relevant to current problems by siting it in the divisions and the difficulty of conducting any research that departed from divisional imperatives in the divisions. As overall strategy emphasised the one or the other so the role of the central department waxed and waned. It is difficult to beat the company's experience that the role of central research must be continually revised with the understanding of the state of technological potential and strategic position.

[22] The first major act was to begin a coordinated move into the life sciences in the 1980s.

[23] In saying this I am expressing a belief that 'other people' might believe in the infamous 'linear model' of innovation, a key component of which is that basic research precedes applied research and development and innovation. In a recent article Edgerton argues that the linear model's principal function is as an academic 'straw man' with which innovation analysts can disagree before setting forth their own ideas on innovation – his exhaustive search for those who purportedly believe in this model turns up no clear believer with a detailed catalogue of their belief (Edgerton 2004).

[24] The Hill et al. research takes particular issue with Whittington's (1990) thesis that R&D is moving from centralisation to fragmentation (Whittington 1990).

[25] The description is from the trust's website www.wellcome.ac.uk/.

[26] Result from Faberberg (1987) cited in Edgerton (1996: 57).

[27] British private sector R&D in the interwar period was also very healthy – however, in this period firms were so secretive about the amounts spent that to many contemporaries it appeared that the comparatively small amounts spent by government were more significant. In Sanderson's paper presenting figures for private R&D expenditure in this period he describes the consequent 'distorted picture among historians of the interwar years that exaggerates the role of bodies whose contribution to industrial research, though honourable, was quite marginal to the total activity and which virtually ignores, belittles or even slanders the vast bulk of industrial research which was carried out privately within the firm' (Sanderson 1972). The prize-winning *Economic History Review* article by Edgerton and Horrocks is able to revise upward the amount known to have been spent by interwar British firms on in-house R&D – the important kind (Edgerton and Horrocks 1994).

This situation has important consequences because those who have taken interwar historians' beliefs about British R&D at face value have tended to assume that interwar British R&D expenditure was in some way inadequate and therefore an explanation of British relative economic decline. Sanderson's article has not stopped authors such as Barnett, Mowery and Chandler from making this argument (Edgerton and Horrocks 1994: 215). Edgerton and Horrocks' review of the evidence and its deficiencies makes a convincing rebuttal of the argument of R&D insufficiency pre-1945.

3 Patterns in Technological Development and the Creation of Technological Standards

Paths of Development, Increasing Returns to Scale and Dominant Designs

When a long time span of technology development is reviewed it reveals what appears to be a significant pattern in that development. The micro political and social processes that were the subject of the last chapter disappear and it is the more durable pattern created by a succession of artefact changes that is visible.

The basic pattern is exemplified by the complex assembled product technologies of the aircraft and car industries. In these, a period of design experimentation and great design variability characterised the beginning of both industries, but this was followed by a reduction in design variety and then a period of decreasing unit costs for this relatively stable design.

Behind the pursuit of the path of technology development in aircraft, electricity supply, mass production in the car industry and continuous flow processes in the chemical and oil processing industries is this objective of achieving increasing economic returns to scale in production – economies of scale. So for some increase in designed plant capacity, if inputs rise more slowly than the outputs an increasing return to scale has been achieved. Of course the achievement of decreased unit costs with increased scale of production creates the possibility of extending the market further through a lower price – the process is a form of positive feedback. However, in practice, diseconomies of scale are eventually encountered that deter further escalation of scale and effectively establish a consensus on 'normal' plant size for a technology.

The observation of the positive feedback process of 'increasing returns' to scale is an old one; Adam Smith discussed the achievement of increasing returns with the progression of the division of labour and with the increasing extent of the market on the first page of his *Wealth of Nations* (Smith 1986: 109). The economist Kenneth Arrow describes the concept of increasing returns as having had 'a long but uneasy presence in economic analysis' (Arrow 1994: ix) but one that in recent times has attracted great interest as an explanation of economic growth. In recent years the economist Brian Arthur has extensively mathematically-modelled the diverse scenarios where increasing returns may prove important (Arthur 1994).

The question is whether we can claim more than that the achievement of economies of scale is a common objective in the management of technology. And before this question is the question of what sense can be made of long-run patterns, or paths of artefact change alone.

Generalising from Patterns of Artefact Design Evolution

MacKenzie has argued (MacKenzie 1992) that some of the alternative economics writing can convey the sense that the existence of paths gives a strong and clear direction to future design, and the following serves as an example:

> A firm's core business...stems from the underlying natural trajectory embedded in the firm's knowledge base. (Teece 1988: 264)

'Natural' usually implies something not controlled by humans, some order not subject to design. 'Trajectory' is a word often used to describe the predictable path of a thrown projectile under the influence of gravity, so 'natural trajectory' can be taken to suggest movement along a preordained path not subject to human design. A practical approach would be to ask what role the existence of such paths plays in design.

The management literature also refers to patterns in *artefact* design change through time (Utterback 1974; Anderson and Tushman 1990); Anderson and Tushman strongly distinguish between an 'era of ferment' in design and the period when a 'dominant design' emerges, so:

> In the early automobile and airplane industries, technological variation between fundamentally different product designs...remained high until industry standards emerged to usher in periods of incremental change elaborating the standards. (Anderson and Tushman 1990: 613)

> A dominant design is the second watershed event in a technology cycle, marking the end of the era of ferment. (Anderson and Tushman 1990: 613)

We have similar problems of interpretation as with 'natural trajectory'. First, 'fundamentally different product designs' could be taken to mean that the significant design variation is between entire products. This appears to be what Anderson and Tushman mean, for example:

> Whether in sewing machines or rifles ... bicycles ... synthetic dyes ... machine tools...reprographic machines or photolithography... *single designs emerge to dominate rival designs* [my italics]. (Anderson and Tushman 1990: 613)

If these are the claims, they deserve a closer look in the case of aircraft and the emergence of the DC-3 as a 'dominant design' – something of an exemplar for these writers (Nelson and Winter 1982; Nelson and Winter 1977; Anderson and Tushman 1990).

The Evolution of Aircraft Design and the DC-3 as a Dominant Design

According to Miller and Sawyer's historical account, the DC-3 was 'not a startling innovation; it was the logical development of an earlier model' (Miller and Sawyers 1968: 102). Had the DC-3 been an innovative synthesis of scattered design elements in many models one could associate its dominance with 'innovation'. However, it was a specific existing model – Boeing's early 1930s' model 247 – that triggered Douglas's commitment to build the DC series. *Aviation* magazine observed that it was the 247 that 'immediately outmoded all passenger airplanes in service' (Miller and Sawyers 1968: 99). The DC series simply copied the 247's all-metal, stressed skin monoplane structure, sound-proofed fuselage, cellular structure wings, retractable undercarriage, cowled engines and variable-pitch propellers (Miller and Sawyers 1968: 98).

New airframes tend to be designed around new model, more powerful air engines. Douglas produced just one 'DC-1' or 247-'clone' when it was decided to stretch the design and build it around a newly available engine to make the DC-2. The DC-3 was a further stretching of this design around another advance in aeroengines and in response to the president of American Airlines' interest in an aircraft that could provide an overnight US coast-to-coast sleeper service – a potentially lucrative target market (Miller and Sawyers 1968: 101).

The key design feature of the DC-3 was the widened fuselage and hence increased payload that gave the DC-3 seat-per-mile costs between one-third and one-half lower than those of rival passenger aircraft, while costs per air mile rose by only 10% (Miller and Sawyers 1968: 101). The concept of 'stretching' was not original, but in the DC-3 it finally raised payload capacity sufficiently that the introduction of the DC-3 into passenger service 'marked the end of profitless air line flying' (Miller and Sawyers 1968: 102). With profitability came the creation of an extended true market and being 'first' meant the DC-3 model became 'dominant'; by 1938, 95% of US commercial traffic was carried by the DC-3 (the total population was then small; by 1941 'dominance' was achieved with only 360 DC-3s in service) (Miller and Sawyers 1968: 102).

First, the 'era of ferment' appears to have been largely over with Boeing's 247. Yet the 247 is not referred to as a 'dominant design', as perhaps it should be. Rather than a clear *watershed* between the periods, we have a muddy transition where the later 'pattern' of innovation is already established; the 'stretching' of the stable design had already begun with the DC-2. No doubt all the design elements copied into the DC-3 were necessary to its success, but it was this otherwise unremarkable step of stretching the design that enabled the final breaching of the great lucrative market of passenger transport that dramatically changed the 'path' of development and rewarded the makers of the DC-3 rather than all the innovators that had gone before. If we want to think of the DC-3 rather than the 247 as 'the' dominant design it would be better to think of the DC-3 as 'made dominant' by its market success to avoid the misleading impression that it represented brilliant 'technical' advance.

It follows from this account that the 'degree' of dominance a design achieves in different technologies will have a context-dependent element. Again, the DC-3 is instructive; the high degree of DC-3 civil dominance was because it was 'first' into the large and profitable market segment of passenger transport. But its dominance was further consolidated because during the Second World War it was selected by the US military as the air transport of choice. Immediately post-war *4000* ex-military DC-3s were sold into commercial use (as stated above, 360 were in civil use in 1941) (Miller and Sawyers 1968: 124).

Ironically, the enormous stocks of spare parts and trained mechanics that the military also brought into existence ensured that refitted DC-3s were being sold long after the war, rather than arguably more advanced replacement models that included, for example, pressurised air cabins (Miller and Sawyers 1968: 126). In other words, the prolonged period of use of the DC-3, when it should have been in rapid decline, was an example of path dependency (see the section on Increasing Returns and 'Path Dependency'); it was the consequence of an extraordinary event in its history.

The problem is that if we seek to interpret the nature of design decisions from an oversimplified 'model path' of 'design ferment' leading to a dominant design, these decisions appear to be almost automatic. The observation that design variety is reduced over time in certain complex component technologies does not tell us how predictable and controllable was the process. The same applies to the observation of the achievement of increasing economies of scale. Nor is the general relationship between the two processes entirely clear: in the case of aircraft, design consolidation occurred prior to the achievement of significant reductions in unit carrying costs and was not a product of these reductions. This is not to dispute that design standardisation and economies of scale (reduced unit carrying costs or reduction in unit production costs) are desirable objectives in the management of technology, but it will prove worthwhile to outline some of the limits and potential obstructions to their achievement.

The Indeterminate Beginnings, Ends and Timing of Stages in a Path of Development

What Constitutes a 'Path'?

The detailed accounts of design contained in Graham's account of RCA's VideoDisc and Hounshell and Smith's account of DuPont's exploitation of chemical technology already provide a basis for tempering the ideas of paths of development and dominant designs as management guides. Managers in RCA were right to think that a consumer video product was a good idea, but that did not tell them which technologies to research and develop or when to do so. In this case, if we must refer to the magnetic tape video recorder as a dominant design, we must acknowledge that market participants, including the developers of magnetic tape technology VCRs, expected a disc technology to be the basis for a dominant design. That certain designs come to dominate – to represent – a product tells us nothing about how the elements of that design were selected and combined or how expectations of success interacted and were modified by market experience.

The VideoDisc case also provides insight into the problem of paths of development. RCA thought of itself as a 'systems innovator' because of the nature of its successes in radio and TV. Many in RCA believed that the company should pioneer and control both the software and hardware of VideoDisc technology in the same way as it had done with radio and TV. The selection of capacitance technology by the R&D department had offered the advantage that *if* achieved, it would have given RCA the desired proprietary control over the videoplayer industry.

So this approach to development of a technology was rooted in RCA's historical experience of paths of development, but in retrospect it contributed to RCA's failure. Such extinguished paths are easily forgotten and with them the fact that in practice alternative conceptions of the future path of development compete. The real virtues of one path over another are not necessarily obvious to the participants at the time.

Similar use could be made of DuPont's excursion into the funding of fundamental research. This was also judged a plausible path of development because of the successful experience with academic research and the discovery of nylon. Although this path yielded returns, it was judged a false path in terms of DuPont's chosen financial criteria.

Of course, these examples raise the issue of what exactly constitutes a 'path'. As in the above examples, I understand it as any recognised or imagined pattern in development that *comes to be used* to guide management decisions – whether the outcome is success or failure.

Indeterminacy in Paths of Artefact Design and Economies of Scale

One might object to the RCA systems-path that it was a different kind of path to that understood by Teece, and Anderson and Tushman. Even if we restrict ourselves to a more repetitive and stable artefact path, we find similar uncertainties in practice to what was found in RCA's case.

The apparent logic of a path appeared positively misleading to Hewlett Packard in the example of computer disc drive standards (Christensen 1997). What usually drove the decreasing size of disc drives was the decreasing size of computers. The minicomputer took 8 inch drives, the personal computer took 5.25 inch then 3.5 inch drives, later there were 2.5 inch drives and 1.8 inch drives. And so there was a regular path of design changes through time and this became the basis for Hewlett Packard's belief that when hand-held, palmtop computers were planned by some of the major computer manufacturers, there would once again be a jump in demand to a new standard disc drive, this time for 1.3 inches.

The demand for the palmtop computer, and therefore for 1.3 inch disc drives, fell far short of expectations and Hewlett Packard was left with excess capacity for an overspecified product. Because its 1.3 inch drive had shock sensors and other design features to suit its intended computer usage it was too expensive for the actual demand that emerged during its development, which was for use in electronic cameras, industrial scanners, miniature cash registers and Japanese language portable word processors. In retrospect, we can say that Hewlett Packard had been misled by a combination of anticipation of the continuation of the historic 'path' in disc drive design change[1] and the desire to capture potential economies of scale ahead of any competitors. The result was misplaced confidence in an excessively narrow conception of the future design-market match. Historic paths of design can mislead management.

Aircraft provide a last example. Both Lockheed and McDonnell Douglas were ruined by failed development gambles on new model civilian aircraft, leaving only Boeing and Airbus Industrie as major civilian producers. Yet at the time of writing, Airbus Industrie has committed itself to building the $12 billion 'super-jumbo' A380 aircraft, capable of carrying 500 to 600 passengers (Daniel 2003: 15). This represents a continuation of the decades-old path of stretching aircraft frames to obtain a lower cost per passenger mile, but the threat that diseconomies of scale will restrict volumes of traffic to below the break-even point for this scale of development are openly discussed in the press. For example, for such a large aircraft, passenger loading and unloading will take longer and it will need longer runways than most airports can currently offer, so it is expected that it will be restricted to long-distance intercontinental traffic. Airbus made the decision to develop the A380 in December 2000 during a period of financial euphoria (Daniel 2003). Although Airbus was careful to acquire provisional orders before committing itself to full development of the design, as boom has turned to recession some of the advance orders are being cancelled. To complicate matters, Boeing has announced that it does not believe there is a large enough market to justify the enormous development cost. So if the participants' public pronouncements are trusted, it is impossible for an observer to know whether the A380 will eventually succeed, or whether it represents a 'step too far' on the well-trodden path of aircraft design development.

A Proliferation of Paths of Development

Once the laser effect had been demonstrated research and development exploded and by 1962 Bromberg cites an estimate that 400 companies had some form of laser research with a dozen large programmes of development (Bromberg 1991: 63). Although there was a degree of hubris surrounding the laser in its early years, the fundamental reason for the large *number* of R&D efforts was the large number of distinct potential uses for the technology, each representing a potential path of development.

These ranged from the obvious and relatively immediate: for instance, xenon arc lamps had been used from the 1940s to provide an intense light to 'photocoagulate' detached retinas and the idea of substituting the more 'focusable' monochromatic light was an obvious one and soon achieved (Bromberg 1991: 130). For optical fibre uses, decades of development and further major 'breakthroughs' were needed in solid state lasers. The military funded early research into laser beam weapons, then lost interest as the obstacles appeared insuperable, then regained interest with the development of the high-power, continuous-wave carbon dioxide laser. Other uses were for battlefield range finders, welding, distance measuring, radar and of course spectroscopy.

The result of this incredible diversity of development paths was an early specialisation of firms into what we can call either laser component manufacturers or laser 'system-builders'. By 1987 there was a worldwide annual market for lasers of $600 million dollars, but an annual market for laser systems of $5 billion (Guenther et al. 1991).

The most striking feature of laser development over many decades is not a transition from design diversity to dominant design in a single path, but design *and use* diversity: the creation of multiple paths of varying fortunes. This distinctive pattern of development stands in contrast to car and aircraft technologies and suggests that they have to an extent been privileged in the literature, perhaps because of their genuine importance and high degree of visibility.[2]

Robotics technology provides an interesting complement to the laser. With the advent of cheaper IT in the 1980s, reprogrammable robots became possible, but there were different conceptions of how the technology would evolve. The engineering company Unimate introduced a general-purpose pick-and-place robot that could be programmed and reprogrammed to suit a range of contexts and that Unimate clearly intended to become the 'dominant design'. The concept was mass production for unit cost reductions with the reprogrammable ability allowing this standard unit to capture a broad range of uses (Fleck and White 1987).

The expected standardisation around one mass-produced design never happened. Instead, programmable robots became increasingly adapted to specific production environment uses. Fleck called this process 'innofusion' to capture the way that diffusion and innovation of robotics occurred together through time (Fleck 1987). Robots could be powerful giants for beer-cask manipulation, or built into CNC machine tools. In other words, significant development of the technology occurred as robots were implemented into working production systems and design variety increased through time. Although at least one company was prepared to gamble that a path of increasing returns had opened with the advent of programmable robotics, instead, and like the laser, there were increasing numbers of distinct paths of robotics development through time.

The example of robotics demonstrates the difficulty of estimating whether a path of increasing returns is about to appear; the example of the A380 and the miniaturising path in disc drives, whether an established path can be made to continue. If these examples show that it can be difficult to know how and when to constitute the next step in a path of increasing returns, they reinforce the conclusion that if anything in the field of the management of technology can be considered a general phenomenon, it is the *pursuit* of increasing returns to scale.

All of these examples exemplify the difficulty of constituting future *steps* in a path for existing organisations, technologies and markets. If a new technology requires a long period of cumulative learning before it can be economic, then there may be a problem of how to constitute the very path itself. The global success of the Danish windmill industry is principally a result of the way the Danish state and society enabled an alternative path of learning early in this technology's development.

The Danish Windmill Industry – Success through the Cultivation of Variety among Paths

The success of the Danish windmill industry is all to do with the early establishment of an alternative path of development to the predominant 'engineering science' approach to windmill development. The story reinforces the potential problems with the 'engineering science' approach to R&D as well as the important role of the state and subsidiary institutions in enabling the exploitation of alternative paths.

Windmill R&D lapsed in the 1950s but revived with the oil price shocks of the 1970s. After a decade of subsidised research, development and implementation in the 1980s:

> California produced scores of unsuccessful turbine designs, poorly performing turbines, and disastrous turbine failures.... Between 1975 and 1988 the United States government spent twenty times (and Germany five times) as much for wind power research and development as did Denmark, yet Danish manufacturers made better turbines – have, indeed, since the early 1980s been the most successful wind turbine producers. Danish wind turbines supplied about 45% of the total worldwide wind turbine capacity in 1990. Most US manufacturers failed in the 1980s and by 1990 only one major manufacturer of commercial turbines (US Windpower) remained. (Heymann 1998: 642)

The early Danish lead in this industry has been maintained; in 2002 Danish companies in this industry controlled half the world turnover of 6 billion euros (Krohn 2002).

Garud and Karnøe characterise the US path as a search for 'breakthrough' in contrast to the Danish path of 'bricolage', which they define as 'resourcefulness and improvisation on the part of involved actors' (Garud and Karnøe 2003: 281). In other words, the Danes employed a craft approach to development.

Those pursuing 'breakthrough' were aware of the Danish approach, so that in 1993 one of the US wind pioneers would look back in time and say:

> We trusted our engineering tools too much, and felt that they could solve the hard problems, and we simply didn't believe that the engineering problems were as hard as they were. We felt bright and able ... to solve anything. We thought in a typical American fashion that there would be inevitable breakthroughs that would make the "pedestrian" Danish approach obsolete overnight. (Stoddard quoted in Garud and Karnøe 2003: 282)

Most government-sponsored R&D programmes, including the Danish one, focused on the development of large-scale turbines capable of high-power outputs in the order of megawatts. This represented an attempt to scale up by a factor of 30 from proven working designs and Heymann points out that scale up by a factor of three is considered reasonable in other technological fields (Heymann 1998: 668). Such large turbines were 'unanimously considered feasible within a development time of a few years' (Heymann 1998: 660), yet they all failed.

The successful craft approach evolved without these direct R&D subsidies as 'enthusiastic amateurs and skilled artisans' (Heymann 1998: 661) began development based on the Danish engineer Johannes Juul's design principles, embodied in his three-blade prototype Gedser turbine.[3] In contrast, it was the two-blade prototype windmill advanced by the German engineer Ulrich Hutter that formed the basis for the engineering science path of development (Heymann 1998).

An important early contribution came from a carpenter, Christian Riisager, who used off-the-shelf parts and 'truck gears, axles and brakes. In spite of his limited theoretical background and experience, by 1976 Riisager had produced a surprisingly reliable 22 kW turbine with an attractive wind to electric energy conversion efficiency of 30% (Heymann 1998: 661). Many new entrants to this path of development were attracted by this relative design success.

It is ironic that 20 years of incremental development beginning with Juul's ideas have finally generated the working 2 MW machines originally envisaged by the 'one-step' breakthrough projects of the 1980s (Krohn 2002). The success of the Danish industrial craft-learning path was greatly aided by the policies of various industry-centred organisations and by the policies of the Danish state (quite apart from the Danish state's direct R&D programmes). These help explain why a similar 'alternative path' did not form elsewhere and can be discussed briefly under the following headings that reflect the structure of Garud and Karnøe's analysis.

The Constitution of the User and the Produce–User Relation

With the 'breakthrough' orientation prevalent in the USA, Garud and Karnøe suggest that engineers in US firms had poor interaction with users. A major reason was that the market concept in the USA was of a wind farm supplying power of the order of megawatts to an electric utility. This impoverished the potential for incremental learning from producer–user interaction over time (Garud and Karnøe 2003: 295).

In contrast the Danish pattern of a physically dispersed set of single installations was itself a manifestation of the large number of owner–users in Denmark: 80% of Danish-installed turbines were owned by farmers or wind energy cooperatives[4] (Krohn 1999). In the late 1970s the cooperatives were linked to a 'criterion of residence' (Tranæs 1997) that stipulated that members should live within 3 km (later extended) of their turbine so that together they would share both the inconvenience and the benefits of ownership. This criterion helped to tie the owner–users closely to the physical operation of their turbine, in contrast to the USA where ownership and usage were separated by the tax-credit nature of government subsidy; 'owners' would invest in a development venture primarily to obtain tax credits and had a weakened interest in selecting and improving windmill models by the installation developer (Garud and Karnøe 2003: 298).

Although supported by a state investment subsidy, Danish owners had their own money invested in windmills situated close to where they lived. They had a strong interest in the performance of their investments and early on they formed their own organisation[5] which in its monthly journal, *Naturlig Energi*, disseminated operational experience of *each separately installed turbine*. According to the chairman of the user organisation, Flemming Tranæs, this created a statistical series of data that greatly aided the functioning of the windmill market (Tranæs 1997). Besides the obvious advantages to producers and users, comparison of operating performance swiftly revealed the importance to economic operation of intelligent siting of even well-designed turbines. The ability to site turbines for maximum performance today constitutes one of the know-how advantages of the Danish industry.[6]

The Danish Windmill Users Organisation also vigorously lobbied the Danish parliament in its many debates prior to the passing or amendment of energy regulations, but it also negotiated grid connection and electricity purchase rates with the utilities on behalf of users. It also pursued through the courts the numerous utilities that found clever ways to renege on their collective agreements (Tranæs 1997).

This collective organisation was therefore effective in limiting the potential *disadvantages* that attach to a highly dispersed, fragmented ownership structure, while that same structure had the advantage that it allowed a gradual learning process based on incremental advance from existing technology.

The Different Assumed Roles of US and Danish Industry Testing Centres

A Danish Wind Turbine Test Centre (DWTS) was established at the Risø National Laboratory in 1978 and adopted the straightforward mission of aiding the advance of existing manufacturer practice. Its role in industry development was modified in 1979 when the Danish government required the testing centre to operate a compulsory licensing system for wind turbines to be eligible for a government investment subsidy (Garud and Karnøe 2003: 300), but, significantly, no absolute criteria were laid down for the good reason that it was unclear what these should be. Instead, the testing centre began with very loose criteria and evolved these as it interacted with the industry, and yet, 'a firm could deviate significantly from target standards ... [if it could] ... convince the DWTS that its wind turbine design would work. In the process, the standards themselves co-evolved' (Garud and Karnøe 2003: 301). In other words, innovative design departures were not blocked by rigid enforcement of standards, something that was itself only possible because the industry and test centre enjoyed a high degree of mutual trust.

The US National Renewable Energy Laboratory never enjoyed the close DWTS relationship with industry. Instead, 'The centre's organisation and set of skills exemplified its government-derived mission, to pursue basic engineering science research in order to establish the theoretical basis for the design of an "ideal" wind turbine ... [and] focused on inducing those in the industry to generate a revolutionary design' (Garud and Karnøe 2003: 302). By one measure, and despite the higher US funding levels, 12 of the 25 most important published turbine design papers came from the Danish test centre and 5 from the US laboratory (Garud and Karnøe 2003: 304).

The Adaptive Role of the Danish State

The Danish style characterised by devolved regulation through mandatory licensing by the test centre, in combination with a percentage state contribution to user investment, better enabled the establishment of the institutions that would enable cumulative learning by *one* of the Danish producer communities.

As Garud and Karnøe argue, the style of intervention was 'hands-on' in Denmark, with regulatory interventions frequent and adapted to the changing industry circumstances; investment subsidies were progressively reduced from 30% in 1979 to zero in 1989 as turbine efficiency increased, so that the 'market' incentive to invest was maintained without excessive waste of public money (Garud and Karnøe 2003: 309, 306).

It would be quite wrong to think of the Danish state as having from the beginning *planned* this global success in windmill technology; it also supported the R&D approach. It is interesting to speculate that the small size and population of the country of Denmark played a role here; unlike the parliaments of Britain, Germany or the USA, with populations more than 10 times greater, the Danish Parliament, the Folketing, proved to have the time as well as the will to legitimate and stabilise a pattern of use that was demanded by a small segment of the small (5 million) Danish population. This ability of the Danish Folketing to intervene and regulate is a feature of many other areas of Danish civic life[7] and it can be thought of as the Danish 'style' of governance.

Increasing Returns and 'Path Dependency' – a Process for 'Lock-in' to Specific Designs

There is no doubt that the pursuit and achievement of increasing returns to scale are important in the management of technology. However, Arthur has suggested a general process by which increasing returns can result in the selection and establishment in use of a *clearly inferior design* (Arthur 1994). His argument is as follows: if there are a number of alternative designs for a developing new technology, it is possible that 'chance events' may give one relatively poor design a lead in the market. Thereafter, increasing returns to scale of production may enable that design to drive out alternatives through a price advantage – even if the alternatives have the potential for *superior* returns to scale if given the chance to develop.

When long-term outcomes depend on early, initial conditions of development in this way the process is described as *path dependent*. Few would argue with this idea in other areas of life, such as political history. It can also be expressed as the idea that history matters because it explains features observable in the present.

Liebowitz and Margolis make one of their many useful contributions by distinguishing between three types of path dependency that may occur in technological development. The first includes standards such as for a clock face, or a metre rule, where alternatives are possible, but the one selected is 'good enough' or even optimal, but remains in use because it becomes established in use. The second type may generate what come to be seen as inefficient outcomes, but only in the light of later knowledge; no better could necessarily have been done at the time. The development of windmill technology appears to be an example of this kind, if the relatively short-lived and dysfunctional two-blade windmills are seen as inefficient outcomes. The third type is when an inefficient outcome is generated by initial conditions, when in principle a superior alternative path exists and is not taken (Liebowitz and Margolis 1995: 207).

Arthur's idea that increasing returns could lock an inferior design into use is of this third type. If this process exists, it would contradict the common-sense norm introduced at the start of this book – that design selected for use represents fitness for some purpose. Arthur does indeed cite what he thinks are important empirical examples of this process and claims that in general 'it appears to be the appropriate theory for understanding modern high technology economies' (Arthur 1994: 2).

Rather than mathematical models of idealised situations, what matters is whether there are any documented empirical examples of the process. There are two objects in the following reviews of Arthur's suggested examples of the 'lock-in' process: to examine the process of management of standard selection as an end in itself, but also to determine if these really are examples of lock-in to *inferior* technological design. For Arthur is right to point out that the issue of competing standards occurs frequently in the development of information technologies.[8] The most frequently cited example of lock-in to an inferior standard is the selection of the QWERTY keyboard design.

Chance Events and the Adoption of the QWERTY Keyboard Design[9]

QWERTY indicates the first six letters of the top row of the standard English typing keyboard of today. It was selected and developed for what would become the earliest commercially successful typewriter in 1873 by the inventor Christoper Latham Sholes, 'the fifty-first man to invent the typewriter' (David 1986: 34). The keyboard design began as an apparently arbitrarily chosen alphabetical layout (David 1986: 36) whose traces remain in the QWERTY layout of today (FGHJKL). The design was bedevilled by typebar clashes that occurred whenever two letters representing two bars situated in proximity were struck in close succession. Through trial and error Sholes moved bars and hence letters on the keyboard around to reduce typebar clashes[10] and the result was close to the QWERTY layout of today.

> QWERTY thus had evolved primarily as the chance solution to an engineering design problem in the construction of a typewriter which would work reliably at a rate significantly faster than a copyist could write. (David 1986: 36)

The marketing rights to this machine were cleverly sold to Remington, the large arms manufacturer of the time (David 1986: 36). Remington continued development and launched and marketed the product giving the QWERTY keyboard, embodied in the Remington–Sholes machine, a lead in the marketplace.

Within a few years mechanical design improvements enabled the inventor James Bartlett Hammond to largely avoid the jamming typebar problems that had propelled Sholes's QWERTY keyboard choice and enabled him in 1881 to offer an early ergonomically 'ideal' keyboard design, the DHIATENSOR (David 1986: 38). Like August Dvorak's 'simplified' keyboard (DSK) of the 1930s, the DHIATENSOR keyboard would position the most used English letters in the home row – the row of characters on a keyboard to which touch-typists always return their fingers. In principle, this would enable higher typing speeds by reducing the number of awkward finger movements a typist must make. In contrast, the home row in QWERTY is ASDFGHJKL and contains only half of the most frequently used letters in English.

By the mid-1890s, despite the availability of apparently superior keyboard designs the QWERTY keyboard was clearly consolidating its position as industry standard. Rival manufacturers, including Hammond, chose to switch to the QWERTY standard. What was the process that consolidated QWERTY as the keyboard standard?

Technical Interrelatedness of QWERTY Skills and QWERTY Keyboard

David points out that typewriter keyboard design became 'technically-interrelated' with the 'software' of typing skills (David 1986: 39) in a way that was not anticipated by the original machine innovators. In 1882 the 'radical innovation' of touch-typing was proposed by the owner of a shorthand and typing school (David 1985: 40). Touch-typing exploited a typist's memorisation of key positions to attain high typing speeds, but this committed the touch-typist to a particular keyboard configuration – and it is significant that touch-typing was first proposed and taught on a QWERTY keyboard.

The decision to invest in skills was taken by individual typists rather than firms because typing is one of those portable, general skills that firms lack an incentive to provide. While individuals tended to choose to train in QWERTY as the most widely used keyboard in order to increase their employment prospects, businesses had an incentive to buy machines with the keyboard that gave access to the largest pool of skilled typists (David 1986: 42). In other words, investors in machines and skills had an interest in joining the larger established market for machines and skills and their decision to join that market strengthened its appeal to others who had yet to make a commitment. This process would transform the initial lead that the QWERTY keyboard had gained in the marketplace through its association with Remington into overwhelming dominance (David 1986: 44).

At this point it is perhaps necessary to point out that the process of 'lock-in' by which QWERTY became a standard was not primarily one of increasing returns to scale in production technology. It is true that in his original 1986 article David discusses this process in terms of the economic advantages that accrue to the individuals making the decisions and that he coins new jargon, 'system economies of scale', to name the process (David 1986: 42). However, the description of the decisions leaves it possible to conclude that these individual advantages amounted to no more than this – that given a choice between markets *that are otherwise equal*, individuals would rather join the larger and better functioning market in machines and skills. The judgement that this was so in this case crucially depends on the relative 'efficiency' of the competing keyboard standards – and I am anticipating here the later conclusion that the QWERTY standard was 'adequate' and not greatly better or worse than rival keyboards. David assumes that QWERTY was clearly worse than its rivals; if this were so, it follows that there must have been powerful forces at work, perhaps economic forces, to enable it to succeed as a standard.

In other words, our judgement of the significance and nature of the *process* by which this design standard emerged as dominant over its competitors depends on whether it could select a standard that was inferior compared with its competitors.

The Relative 'Efficiency' of the QWERTY Keyboard

There are three types of evidence that support the idea that the QWERTY keyboard is an inferior standard. The first is the history of chance events that led to its selection. This gives us a kind of negative confidence that there was nothing advantageous in the standard – it was simply 'lucky' in its history. However, this does not in itself demonstrate that it was inferior to other keyboards.

The second kind of evidence is that participants in keyboard design sometimes make the appealing 'ergonomic design' argument for their alternative keyboards; that is, those designs that claim that because they have the frequently used letters in the typist's home row they are more speed 'efficient'. This is an *a priori* argument and it should itself be subject to testing through the third type of evidence – controlled tests of typing efficiency on the different keyboard designs.

It is, then, the empirical test of speed efficiency on different keyboard designs that really matters. The economists Liebowitz and Margolis are dogged critics of the idea that increasing returns – of some kind – can 'lock in' inferior technological design to use and they pay particular attention to the experimental evidence in support of the QWERTY story.

In their review of the experimental research, Liebowitz and Margolis quickly raise the issue of the trustworthiness of data associated with the man who had most to gain by the discredit of QWERTY – August Dvorak, the inventor of the Dvorak simplified keyboard (DSK). Dvorak patented his keyboard in 1936 and so acquired a financial interest in its success.

David's only cited experimental test evidence against QWERTY is an anonymously authored wartime naval research project. This found that after retraining a group of QWERTY typists on DSK to reach their former (QWERTY) typing speed, for a standard time of further speed training, the efficiency gain for the DSK typists was a staggering 75%, while the same time of speed training given to a control group of 18 QWERTY typists on QWERTY machines produced only a 28% increase in speed (Liebowitz and Margolis 1990: 11).

Liebowitz and Margolis throw doubt on the value of this research, first because they find that Dvorak was the US Navy's expert in time and motion studies during the war and was presumably linked in some now unknown way with the organisation of the naval research (Liebowitz and Margolis 1990: 12). They also fault the method of the research[11] in various ways, but the strongest evidence that something was wrong with this research is that when Liebowitz and Margolis look for research reviewed by, or conducted by, independent researchers rather than Dvorak himself, they are able to cite nine studies (my count) that find no advantage or an advantage of a few per cent for the DSK

(Liebowitz and Margolis 1990: 15).[12] Liebowitz and Margolis have conducted a thorough review of the evidence and I am persuaded when they conclude:

> At the very least the studies indicate that the speed advantage of Dvorak is not anything like the 20–40% that is claimed in the Apple advertising copy that David cites. (Liebowitz and Margolis 1990: 17)

If Dvorak has a speed advantage, it is a very small one, so small that it is unlikely to have played a role in real keyboard selection decisions.

The plausible 'ergonomic' keyboard design argument has proved misleading, at least as concerns speed. It is now reasonable to view the skill of typing as more independent of keyboard layout than previously assumed.

QWERTY as Fable and the Real Significance of QWERTY

Liebowitz and Margolis call the story of QWERTY-as-inferior-standard a 'fable' and catalogue its uncritical repetition in the *Washington Post, New York Times, Newsweek, Fortune* and other popular US news magazines. Further, in 1994 the *Social Science Citation Index* recorded 24 academic journal citations for David's *American Economic Review* version of the story 'the very large majority of these are uncritical uses of the QWERTY story' (Liebowitz and Margolis 1996). It has also entered popularising economics texts, for example Paul Krugman's book *Peddling Prosperity* has a chapter entitled 'QWERTYnomics' devoted to this 'parable that opens our eyes to a whole different way of thinking about economics' (Krugman 1994: 223).

Liebowitz and Margolis make no attack on David's 'technical interrelatedness' argument and that still stands as the process that eliminated variety among many competing and adequate standards in favour of a single standard. This selection did not occur *despite* the inferiority of that standard, but because that standard was 'good enough' and users could express their preference to participate in the more effective, and thus larger, market in machines and skills. We need no new jargon to describe the individual preference to join a larger and more reliable market.

A last irony in this story of a fable is how the fable has benefited from the extraordinarily wide diffusion of personal computers with their QWERTY keyboards as work tools. This lends the story a popular accessibility and immediacy it might not otherwise have. However, in the hundred years of their existence there have been some obvious changes in the *uses* of keyboard interface machines.

As stated before, the object of the first typewriters was to 'work reliably at a rate significantly faster than a copyist could write' (David 1986: 36). While the typewriter had this original major use of copying documents, speed typing had great significance, but the photocopier and the word processor have long ago eliminated the tedious job of document replication as the major use of keyboard machines. With their elimination the significance of speed typing and 'efficiency' defined in terms of percentage improvement in 'words per minute' lose much of their relevance. The QWERTY keyboard today is more than ever an *adequate* keyboard for most of the uses to which it is put.

Managing Neutral Standards through the Recruitment of Competitors – VHS Versus Betamax Video Recorders

After QWERTY, the competition between VHS and Betamax videocassette recorder (VCR) standards is 'probably the second most popular example' of how inferior design can become 'locked in' to a technology (Liebowitz and Margolis 1994: 147) (see also for example, Arthur 1994: 2). This case, like the keyboard case, will also prove to concern standards that are better considered similar – the competition is between nearly *neutral* standards. The irony here is that this is the more difficult circumstance to manage and the VHS versus Betamax case illustrates these difficulties well. The management problem centres on how to recruit past and future product competitors to cooperate on the adoption of one standard, while also seeking maximum individual advantage from the arrangement. First, we must know why the standards would remain technically similar.

By December 1970, 'seeking to strengthen its hand against foreign rivals' (Rosenbloom and Cusumano 1987: 61), Sony had succeeded in an effort to agree a common standard for magnetic tape home video with its two major prospective Japanese competitors, JVC and Matsushita, at the cost of transferring to these rivals its leading-edge technology. Products launched with this 'U-matic' standard for $^3/_4$ inch tape failed to attract enough buyers to be viable, but the standard became important as the common basis for development of the VHS and Betamax standards and therefore ensured that these standards would be technologically very similar (Cusumano, et al. 1992: 82). VHS and Betamax variation would indeed be limited to different cassette sizes, means of threading the tape and tape speed (Liebowitz and Margolis 1994: 148). The similarity in technology would mean that no permanent significant lead in quality could be established within this group of producers. However, the choice of different cassette sizes was enough to ensure the incompatibility of the VHS and Betamax standards and how this came about deserves some explanation.

In the further development of the U-matic standard, different beliefs were adopted about consumer preferences. Sony prioritised portability and so introduced a small cassette with an initial one-hour playing time. Despite many extensions to playing time, the smaller Betamax cassette would always contain less tape and therefore have a shorter playing time than the VHS rival. It would be reasonable to imagine that it would matter to consumers if they could not record a normal length film, but Sony was first to market with Betamax and quickly doubled playing time. Although Betamax would always have lower playing time than VHS, as playing time increased for both standards the difference between them would matter less and less to consumers. But as far as playing time mattered, it would always be to the advantage of VHS, the eventually successful standard.

According to the sources, this early difference mattered most because it affected standard recruitment. Matsushita refused Sony's 1974 overtures to join the Betamax standard, not only because 'top managers did not want to acknowledge Sony's leading position' (Rosenbloom and Cusumano 1987: 64), but also because they believed consumers would prefer a higher capacity tape (Rosenbloom and Cusumano 1987: 64; Liebowitz and Margolis 1994: 148).

Sony only began active recruitment to the Betamax standard again when it had doubled Beta playing time, in March 1976 (Cusumano et al. 1992: 58), but by then it had also begun manufacture (in April 1975), a step which probably prejudiced the chances of further recruitment as it now appeared that Sony was attempting to steal a march on its rivals (as it surely was). Yet it is not certain which actions proved decisive; Sony knew that Matsushita was its 'most formidable rival' (Cusumano et al. 1992: 57) and the most important potential recruit (Cusumano et al. 1992: 56) because of its larger size, prowess in mass production and spare VCR manufacturing capacity (Matsushita's own design had failed). Cusumano et al. also report that by mid-1975 Sony knew 'that JVC was working on a competing format, which, because of JVC's position as a Matsushita subsidiary, Matsushita was likely to support if Sony did not make a special effort to persuade them to adopt the Beta format' (Cusumano et al. 1992: 58). Even the Japanese government, in the usual form of MITI and the most likely source of additional pressure, failed to persuade Matsushita to back Beta (Cusumano et al. 1992: 58).

Matsushita's adoption of the VHS standard in early 1977 was probably the decisive event for the future battle of the standards; as early as 1978 Matsushita alone produced just over half of all Japanese VCRs (Cusumano et al. 1992: 88). In addition to its size and scale of VCR production, Rosenbloom and Cusumano write of Matsushita's 'astute actions' in being willing to supply other Original Equipment Manufacturers (OEMs) in addition to its own brands (that is, Matsushita would manufacture VCRs, then allow others to sell the machines under their brands) (Rosenbloom and Cusumano 1987: 64). Matsushita's alliance with RCA in the USA also ensured that most US manufacturers adopted VHS. The number of firms adopting VHS finally settled at 42 compared with 11 for Beta (Rosenbloom and Cusumano 1987: 42).

However, the superior recruitment to VHS, as well as the success of VHS in the market, continued to be influenced by the one qualitative difference between the formats: the longer playing time of the VHS cassette (Liebowitz and Margolis 1995: 222). For example, when RCA declared that it wanted three hours of cassette playing time, Matsushita developed a four-hour VHS cassette to meet this requirement and RCA joined the VHS standard. And one of the sources cites an executive who ascribed VHS market success when released in the USA to its four-hour playing time compared with Betamax's two hours (Liebowitz and Margolis 1995: 222).[13]

Lessons and Non-lessons from VHS Versus Betamax

Several sources acknowledge that there was (and is) a popular *perception* that Betamax was better than the VHS format (Klopfenstein 1989: 28; Liebowitz and Margolis 1994: 147).

Yet the evidence is against any fundamental quality difference *associated with a particular standard* – for the good reason that the two standards were developments of the earlier U-matic standard. There is plenty of confirmation of this view: an analysis of the different VCR models introduced between 1975 and 1985 showed that 'at no time did either format establish more than a transient advantage in features, prices, or picture quality' (Cusumano et al. 1992: 60). Klopfenstein cites two experts and five reviews from the US consumer organisation 'Consumers Reports' that, overall, could find no clear superiority in picture quality between the two formats (Klopfenstein 1989: 28). Cusumano et al. plausibly suggest that the fuzzy and incorrect general perception of Betamax as 'better' is probably linked to Sony's positive reputation as an innovator, rather than demonstrable quality differences between the *standards*.

As with the keyboard case, we have the late onset of the 'lock-in' effect in favour of the dominant, but qualitatively merely adequate, standard. Unlike the keyboard case, there was the possibility of managing the recruitment to the standards.

With hindsight it is tempting to fault Sony for attempting to gain a 'first-mover' advantage before settling the standards issue. However, if Sony had restrained its activity in this way it would have meant throwing away a development lead-time of two years. Sony had been 'burnt' in this way before, at the time of the U-matic standard development, when Sony's engineers judged that they had given away a technology lead to its competitors in exchange for the dubious value of recruiting Matsushita and JVC (Cusumano et al. 1992).

Although the Sony chairman Akio Morita is reported as having later acknowledged that Sony should have worked harder at standards recruitment (Cusumano et al. 1992: 56), in the early years the two standards coexisted and prospered in a rapidly expanding market, albeit with VHS dominant. Sony *did* benefit from its two-year development lead and the period of standard coexistence; the VCR was hugely successful in both formats for years, to the extent that for JVC and Sony, video products became the 'dominant source of revenue in the late 1970s and early 1980s' (Rosenbloom and Cusumano 1987: 64) and this success allowed expansion of both firms into other areas. As long as the major use of the VCR remained home recording, as discussed above, it is reasonable to conclude that there was little to choose between VHS and Betamax – they were competing neutral standards.

This would alter with a change in VCR use, but the rise in importance of pre-recorded cassettes was apparently not foreseen by market research until as late as 1980–1 – because, when asked, consumers indicated no interest in the possibility of such tapes (Cusumano et al. 1992: 90). When this change began, the effects were rapid. Sales of pre-recorded cassettes doubled every year between 1982 and 1986 and as the available range of pre-recorded cassettes began to feature in choice of machines, Beta went into rapid decline; sales of Beta increased by 50% in 1983–4, but decreased 50% in 1984–5 with effective extinction a few years later (Cusumano, et al. 1992: 90). As in the QWERTY keyboard case, the lock-in effect that favours the larger working market began to operate late in the competition between standards.

The value of this case lies in its illustration of the trade-offs involved in the management of cooperation for standards with your competitors. Sony's position was weak because of the company's smaller size, but strong in terms of its past record of VCR technology development. If the intention was to grow and become more powerful through innovation, standards collaboration with its concomitant loss of technological advantage was not an ideal strategy. Sony faced difficult decisions, but the company's choice was reasonable enough given the contemporary uncertainty. A last point would be that at some unknown but not catastrophic cost, Sony was able to switch to the VHS standard as Betamax went into decline.

The Personal Computer Seen as Variation on the Standards Stories

The same issues arise with variations in other technologies characterised by a hardware–software interaction in the marketplace. An interesting variation comes from Cringeley's popular history of the personal computer (PC) and IBM's entry into the $1 billion PC market of 1981 (Cringeley 1996). Before the IBM PC there had been 3000 Apple computer applications and over 5000 CP/M operating system applications on a multitude of hardware platforms (Cringeley 1996: 164). According to Cringeley, the early development of a 'compelling application' in the form of a spreadsheet designed to run on the IBM PC together with the IBM brand name, gave the IBM PC and Microsoft DOS operating system one a combination 40% of the market (Cringeley 1996: 163). Here we have an 'accidental event', thereafter, software applications writers, by working to this dominant standard, favoured its further consolidation.

IBM's proprietary control of the hardware standard was weak, as it had been forced to largely outsource the design of the components of its PC. Other hardware manufacturers soon managed to produce IBM hardware design 'clones', so that the legacy of IBM's short-lived leadership of the PC market would prove to be the establishment of a PC hardware standard and the growth of Microsoft, IBM's somewhat arbitrary choice to be sole supplier of its PC operating system, PC-DOS.

The PC has its own variation of the inferior design lock-in story. Each semiconductor chip advance created an opportunity for an operating system upgrade, and therefore for potential new entrants to dislodge Microsoft's domination of the supply of operating systems. With successive operating system upgrades, Microsoft developed a strategy of maintaining 'backward compatibility' with existing software applications. Millions of old users found it attractive to upgrade to each new Windows edition rather than risk switching to a possibly technically superior, but certainly application-poor, rival operating system. The achievement of successful backward compatibility allowed Microsoft to overcome its vulnerability to new operating system entrants (like OS/2). By 1996 more than 85% of computers by number ran Microsoft's operating system and more than half of the money spent on applications went to Microsoft (Cringeley 1996: 318).

If Microsoft's DOS operating system was compared directly with Apple's Macintosh operating system, then DOS did appear to be the inferior product. Apple had the more robust operating environment compared with the crash-prone and awkward 'command-line' Windows–DOS operating system.

A detailed comparison of advantages and disadvantages does not solely favour Apple, but as Liebowitz and Margolis point out, a purely operational comparison misses the role of Microsoft's strategy of 'backward compatibility'. In contrast to Microsoft, the Apple Macintosh represented the second time that Apple had introduced a generation of operating system incompatible with the stock of existing applications.

> Apple's behaviour signalled to customers that the company would not seek continuity in its operating systems in the future.... Computer users, particularly commercial users, who found such abrupt changes disruptive took note of these policies. (Liebowitz and Margolis 1999: 129)

In other words, if user expectations (and suspicions) about future producer behaviour are taken into account, it is possible to recognise the quality and efficiency of the Apple Macintosh OS – *machine* combination, yet still believe that when buying IBM-Microsoft customers were buying what they preferred and what was better for them.

Conclusions on Increasing Returns and Lock-in to Inferior Technologies

Arthur does suggest other technologies where increasing returns may have 'locked in' clearly inferior designs to use, but none of these receive more than one or at most two paragraphs of discussion.[14] The QWERTY and VCR cases warn us that a thorough historical analysis will be necessary to establish the effect. As Liebowitz and Margolis write:

> The theoretical literature establishes only that within certain models that incorporate particular abstractions, market failure of the type that causes the wrong network to be chosen is possible.... It is essential that the literature present real examples of demonstrable market failure. (Liebowitz and Margolis 1994: 148–9)

It is possible to suggest speculatively that the effect is likely to be rare for several reasons. First, the alternative design conceptions for a developing technology are not 'randomly' proposed, but represent attempts to improve on rival designs by their designers; and here there is a degree of selection operating that weeds out the truly dreadful from ever being developed or from reaching the marketplace. Second, while the assemblage of design components of a major new technology are understood to be uncertain and in a process of development, one might expect entrepreneurs to defer the high capital investments in plant development that are implied by a commitment to realise increasing returns (in the material production technologies). In other words, when the attempt to obtain increasing returns is made, it will be when there is some reason to expect design stability, or as in the QWERTY and VCR cases, design neutrality. Third, we already know that lower unit prices achievable by scale production routinely do *not* block the development of clearly superior, rival designs in certain industries; industries like the chemical or semiconductor industry were or are characterised by cycles of research, development and design obsolescence and the ability to build plants with scale economies as a condition of survival has long been taken for granted. In other words, if an inferior design ever became 'locked in' to use by the strategy of obtaining increasing returns to scale, a rival or later cycle of capital investment would renew the opportunity to reverse that design commitment. The achievement of economies of scale does not prevent imitation of the strategy and continuing competition in the designs that embody those economies of scale. This is possible for it is quite normal for the extent of the market to be far greater than the achieved economies of scale in a single plant.

Tit-for-tat Games in the International Management of Neutral Standards – the French State and Colour TV Broadcasting Standards

This leaves us with the conclusion that the problem of managing industry standards is likely to be greatest when there is a range of neutral standards available for adoption, with none having a clear superiority. If the VCR and PC cases are a guide, when neutral standards are available the ability to recruit to one standard is likely to depend on many aspects of strategic context, especially the history of collaboration and competition within the industry and actual and estimated ambitions and abilities within individual companies.

The management of standards at the international level throws an interesting light on the significance of the choice to compete either within an agreed standard or through different neutral standards.

There is no international authority to take sanctions against states that actively promote a standard to their national advantage. However, where states are wary and similar in power it is difficult for one state to force its preferences on others and the attempt to do so may backfire. A nice example of the attempted use of 'force' to advantage national interests is the French state's sponsorship of its national SECAM colour TV broadcasting standard (Crane 1979).

Black and white TV broadcasting standards had been highly fragmented and the costs in terms of limited markets for national European TV programmes were painfully evident. As a result, most European states were determined that there should be a 'European' standard for colour TV.

However, the French government decided to promote aggressively a French 'national' standard – SECAM – against all other standards, despite the opposition of French standards authorities, segments of the French TV manufacturing industry and other countries. Crane writes that the French government saw the advancement of SECAM and the consequent promotion of French TV manufacturing as partial compensation for the loss of national 'technological face' perceived to have occurred at this time because the US firm General Electric had bought the major French computer manufacturer.

Once again we have standards which are 'neutral' in terms of intrinsic qualities. Crane writes that tests on the three competing standards eventually developed did not reveal any one to have a clear advantage over the others. Although the three standards were incompatible, like the VCR standards, they were related in development terms – 95% of SECAM and PAL were based on the same US patents as the US NTSC standard. PAL was so similar to SECAM, that French technicians even referred to it as a SECAM variation (Crane 1979: 57).

The French state nevertheless portrayed the US standard NTSC as inferior by stressing its weaknesses and ignoring its strengths and there was an increasingly acrimonious and public row with the USA.

The French state also sought to promote SECAM by assuming responsibility for its marketing. So countries without a scientific/technical infrastructure were offered free technical training in SECAM broadcasting. Those that accepted, the Soviet bloc and various Third World countries, did indeed become the countries where French SECAM-based products (programmes and equipment) sold well.

Countries that possessed a developed TV manufacturing sector were more difficult to deal with. The French state deliberately broke a long-standing working agreement to exchange patents for free between the major French electronics firm CSF and the German firm AEG-Telefunken. If this had remained in force, little advantage could have accrued to French industry within the developed markets of Europe. On the other hand, the strategy was transparent to the Germans. When AEG-Telefunken was asked to pay for access to the SECAM standard the Germans saw this as unreasonable, chauvinistic behaviour. The German company reacted by developing the PAL standard and ensured that this was fostered and developed as a national rival to SECAM. The result was that most of Europe adopted PAL, the US NTSC, and France SECAM.

The French achievement was to trigger a competitive reaction that ensured there would be no common broadcasting standard in Europe. Although the French government appears to have thought in terms of hardware advantages, Crane suggests the real economic loss may have been to the developing European programme industries, because a common European broadcasting standard would have allowed early interchangeability of successful programme material. However, an ironic twist to the story is that when the Japanese TV manufacturers swept away European manufacture in the 1980s, they first targeted the larger PAL market. The French TV makers survived a little longer only because their government had *failed* to 'lever' SECAM onto the other European states.

The VCR and, even more so, the colour TV case demonstrate the dilemma of whether to cooperate within or compete through technological standards. The situations are similar to those discussed through various game-playing scenarios in a book like Axelrod's *Evolution of Cooperation* (Axelrod 1984). The differences are primarily to do with certainty; a computer game scenario is exactly defined beforehand, whereas in our cases the abilities and strategies of the participants were uncertain, or in the French case, not recognised or grossly underestimated. On the other hand, one of the general findings of Axelrod's games seems relevant here: when there is a population of independent decision-makers and repeated rounds of gaming interaction, the simple strategy of 'TIT for TAT' often proves most successful as a way of disciplining cheats on a cooperative strategy (Axelrod 1984: 20). When the French government unilaterally decided to 'defect' from the European cooperative game, a German company was able to administer a TIT for the French TAT.

Notwithstanding the language of game theory, one can hope that such experiences, and the fact that many European states have similar technological abilities and long memories, will prove to have a disciplinary effect on would-be European national 'cheats'. The ability to agree common European technological standards is a necessity for a properly functioning single European market. This example reminds us that in a Europe of independent sovereign states it is the *memory* of the fate of cheats on common standards that underpins the continuing readiness of states to abide by common standards.

Conclusions on Standards and Paths of Development

In its broadest sense a path of development is simply a sequence of design and development events with some kind of relationship between them. A readily and retrospectively identifiable type of path consists of regularity in an artefact lineage through time. Of course, the synoptic, retrospective identification of apparently clear beginnings and ends of a path does not mean that these beginnings and ends were or *could be* clearly identified by the participants in the technology when it most mattered to them; the clarity is partly a product of hindsight.

Laser and robotics were used to illustrate the proliferation of paths as multiple evolving artefact–use relationships. Danish windmills were used to illustrate the problem of constituting the path of learning itself, but the usual context for the discussion of paths is when there is the object of increasing returns.

The attempt to exploit increasing returns to obtain economies of scale remains one of the general objects of the management of technology. However, the attractively simple model that in the growth of complex product technology, design variety is succeeded by a dominant design *achieved through increasing returns* was faulted in the classic case of the DC-3. Design stability was achieved before increasing returns allowed passenger transport to become economic.

Much popular attention has been directed at the relationship between a form of increasing returns and the selection of allegedly inferior designs for use. It was argued that the most cited examples of this effect were in fact examples of lock-in to adequate or 'neutral' design standards. In the absence of compelling examples of 'lock-in to inferior design' one can conclude that increasing return effects are interesting in their own right, but not because of unsubstantiated theoretical speculation about their role in inferior design selection.

Notes

[1] For the account of Hewlett Packard's involvement in the Kitty Hawk 1.3 inch disc drives see Christensen (1997: 150). Christensen draws a different lesson to the one here.

[2] In contrast barbed wire 'plugged' the 'component gap' in an existing technological system.

³ The era of low-cost energy that began in the 1950s led to the discontinuation of research into wind power. When it restarted in the 1970s recourse was made to the earlier work, but this already had two quite distinct design paths represented by the work of Juul and the German engineer Hütte (Heymann 1998).

⁴ In 1999, over 100 000 Danish families owned shares in wind energy cooperatives (Krohn 1999).

⁵ The Danish Wind Turbine Owners' Association; see the articles by its chairman Flemming Tranæs at http://www.windpower.dk/articles/coop2.htm.

⁶ Talk given by Aidan Cronin, manager, at Vestas AS, Ringkobing, Denmark.

⁷ For example, the Danish housing market has many varieties of communal housing termed *bofaelleskaber*. The law governing the creation and operation of the many varieties of communal housing has been changed repeatedly in recent years.

⁸ *Collins Dictionary* definition of a standard is 'an accepted or approved example of something against which others are judged or measured'; such a general definition of course applies as much to regimes of intellectual property rights as to the design of artefacts. The DC-3 became a standard in this sense, despite many of its constituent design elements having been demonstrated in other aircraft models; that a design is taken as a standard does not tell us how it became a standard.

⁹ The following account draws on David's article (David 1986).

¹⁰ Unlike some of the derivative authors, David never claims these changes were made to *reduce* typing speed.

¹¹ The major fault with the research was that three members of this second group had no typing experience so their initial and final word-per-minute scores were created by averaging early and late typing tests; this would reduce the QWERTY-trained, QWERTY group 'efficiency gain' by an unknown amount. More seriously, the high efficiency gain is dependent on an entry speed of around 30 wpm known to be subefficient; the US Navy defined competence even then as 50 wpm (Liebowitz and Margolis 1990: 11).

¹² There is a DSK 'advocacy' webpage that attempts to attack Liebowitz and Margolis' arguments ('the fable of the fable' (Brooks 2000)). It is not convincing. The webpage depends heavily on another DSK-advocacy publication by Cassingham whose book gives no sources so a reader is disabled from checking some of the statements. Some are known to be wrong and so cast doubt on the quality of the work; for example, Cassingham makes the incorrect claim that the QWERTY keyboard 'was designed to slow their [typists] speed' (Cassingham 1986: 11) but gives no source. It is clear from David's account that since there was no touch-typing at this time, Sholes object was not to slow typists down.

¹³ Lardner cited by Liebowitz and Margolis (1995). As Betamax playtime passed three hours, this quality probably became unimportant – but by this time Betamax was the secondary standard (Liebowitz and Margolis 1995).

¹⁴ For example, he suggests the pressurised water reactor was selected over the theoretically more efficient gas-cooled reactor for chance reasons, that the steam car may have had superior long-run development potential to the internal combustion engine (Arthur 1994: 10–11). A serious discussion of nuclear reactor design would consider the problem-ridden development of just such an 'efficient' and 'advanced' gas-cooled reactor (AGR) by the British state. As for the steam car, in the year 1900, 4192 cars were manufactured in the USA, of which 1681 were steam, 1575 electric and only 936 gasoline. As early as 1905 at the New York automobile exhibition there were five times as many gasoline as steam and electric cars and the dominance of this power source choice was soon consolidated (Basalla 1988: 198). The theoretical and practical thermodynamic advantage lies with the internal combustion engine – there is no water tank to heat and cool uselessly every time the engine is started and stopped. Drivers need not search for sources of water to top up their boilers as well as search for fuel sources. The initial advantage in number of steam cars almost certainly derives from the mature development of this technology for many uses. In contradiction to Arthur's use of the example, the thermodynamically inferior steam car had the initial advantage of established infrastructure over the superior gasoline internal combustion engine, but if there were any increasing returns they quite obviously failed to consolidate steam over the superior internal combustion engine.

4 Competition and Innovation as Substitution Threat

Competitive Scenarios of Innovation

The simplest way to express the role of competition in the management of RCA's and DuPont's innovative activity is that it appeared as a perceived sense of threat to future projects and revenues. The sense of threat felt in the present was important as a mediator of actions, but it derived from a variable mental construction of possible future scenarios. These were built from estimates of the abilities and intentions, declared or imputed, of established competitors. Innovation is a future-directed activity and it is as the mentally constructed shadow of future competitive scenarios that competition-as-threat weighs on innovative plans in the present. A detailed analysis of the management of innovation in a market economy is always likely to include the analysis of such constructions, whatever the technological context.

In the analysis of the evolution of technologies over many decades it often makes sense to think of two general competitive scenarios for innovation and the related perception and experience of threat. The first is that of the innovating firm and its relationship to other innovators, the second is that of the established firm experiencing substitution from the innovating firm.

Of course innovating firms face potential imitation and substitution from their known rivals, but when the scenario is the exploitation of a major technology, in some degree there exists the option of pursuing projects that avoid immediate competition with rivals. It is impossible to be generally determinate, because with uncertainty concerning future innovative opportunities and innovative activity depending on such potentially scarce inputs as technical skills, the actual 'population' of innovating firms may be variously matched to innovation opportunities. The idea that a field of innovation may be more or less crowded and that profits within that field are to a first approximation an indication of the degree of crowdedness represents the DuPont experience and remains a useful qualitative schema of application elsewhere.[1]

There is no *necessity* for innovation to substitute for existing goods and services – in principle, an innovation may have an entirely novel use. In practice substitution is a common phenomenon and it will make sense to think of innovations as having different degrees of substitution for technologies that already exist. As regards competition, if the substituting and established technologies are sufficiently different, innovating firms may not need to treat established technology firms as serious competitors, despite trading in the same market, for such firms may lack the ability to respond to the substitution threat. This being a possibility, it again makes sense to consider the experience of substitution separately. Because innovating firms are the subject of this book, the first scenario is often referred to: particular examples are RCA and DuPont in

Chapter 3 and radio and Dyson's vacuum cleaner in Chapter 5. The current chapter is principally concerned with the second competitive scenario, of substitution threat and possible firm response. However, in this chapter as in the next three, there is introduced the operation and reform of the institution most relevant to the practice described. In this chapter that will be competition policy, its changing intellectual basis in economic thought and how these changes relate to the practice of innovation.

Creative Destruction and Firm Response

It is almost conventional to cite Schumpeter on the nature of substitution threats because he nicely dramatises the possible transformative effects of this kind of competition from innovation:

> in capitalist reality, as distinguished from its textbook picture, it is not that kind of competition which counts, but the competition from the new commodity, the new technology, the new source of supply, the new type of organisation – competition which commands a decisive cost or quality advantage and which strikes not at the margins of the profits and the outputs of the existing firms but at their foundations and very lives. This kind of competition is as much more effective than the other as a bombardment is in comparison with forcing a door, and so much more important that it becomes a matter of comparative indifference whether competition in the ordinary sense functions more or less promptly. (Schumpeter 1943: 84)

For Schumpeter, this is nothing less than the central dynamic process in capitalist change. This sense of the destructive power of innovation on old industries and technologies is captured by the idea of 'creative destruction', a term which has gained widespread use and refers to the substituting impact of innovation on established businesses.

This is potentially the strongest form of threat a firm can face and for this reason alone, the 'reaction' of firms is of particular interest, but an investigation of this reaction is also an investigation of economic transformation in action.

There are a limited number of possible reactions to the threat of full technological substitution for a given use or market:

1 Switch to the new technology.
2 Accelerate improvement of the established technology (the 'sailing ship effect').
3 Exit from the market.

Since it will be argued that despite frequent reference to the sailing ship effect in the literature, it is in fact rare or non-existent, the switch decision is the most interesting and the only 'true' response, in the sense that if it happens at all it must be by deliberate strategic choice (unlike exit).

In Search of the 'Sailing Ship Effect' – Acceleration of Innovation in the Old Technology as a Response to Substitution Threat

First, the eminent historian Nathan Rosenberg nicely summarises what can be shown to happen when established firms face radical substitution threats:

> The imminent threat to a firm's profit margins which are presented by the rise of a new competing technology seems often in history to have served as a more effective agent in generating improvements in efficiency than the more diffuse pressures of intra-industry competition. Indeed, such asymmetries may be an important key to a better understanding of the workings of the competitive process, even though they find no place at present in formal economic theory. (Rosenberg 1976: 205)

But many writers have gone further and claimed that established firms react to a substitution threat by *accelerating innovation in the old technology*, a response sometimes referred to as the 'sailing ship effect' (Foster 1988; Cooper and Schendel 1988; Utterback 1996). So in a popular recent book by Utterback:

> Of course the established players do not always sit back and watch their markets disappear. Most fight back. The gas companies came back against the Edison lamp ... with the Welsbach mantle ... there was nothing incremental about that. Purveyors of established technologies often respond to an invasion of their product market with redoubled creative effort that may lead to substantial product improvement based on the same product architecture. (Utterback 1996: 159)

Utterback's view depends on two more nineteenth-century cases: gas versus electric lighting and mechanical versus harvested ice. The gas industry appears to be a particularly promising candidate because the Welsbach gas mantle was certainly a radical innovation; it improved gas lighting efficiency by five times (per unit of gas burnt) and for a while it was not clear that electric lighting would prove the more effective and efficient lighting source (Bright 1949: 126).

However, while the Welsbach mantle certainly aided the gas lighting industry, it is not an example of the sailing ship effect. Welsbach was an independent inventor–entrepreneur, an Austrian who founded and ran his own electric lighting company in Vienna. Nowhere does the historian Arthur Bright claim that he developed his mantle in the USA or with gas company funds or in 'reaction' to the threat from electric lighting.

This story has a further bizarre twist. The thorium and cerium oxides whose superior fluorescence was responsible for the gas mantle's luminous efficiency could only be obtained by extraction from ores rich in other rare earth elements such as tungsten, iridium and osmium. As the gas mantle became standard, and the production of thorium and cerium compounds grew, the other rare earth metals became cheap and available for experimentation. So we have the paradox that the success of the Welsbach gas mantle helped make research possible into rare earth metal filaments, research that would eventually result in the elimination of gas lighting by the tungsten metal filament (Bright 1949: 168). Far from Welsbach having any commitment to the 'gas industry', Welsbach himself would contribute to its final destruction as a source of lighting. He developed his work on the gas mantle by attempting to pass current through filaments of platinum coated with thorium oxide. When this failed (because of the different coefficients of thermal expansion) he began experiments that resulted in the use of osmium metal in the first commercially successful metal filament lamp (Bright 1949: 174). Millions of these osmium filament lamps were sold because it was a clearly superior light source to Edison's carbon filament lamp, and a new path of research was opened.

Welsbach helped to prolong the gas industry's hold over the lighting market, but he then helped to destroy it. He is better characterised as a leading 'lighting' innovator, without particular commitment to the vested gas or electric lighting technologies. The coincidence in timing between the advent of electric lighting and the Welsbach gas mantle is just that – a coincidence, not a 'strategic response'.[2]

This example shows that *continuing innovation* in an old technology that occurs for normal competitive or other reasons must be distinguished from *innovation as a response to a novel technological substitution threat.*

Solvay Versus Leblanc – Substitution in the Early Chemical Industry

The late nineteenth-century chemical industry furnishes us with an example of the difficulty of interpretation of evidence *and* an extraordinary reaction to a strong substitution threat worth examination in its own right: the established Leblanc process firms managed the extraordinary feat of delaying, although not halting, the creative destruction of their technology by the Solvay producer. This justifies some detailed analysis, but the case has also been misinterpreted as an example of the sailing ship effect:

> the alkali industry case also demonstrates the so-called 'sailing ship effect' … it illustrates how established companies can become locked into existing technological trajectories. Rather than attempting to capitalise on the possibilities offered by the emergence of a superior new substitute technology, they vigorously defend their position through the accelerated improvement of the old technology. (Rothwell and Zegveld 1985: 41)

This statement asserts a *general* preference on the part of the established firms for increased old technology development, *despite recognition of a superior substituting technology* – in other words, economically irrational behaviour, and very strange if true.

One of the most sophisticated economic historical analyses also finds the behaviour of the old technology management difficult to explain, first allowing that:

the Leblanc-Solvay case has been viewed by many as a prime example of entrepreneurial failure. (Lindert and Trace 1971: 263)

Lindert and Trace calculate that if the Leblanc producers had switched whole-sale into the Solvay process, it would have been profitable *at the product prices actually pertaining in the 1880s to 1900s.*

> Our hindsight calculations show that a wholesale conversion to Solvay plant, if ordered in 1888 and fully installed by the start of 1890, would probably have brought the new United Alkali Company extra profits in every remaining pre-war year except 1893. (Lindert and Trace 1971: 260)

This is probably true given their assumptions and might be thought to 'prove' the Leblanc producers to have been economically irrational in not switching to the Solvay process. However, as will be shown below, it is likely that different actions would have generated a quite different set of historical prices. Although for Lindert and Trace the mystery is why the Leblanc producers did not adopt a switch strategy, they rightly point out that all depends on how the Leblanc producers *then perceived possible actions and their consequences.* They also emphasise that direct evidence for such intentions is poor. Nevertheless, they end by plumping for a kind of 'cultural' explanation of entrepreneurial failure:

> Explaining the costly conservatism of the Leblanc producers requires a certain amount of conjecture ... Our tentative inference is that the Leblanc management exhibited the early-start mentality, with its profit-losing attach-ment to continuity and its reluctance to admit a major mistake. (Lindert and Trace 1971: 264)

However, this conclusion was made without reference to a major publication that contains details of the thinking of the Leblanc producers (Reader 1970).[3] Reader's book makes it possible to provide an explanation of the Leblanc management behaviour that does not need an 'early-start mentality', or economic irrationality, or the sailing ship effect.

Solvay Versus Leblanc – from Creative Destruction in a Free Market to Technological Stasis in a Rigged Market

The Solvay brothers preferentially licensed their superior ammonia–soda process (other ammonia–soda processes existed) for making sodium carbonate to Brunner Mond in Britain, and this company began production in a halting fashion in 1874 (Reader 1970: 53). Given that whatever the production tech-nology sodium carbonate was a standard product sold into a national market, one might expect that competition between the Solvay and the Leblanc processes should be intense, resolved relatively quickly and only limited by the time taken to build a new Solvay plant. However, it would take until the First World War for the last Leblanc plant to close, over 42 years after the start of Solvay production (Lindert and Trace 1971).

At first Leblanc sodium carbonate production did rapidly decline with rising Solvay production, despite a growing total market for sodium carbonate in the second half of the nineteenth century. Figures compiled in Table 4.1 from a standard chemical industry history (Haber 1958) show creative destruction well underway in the 1880s, and crude extrapolation of trends suggests that it would have been possible for it to have been completed before 1900. Instead, there is a dramatic cessation of expansion of Solvay production by Brunner Mond around 1890 and static output for approximately 10 years.

The immediate reason is that Brunner Mond entered a market-fixing agreement with the Leblanc producers despite the very evident historic success of the Solvay process in the carbonate market. To understand why the owner of the successful and superior Solvay process should want to enter such an agreement, we need to examine the evolution of prices prior to the agreement and the detail of the agreement itself.

Market Prices and the Logic of the Market Agreement

Prices for sodium carbonate in Europe fell over 75% between the 1860s and late 1880s (Reader 1970: 54), a scale of decline greater than any estimated cost difference between the rival production processes. With a general trade depression between 1884 and 1889 the decline accelerated and while this aggravated losses for the many Leblanc producers, it also brought near zero profit for Brunner Mond by 1887 (Lindert and Trace 1971: 257). It is this collapse of profitability *for all producers* that induced the restructuring of the alkali industry and the rigging of output and prices of the main alkali products.

The independent Leblanc producers (more than 40) reacted to the slump in prices by cartelising in 1891 into the United Alkali Company (UAC) (Reader 1970: 106). Haber writes of the promoters of the UAC that they 'made it clear they intended to effect far-reaching reforms' and that they were motivated to

Table 4.1 Solvay versus Leblanc production of sodium carbonate in Britain in thousands of tons

Solvay production (Haber 1958: 158)		Leblanc production (Haber 1958: 152)	
Year	Tons (000)	Year	Tons (000)
1878	4.0	1878	196.9
1880	18.8	1880	266.1
		1882	233.2
		1884	204.1
1885	77.5	1886	165.9
1890	179.5		
1898	181.0		
1900	225.0		
1903	240.0		

Note: Data derived from (Haber 1958: 152, 158). Additional data: 1878 Solvay figure from Reader (1970: 53) and 1900 figure from Reader (1970: 174). The United Alkali Company refused to publish precise Leblanc sodium carbonate figures from 1893 (Lindert and Trace 1971: 282).

'establish friendly relations with the Solvay manufacturers' (Haber 1958: 182). These 'friendly relations' would eventually comprise an 'agreement on price levels and market shares for each soda[4] product' (Lindert and Trace 1971: 253) and by its periodic renegotiation and renewal the two parties maintained a rigged market into the First World War.

Irrespective of the details of the agreement or its effective prosecution, on circulation of rumours of the formation of the UAC, alkali prices began a speculative rise that restored and maintained overall profitability for both processes. Prices were therefore clearly detached from production capacity and costs and driven by fear of the new cartel's behaviour. This speculative rise in prices cannot be properly distinguished from the later effects of the agreement; the rise in prices and restoration of profits were maintained through to around 1897 (Lindert and Trace 1971). This circumstance, whereby the existence of a cartel alone influenced prices regardless of actual output controls, likely influenced the approach of the two parties to their series of agreements.

The Significance for the Market Agreement of Solvay as a Partial Technological Substitute for Leblanc Output

The key to understanding this case and the details of the market-rigging agreement is that it is not purely substitution of one process for sodium carbonate production by another. The primary valuable product from the Solvay process was sodium carbonate, and carbonate was the original and most important product of the Leblanc process, but the Leblanc process also produced two byproducts,[5] caustic soda (sodium hydroxide) and bleaching powder, that had growing and lucrative markets in the late nineteenth century. Because caustic soda could be made from an *intermediate* Leblanc product (sodium sulphate) whereas the Solvay process required an additional process to convert its carbonate product into caustic soda, the Leblanc process made cheaper caustic (Lindert and Trace 1971: 251). And as far as the Leblanc process was used to make bleaching powder, it necessarily produced either caustic or carbonate so that it would always retain an interest in the state of these markets.

These distinctive features of the two processes at once gave the Leblanc producers some basis for hope in the future (they could hope, and bargain, for a secure monopoly in the expanding caustic and bleaching powder markets) and Brunner Mond some reason to fear retaliatory entry into its process technology by the UAC. For although Brunner Mond could threaten entry into the caustic market, it could not do so with the price advantage it had in the carbonate market, nor had it a price advantage in UAC's chlorine-related markets.

Detail and Strategic Intent of the Market Agreement

The periodically renegotiated details of the agreement reveal the changing strategic concerns of the two producers and in particular provide further explanation for the otherwise astonishing entry of Brunner Mond into the market agreement, which was not only about fixing output and prices, but also about restricting entry into each party's technology.

In the 1891 agreement, Brunner Mond obtained an upper limit for UAC ammonia–soda production of 15 000 tons/year. In return, Brunner Mond agreed an upper limit to its carbonate production of 165 000 tons/year, to limit production of chlorine and not to enter the market for caustic (Reader 1970: 109).

Brunner Mond's great vulnerability was the expiry of its major Solvay patent in 1886 (Reader 1970: 106). With the creation of the giant UAC (1891) a coordinated switch from Leblanc into ammonia–soda production became a major concern. The UAC did subsequently make a limited entry into ammonia–soda technology (not the Solvay version) within the terms of the agreement, with the purchase of two ammonia–soda plants in 1893 (Reader 1970: 108).

There may have been further reason for Brunner Mond to enter an agreement given that Solvay sodium carbonate was only a *partial technological substitute* for the Leblanc process. Brunner Mond could not be sure of the full creative destruction of the Leblanc producers through carbonate output expansion. There was some danger of obtaining the worst of all worlds: not only a low sodium carbonate price, but the danger of retaliatory entry into Solvay technology financed by the premium prices obtainable in UAC's monopolies in the bleaching powder and caustic soda markets.

Nor was an aggressive switch to ammonia–soda necessarily an obvious and profitable strategy for the UAC. By 1891 Brunner Mond held over 50% of British carbonate production (Table 4.1) and capital reserves while the UAC was loaded with debt (Haber 1958: 159). Major entry into ammonia–soda risked a return to collapsing prices and war with Brunner Mond. This is why the actual historical prices for alkali products that Lindert and Trace use to calculate the profitability of the switch decision for UAC are misleading; had UAC attempted to switch, a different set of prices would have obtained. And there was every reason to believe from experience that these prices would have been lower.

One last deterrent faced a UAC strategy of massive entry into ammonia–soda – the technical difficulty of operating ammonia–soda plant without Solvay group assistance. The Solvay group collected details on individual plant performance at fortnightly intervals and, in return, dispersed technical assistance and new patents within the group. Reader comments on the development of the Solvay process that

> it will be evident that the Solvays' success, when at length it came, depended comparatively little on the kind of knowledge which is revealed in a patent specification or expounded in a textbook. It depended far more on detailed experience of the process in action and on trade secrets. These secrets were and are very carefully guarded indeed. (Reader 1970: 45)

One of the novel features of the Solvay process was that it operated continuously and much of the difficulty of management was connected to this feature.

Given the years it had taken the Solvay brothers to build up this knowledge and given Brunner Mond's powerful lead in *operating* the technology, UAC's strategy of negotiating a set of rigged markets, maintaining a division of control of the technologies and placing its hopes on the future of the chlorine and caustic markets looks reasonable. In essence, by the time UAC was created and could threaten Brunner Mond, it was too late for a strategy of wholesale switch into ammonia–soda.

The Reaction of the Leblanc Producers to Solvay Production – No Evidence of the Sailing Ship Effect

The significance of these events for the Leblanc system is that it bought *time and increased profits* with which the Leblanc producers might have sought to improve their position. What did they in fact do?

There *were* improvements to the Leblanc system, but the greatest *technological* innovation during the period of substitution came *before* the formation of the UAC and was the Claus–Chance process of 1882, which made it possible to recover the sulphur previously lost in the form of calcium sulphide[6] (Haber 1958: 100).

We have no clear evidence that the development of the Claus–Chance process was driven by the threat from Solvay production. Rather, the chemical histories tell us that there had been many attempts over decades, and before the arrival of the Solvay process, to recover the valuable sulphur lost in the Leblanc process. Texts like Haber's comment only that theoretical cost savings and potential new markets always provided the incentives to innovate and improve the Leblanc system.

However, the main sources cite UAC's claims at the time of its formation that it would 'modernise' and 'reform' its processes, a promising indicator that there might be a sailing ship effect. So the UAC claimed that its creation subjected the constituent Leblanc manufacturing plant to central comparison and control and this allowed it to establish best operating practice in every works (United Alkali Company 1907: 41). This can be taken as another example of that 'tightening' of operational efficiency that Rosenberg noted as common in substitution cases, but it does not count as the sailing ship effect, which is properly the induction of technological change involving change in the old artefact forms. And minor efficiency savings cannot be expected to save an obsolescent technology – as they did not in this case.

The UAC did found a 'Central Research Laboratory' that it claimed produced (unspecified) improvements in old processes (United Alkali Company 1907: 41). If we knew what these were, we would have candidates for the sailing ship effect, but they are nowhere specified in published company documents (for example, United Alkali Company 1907) or the chemical industry histories. If they existed, they are unlikely to have been of great economic importance, as subsequent events will show.

Other UAC activities that are described are vertical integration into salt and pyrites mines, the latter to secure sulphur supplies. There was also some diversification – UAC bought a railway in Spain (Reader 1970), which if it suggests any strategy, suggests increased effort to *exit* Leblanc production.

The problem with the actions listed above is that even when, like the Claus–Chance process, they are significant improvements to the Leblanc system of production, they are not necessarily efforts focused on the *improvement of the economics of sodium carbonate production*.

We defined the sailing ship effect as if two substitutable technologies would compete for a single market and, on this basis, for the sailing ship effect to exist we would require evidence of efforts to accelerate innovation specifically in Leblanc sodium carbonate production. It is more likely that the continuing efforts at improving operational performance were primarily intended to benefit the development of the bleaching powder, caustic and sulphuric acid markets and to appear more as strategies of 'exit' rather than the sailing ship effect. Improvements in the economics of sodium carbonate production did occur, but they became *incidental* to the main strategy of exit from sodium carbonate production – they were never expected to enable serious price 'competition' with the Solvay process.

The UAC did not publish output figures after 1893 (Lindert and Trace 1971: 282), but by its own published comments it appears that by 1902 at the latest the UAC had ceased to make carbonate altogether (Lindert and Trace 1971: 249; United Alkali Company 1907: 33). It is reasonable to assume that some time earlier, perhaps as early as 1890, UAC came to define itself as primarily a caustic and bleaching powder producer, which happened to make some high-cost sodium carbonate as a byproduct of these operations.

It is now clear that the 'paradox' of the long period of coexistence of Solvay and Leblanc production in Britain is no paradox at all. The Leblanc system of the 1890s was not primarily a sodium carbonate production technology as it had once been, but a chlorine and caustic soda technology. The final judgement on the UAC is that the company made a reasonable decision to manage an exit from carbonate production, rather than attempt a high-risk, major switch into the Solvay process.[7]

The Alkali Industry and the Hypothesis of Entrepreneurial Failure

There is a larger significance to this result. The Lindert and Trace paper was originally presented at a conference of economic historians intent on applying quantitative methods to the historical analysis of the British economy (McCloskey 1971b). The theme of the conference became the revision of the notion of Victorian entrepreneurial failure, for in paper after paper, late Victorian industrial performance was shown to be better than had been imagined when the complaints of contemporaries formed the basis for economic history. The editor of the collected conference papers, Donald McCloskey, commented on the idea of Victorian entrepreneurial failure that:

> Only Lindert and Trace brought in a clear verdict of guilty … again … the conclusion was modified in discussion. (McCloskey 1971a)

If, as I have argued, Lindert and Trace's result is understandable given their sources, but not correct, then the Solvay–Leblanc case becomes further confirmation that Victorian entrepreneurs behaved 'rationally', in the sense that they exploited what opportunities they were able to perceive as opportunities. Even where they made mistakes, those mistakes were understandable given the information they had at the time.

This historical revision contradicts the still popular idea that British 'anti-industrial culture' was responsible for British relative economic decline, an idea popularised by Wiener's influential book *English Culture and the Decline of the Industrial Spirit* (Wiener 1987). Edgerton and Sanderson have argued that Wiener's book is thoroughly misleading in its thesis that anti-industrial attitudes dampened entrepreneurial behaviour and obstructed the provision of technical skills by British universities to the point that economic opportunities in Britain were not exploited (Sanderson 1988; Edgerton 1991). In short, the counter-argument is that as Wiener shows, anti-industrial attitudes may have dogged Oxford University, but this is beside the point when industrialists were busy founding and endowing a whole class of city-based universities to serve their needs. In parallel vein, some of the sons of the sons of company owner–entrepreneurs may not have been entrepreneurial as owner–managers, but one should expect this behaviour to create an opportunity for a new class of entrepreneurs to prosper.

This thesis of a 'failure of British entrepreneurship' became popular in British government circles in the 1980s, in part because of Wiener's and Barnett's (Barnett 1986) books.[8] Some of the drive to encourage teaching and research in entrepreneurship in management schools seems to derive from a continuing fear that a major 'failure of entrepreneurship' is possible. From this brief review I take the positive conclusion that rather than 'anti-industrial attitudes' among select groups being a major problem, it is probably more important to investigate the construction of of the institutional and technological environment within which entrepreneurship could be expected to flourish.

The Strategy of 'Switch' to the Substituting Technology

There is no evidence of the sailing ship effect in the cases discussed here, of electric versus gas lighting or of Solvay versus Leblanc soda production. There is no evidence in other cases sometimes cited of mechanical versus natural ice harvesting, and of steam versus sail in ships.[9] The general problem with the 'sailing ship effect' is that for an *acceleration* of old technology innovation as response to threat, we need some reason why the rate was previously *below* its potential rate. And it is reasonable to expect potential innovations to be developed and adopted for normal competitive reasons within a group of established firms.

That leaves 'switch' or exit as the major responses to substitution threat, with switch the more interesting, because it must be deliberate, and managed.

Although the alkali case was really an analysis of a decision *not* to switch, the presumption was that the merger of the fragmented Leblanc 'industry' into one firm created a management *opportunity* to switch. And in general, if established firms are fragmented, competitive within their technology and therefore individually lacking in resources, we can expect them to have trouble reacting to an innovative threat on an individual basis. The greater their control over the market – that is the more they succeed in deviating from the impossible ideal of perfect competition – and the greater their command over resources, the greater the *possibility* of a successful strategy of 'switch'.

This expected dependence of behaviour on the scale and nature of the threat *and* on old technology market structure makes it difficult to generalise about incumbent behaviour. Nevertheless, Cooper and Smith have reviewed eight distinct technologies that experienced substitution and found that in all of them there are successful attempts to switch into the new technology from the old. In their tentative conclusions they note that in *none* of their examples do the leading, established technology firms succeed in transferring this leading position to the new technology (Cooper and Smith 1992). They also concluded that successful switch was associated with early recognition and response to the threat.

So while it is obvious that a successful 'response' requires access to resources of expertise and finance, less clear is how and when firms perceive a threat and why and how they decide to enter, or manage, the new technology. It therefore makes sense to examine how firms rich in resources that faced major potential threats perceived and acted upon those threats. Watson Jnr's early introduction of computer technology into IBM is something of a classic story of successful technology 'switch'.

Successful Switch – IBM's Early Move into Computer Technology

The relationship between Watsons senior and junior dominates Watson Jnr's autobiography, significantly titled *Father, Son and Co* (Watson and Petre 1990). It is fair to say that the junior Watson lived under the shadow of his father's example and success and struggled with the issue of establishing his own self and identity in the face of that example, long before his father began to groom him to take over the company. As he began to be moved around the company for the sake of acquiring experience he had a great psychological need to acquire successful projects of his own that could help justify his rise to power – to himself, as much as to others. This personal motivation helped an early, amateur interest in computer technology to become a passion as it became tacitly understood by father and son that it would be the son who would decide the role this technology would play in IBM. But if Watson Jnr acquired a personal commitment to promote computer technology, the rest of the company did not necessarily agree with him and he provides us with a first-hand account of the difficulties of managing the 'switch' decision.

The initial problem was that although everyone in IBM was aware of the first US computers such as ENIAC in the late 1940s, it was not obvious that they were relevant to IBM's punch card business. Watson Jnr described his father as thinking of computers as giant scientific calculators, in a separate market realm to IBM's punch card accounting systems (Watson and Petre 1990: 189).

But it was difficult to ignore the potential threat when the university inventors of the ENIAC computer launched a new company, UNIVAC, with backing from some of IBM's most important customers and with a declared mission to replace punch card machines in the accounting office. In reaction to UNIVAC, Watson Snr had his engineers design and cost a magnetic storage device to rival the projected UNIVAC machines. According to Watson Jnr he would eventually reject their proposals because he distrusted the apparently more ephemeral nature of magnetic tape compared with the permanent, unerasable record provided by punch cards, a record that could be checked manually by clerks if necessary (Watson and Petre 1990: 194). So the first opportunity of entry was lost.

By 1948 three circumstances buttressed Watson Jnr's determination to do more in electronics. First, the only electronic device IBM sold, sold very well. Second, when he assigned a man to attend US engineering conferences, he was able to report a large increase in significant computer ventures, usually involving magnetic tape. Finally, some of IBM's largest customers were warning Watson that they had been told magnetic tape would wipe out the need for expensive punch card storage (Watson and Petre 1990: 195).

Watson Jnr knew that his father had to be persuaded to accept magnetic tape technology into the company, but would not accept a rationale based on the prospective death of the beloved punch card. So he set about finding such another rationale. In 1949 he organised a task force of 18 systems experts to study magnetic tape and decide whether to add it to IBM's product line. After three months they decided magnetic tape was *not* needed, so this attempt to smuggle electronics technology into the company failed. Watson then tried to persuade IBM's top salesmen of the advantages of magnetic tape, so that he could take their 'opinions' to his father, but they also insisted punch cards were better. So both these attempts to politically engineer a major shift in core technology failed.

Nor did Watson find an ally in IBM's major R&D laboratories at Endicott. The problem with the R&D laboratories was that its core expertise was in metal engineering, suitable for punch card machines, but irrelevant to the new magnetic tape technology. However, there was a small group of electronics experts within the then-minor Poughkeepsie Laboratory. They had also had their attempts to move into magnetic tape and computers squashed, this time by the vice-president of engineering, who had favoured instead a variety of advanced punch card projects (Watson and Petre 1990: 198). Nevertheless, the Poughkeepsie team did produce an electronic calculator, the 604 machine, that sold at 10 times the rate expected, and prompted Watson Jnr to rerate electronics technology as likely to grow faster and be more important than anyone had previously estimated (Watson and Petre 1990: 199). Once again, a direct approach to his father to authorise a switch to electronics technology failed, because his father called in the head of R&D who defended the laboratories' status quo projects (Watson and Petre 1990: 199).

The opportunity to act came when a finance officer showed that IBM's R&D spending as a percentage of turnover was falling behind the 3% of sales turnover established by RCA and GE. When Watson Snr authorised an increase in research spending, Watson Jnr took control of the appointment, had his favoured man installed and authorised him to begin hiring hundreds of electronics engineers – to create a new R&D competency. Within six years IBM had gone from 500 to 4000 engineers and technicians and had created a first-class R&D competence in the new technology (Watson and Petre 1990: 202).

The import into the company of new expertise was not the end of the battle. Core IBM marketing and sales managers remained sceptical and Watson Snr surveyed the effort closely but discretely. Although the new technology had an internal profile, until it generated successful products there was always the danger of a weakening internal political position and subsequent retreat. So when in 1950 Watson Jnr authorised investment in the largest single develop-ment project in IBM's history, he broke convention by taking no advice from sales and marketing people and patriotically naming the project the 'Defence Calculator' (later the IBM 701), a name that would forestall criticism of its commercial viability from these groups (Watson and Petre 1990: 205).

This large investment in R&D capability allowed IBM to have three models of computer on the market by the mid-1950s, the 701, 702 and 650. Watson called the latter, lower price machine the 'Model-T' of the computer industry and the combination of superior technology and a technically updated sales-force enabled IBM to move ahead of its competitors in computers (Watson and Petre 1990: 244). Only at this point could Watson Jnr write that it 'finally became obvious that we were giving birth to a new industry' (Watson and Petre 1990: 244).

By this account, Watson Jnr's contribution to IBM's success was his very early commitment to the new technology and his persistence in seeking to legitimate it within the organisation – his role was one of 'political engineer'. Although his motivation appears idiosyncratic in that it involves the relation-ship with his father, this makes the account of internal change more valuable because it occurred in advance of any general realisation of the significance of the technology, certainly in advance of any damage done by UNIVAC to IBM's established markets.

Transistor Versus Valves – Compromised Management Motivation for Switch Strategy

Recognition of a threat does not necessarily mean that a policy of switch can be managed. Perhaps the 'classic' case of established firms recognising a potential threat but *failing* to develop the substituting technology themselves is the electronic valve manufacturers' reaction to the advent of the semiconductor transistor as the basic component of the electronics industry. Perceptions of the new technology's significance and the motivation of those charged with its development are once again at issue, but this time intertwined with an industry dynamic that weakened the established firms' ability to confront the switch decision – the tendency for electronics experts to leave the established companies.

Bell had developed the transistor in part because of the high rates of burn-out and malfunction of valves, and when Bell announced the transistor to the world in 1947[10] Bell pointed strongly to its substitution potential for valves (Braun and Macdonald 1982). This proved premature: the development of the transistor would prove long and costly; nevertheless, there is no question here of the 'threatened' firms being unaware of the *potential* threat and many established valve electronics firms did establish research on the transistor. As late as 1959 Braun and Macdonald showed that Bell and eight valve manufacturers accounted for 57% of the total R&D expenditure on transistors and 63% of the major innovations (most of the latter came from Bell (Braun and Macdonald 1982: 60)). Braun and Macdonald remark:

> Despite all the money and effort put into R&D and the fine record of patent awards and innovation of the valve companies, the semiconductor market was early and rapidly falling into the hands of the new companies. This obviously requires some explanation. (Braun and Macdonald 1982: 60)

According to Braun and Macdonald, the two major 'managerial' reasons for this failure appear to be, first, that most valve companies gave responsibility for semiconductor transistor research to the valve departments run by engineers experienced in valve research and, second, that established companies could not retain scarce expertise in transistor research, which drained continually out of the established firms to new start-ups.

The problem with engineers experienced in valve research was that they saw the semiconductor transistor *too much as a valve substitute* rather than as a device that could be developed in its own right, that is as a basis for various novel logic circuits that would create the possibility of entirely new products. The basis for the valve engineers' limited view appears to be that they had learnt to improve valves through the application of technology gained through manufacturing experience. In contrast, the problems of development of semiconductor (sc) transistors required a thorough knowledge of the science of their operation and their exploitation for product innovation, rather than cost reduction.

As for deliberate management resistance, Braun and Macdonald do say that in the valve departments 'the transistor was often regarded, perhaps subconsciously, as the enemy, a threat to the status quo, and a challenge to personal, professional and corporate accomplishment' (Braun and Macdonald 1982: 51). But they do not give supporting empirical instances of this attitude leading to actions that blocked development.

What is clear is that the established firms' bureaucratically determined 'fair' pay and promotion structures in combination with uninspiring leadership of the development of the new technology drove semiconductor specialists into start-up companies. At the risk of repetition:

> Top management…frequently put electronic engineer managers in charge of their sc operations that didn't know a thing about the chemistry and metallurgy of semiconductor devices. … For about a year we had a going-away party every Friday. (Manager quoted in Braun and Macdonald 1982: 65)

The new companies often sought to recruit the best talent at high salaries to run R&D projects focused on radical transformative change in the new technology. When this worked, it redefined leading-edge practice and left other research programmes trying to catch up. Few examples are more stunning than Fairchild Semiconductor, formed to develop a new manufacturing process, the planar process, for superior quality transistors. This process became the basis for *all* modern production and led to a large rise in transistor demand and output in the 1960s as quality and productivity jumped. The importance of this innovation can be gauged from the fact that 41 companies had been generated by former Fairchild employees by the early 1970s (Braun and Macdonald 1982: 71).

There were other 'transformative' innovations, like the move from germanium to silicon as the major semiconductor material. By 1959, despite the established firms continuing to receive 78% of government R&D subsidies, the new firms had captured 63% of the market (Braun and Macdonald 1982: 71).

As explanations for the failure of the established firms to maintain their position in the new technology, despite early entry, it is difficult to separate out the relative importance of bureaucratically levelled reward systems, uninspiring technological development leadership and the continued loss of the people most committed to developing the new technology.

However, these features only mattered because semiconductor technology repeatedly yielded radical changes in the path of development and these made established gains obsolescent. A glance at Table 4.2 which lists the changes in ranking among the major semiconductor manufacturers over several decades shows that many of the 'new' firms also had trouble maintaining their market position. This was a technological game that any one management hierarchy

Table 4.2 The 10 leading US transistor-based device manufacturers during successive waves of technological change ranked by share of world market

Company	Valves	Transistors (1955)	Semiconductors (1960)	Semiconductors (1965)	Integrated circuits (1975)	Integrated circuits (1979)
RCA	1	7	5	6	8	9
Sylvania	2	4	10			
General Electric	3	6	4	5		
Raytheon	4			10		
Westinghouse	5	8				
Amperex	6					
National Video	7					
Ranland	8					
Eimac	9					
Landsdale Tube	10					
Hughes		1	9			
Transitron		2	2	9		
Philco		3	3	8		
Texas Instruments		5	1	1	1	1
Motorola		9	6	2	5	3
Clevite		10	7			
Fairchild			8	3	2	5
General Instrument				4	7	10
Sprague				7		
National Semiconductor					3	2
Intel					4	4
Rockwell					6	
Signetics					9	6
American Microsystems					10	
Mostek						7
American Micro Devices						8

Note: Data abstracted from Braun and Macdonald (1982: 123).

found difficult to play and the problems of this technology's development helped generate the venture capital industry we know today (Wilson 1985).

The Difficulty of Unravelling Valuable from Obsolescing Competences

Because the injunction to 'recognise and act early on a threat' sounds so much like good simple advice, it continues to be worthwhile to examine diverse sources of error in the process. Another classic source of error lies in the correct evaluation of existing competitive strengths, or 'competencies' to use the vogue term of today.

The president of Monroe, a leading electromechanical calculator company in the 1960s, made a 'strategic' commitment to enter the electronic calculator technology, but had this to say about the existing business:

> Our effort was not to create competition for these electromechanical machines. It was not to take away the established base but to seek new business over that base. (Majumdar 1977: 87)

As in the other cases presented here, market reports were misleading, because they failed to anticipate changes in the substituting technology that would transform relative market shares; Majumdar shows that an industry report of the time showed electronic calculators had higher prices and required full-time working to be economic over electromechanical calculators. The common interpretation was that they were therefore unlikely to penetrate the 'market' for calculators bought to make occasional simple calculations (Majumdar 1977: 89).

But the significant additional feature of the case is that the established firms were aware that they competed with one another by their brand names and distribution networks. The first represented the quality of production and the second the quality of service, or maintenance of the product, which like any complex mechanical product was prone to breakdown and to need repair. It was clear to the established firms that the new entrants would have trouble establishing ability in these areas – it was not clear to them that they would not have to. With the development of the single-chip microprocessor the reliability of electronic calculators surpassed that of mechanical calculators and the expensive distribution network turned from a 'competitive advantage' and a 'barrier to entry' into a desperate liability.

Recognition of a threat was not enough – what was missing was the detail of how that threat would evolve and impact on current abilities. However, in this case it is not clear that accurate prediction would have been any use to the electromechanical companies. The microprocessor that would transform the electronics industry was first designed as a 'calculator chip' and its success for that purpose commodified the calculator business so that anyone could manufacture a calculator.[11] The 'industry' characterised by independent specialised calculator firms had been permanently destroyed.

The continued management of a slow, or in this case rapid, decline and self-liquidation is not very exciting, but when there are no convincing alternatives, it accomplishes the extraction of maximum value from the depreciating assets of the established technology and the release of the remaining assets on to the market in a controlled manner.

Wilful Misinterpretation? Pascale on General Motors and the Japanese Manufacturing Threat

It should appear reasonable that the wrong technological expertise contributes to problems in the estimation and reaction of a threat, but other features of the organisation can block the ability to analyse and act – to learn. Pascale makes a strong case that the development of General Motors' management hierarchy during a long period of relative stability in the US car industry disabled its ability to analyse and respond to the threat of rising Japanese penetration of US car markets in the 1980s.

There is no doubt about the scale of the 'revealed' threat. US car producers lost approximately 30% of the US car market to mostly Japanese producers between 1955 and 1990 (Womack et al. 1990: 45). General Motors alone lost 15% of the US car market (Pascale 1990: 237). We know from the unusually large amount of comparative data on this industry (collected as part of the response to the problem) that by the 1980s the three US firms had fallen behind leading Japanese car producer productivity and quality levels (Womack et al. 1990).

Like Ford and Chrysler, General Motors (GM) sent teams of executives on Japanese plant visits; like Ford, GM formed a joint venture with a Japanese producer to build a plant to enable GM to learn directly from Japanese manu-facturing practices (with Toyota, at Fremont).

But the 16 GM managers who had worked directly at Fremont were dispersed around other plants to become ineffective, rather than being deployed as a team to attempt the transformation of existing plants. GM had also authorised two 'in-depth' studies of the Fremont plant that

> tended to pull their punches and reflect GM's biases in their conclusions...
> the effort to make the findings fit with GM's traditional worldview sufficed
> to allay any fears among top management that dramatic change was needed.
> (Pascale 1990: 76)

So the recognition of an external threat and its formal investigation and analysis did not lead to appropriate action. Whereas Ford began work on the imple-mentation of Japanese approaches to production in the early 1980s, GM persisted through the 1980s with an orthodox approach – that of transfer of plant to cheap labour countries and an increased pace of automation.

> How does one explain this persistent tendency of GM managers to ignore
> compelling evidence and the testimony of their own managers on the scene
> at Fremont? The answer lies in its collective identity and deeply etched
> social rules. It is almost beyond comprehension for GM management to
> contemplate the full-blown changes in status, power and worker relations
> that adaptation of the Toyota formula would entail. It is easier to install
> robots. (Pascale 1990: 76)

As explanation of this behaviour, Pascale points to the domination of senior GM positions by finance managers that stifled healthy debate and consolidated a low regard for manufacturing executives (Pascale 1990). Nor has Pascale a problem in identifying worker–manager salary differentials as a significant source of internal tension in GM; at the end of the 1980s the ratio of senior Japanese executives' compensation to that of their hourly workers was 10, but in GM it was 50 (Pascale 1990: 241). The result was an understandable alienation of the workforce from their senior management and a senior management reluctant to seek the forms of cooperation from their workforce that characterised the implementation of the Japanese manufacturing techniques. In this analysis there was no determinate point at which GM would necessarily change its behaviour, because senior management interests had become divorced from the interests of the company. And although this has been a very brief excursion into internal organisational behaviour, it has been sufficient to ruin any simple ideas about how companies *must* react to strong technological threats.

Extension of the Argument to the US Car Industry

Many observers have criticised US management and the management of US car firms in particular for their failure to manage innovation appropriately in the 1980s.[12] Pascale criticises how 'competition' became centred around styling (Pascale 1990) rather than the innovative efforts to transform operations that characterised many of the Japanese firms. The US car industry becomes interesting as an example of an industry that had failed to maintain a rate of innovation revealed to be possible by the entry of Japanese car makers.

From the DuPont history we already know that there is a degree of managerial discretion over the quantity and form of innovative effort. An effort to exploit this idea and generalise about competition is made by Lawrence when he describes the US car oligopoly as suffering from too little competition to maintain adequate innovation (Lawrence 1987). This view is derived from Lawrence and Dyer's detailed review of the role of competition in seven US industries (Lawrence and Dyer 1983), and Lawrence tempts us with a general 'competitive principle':

> An industry needs to experience vigorous competition if it is to be economically strong, either too little or too much competitive pressure can lead an industry to a predictably weak economic performance characterised by its becoming inefficient and/or non-innovative. (Lawrence 1987: 102)

It is implicit in this principle that 'competition' means 'number of firms', and 'competitive pressure' loosely translates as 'pressure of numbers'; it is not a reference to the competition provided by innovation. The problem is that industrial structure is not independent of the kind of innovative opportunity available; DuPont transformed industries composed of small firms into a large firm because that way it could support the costs and own the benefits of organised R&D. And if innovative opportunities remain small scale then innovation can continue in an industry populated by many firms.

So although in Lawrence's terms, the US car industry was non-innovative because there were insufficient firms 'competing', one can suspect that 'number' is not enough and wonder whether the addition of several more car manufacturers, structured fairly alike and with a similar dependency on the US context, for example the form of US industrial relations and supplies of worker-level skills, would have generated more innovative behaviour.

A better explanation of the US car industry's failure to innovate could be called 'institutional'; past events and institutions largely outside the direct control of the industry were originally responsible for the industry's reduced ability to control innovation. This is essentially the form of Lazonick's explanation (Lazonick 1990). In particular it is the specific event of the Depression of the 1930s that led to the resurgence of organised labour in the form of industrial unions and the ensuing entrenched loss of management control over the shop floor workforce from this period that explains the poorer innovative performance *on* the shop floor. This issue of control and workforce will be explored more thoroughly in Chapter 7.

Discussion of the Management of Substitution Threats

Whatever the context, the management of substitution threats is fraught with uncertainty at every stage. There is the problem of recognising the scale, extent and timing of the developing substitution threat when the substituting technology is alien to established firms' experience and is evolving in a manner not deterministically understood by the innovating firms. And yet this issue of 'perception' is not fully distinct in practice from the issue of the motivation of management to act in order to thwart the substitution threat. Internal structures and established patterns of innovation befogged perception of the threat and addled the response to it.

Compromised management motivation could itself be managed by developing the new technology in a 'spin-off' firm, and this is the advice of commentators like Christensen (1997). This is certainly no magic solution for two reasons: it still requires the established firm to have early recognition of the significance of the potential innovation, something that we have seen is no mean feat; and it doesn't save the spin-off company from all the possible problems of developing a new technology, such as knowing how to recruit the better people, avoiding false development paths and new waves of substituting technology. The systematic cultivation of spin-offs by established firms is known as 'corporate venturing' and does not have a very wonderful track record from the evidence of the 1980s.[13]

The real 'general solution' to this problem is not 'firm strategic', but an institutional innovation – the development of the early venture capital industry is testament to the inability of existing firms to recognise and develop all the good ideas their people could generate. From the perspective of the public good, the existence of venture capital ensures that the variable development ability of established firms does not matter – good ideas will be developed *somehow*.

Yet *some* large multidivisional firms have been able to manage systematically the substitution threat in their core technologies. DuPont management remained committed to innovation regardless of substitution problems, so that the company historians go so far as to comment that:

> There has probably never been a clearer case of the Schumpeterian principle of technological innovation bringing about the destruction of capital than with nylon's, Orlon's and Dacron's impact on rayon and acetate. In 1941 rayon and acetate plants represented DuPont's largest investments. From 1954 to 1963 DuPont progressively closed down its rayon plants, and the company made no new major investments in its acetate business after 1950. While painfully aware of this massive destruction of capital, DuPont's executives could appreciate that they, not their competitors, could fully exploit the inevitable product cycle. (Hounshell and Smith 1988: 387)

Significant in this case is the confidence, inevitably linked to the use of the patent system, that DuPont would be the company to reap the benefits of substitution.

DuPont did experience internal attempts to take control and subvert substituting technologies,[14] but senior management remained loyal to the company rather than specific businesses and never made the mistake of yielding to internal interests.

It is time to attempt a conclusion on this issue of substitution threat and the large multidivisional company. We know from the US valve and car industries that organisational form and size alone do not magically convey an ability to manage such threats. The potential great advantage of a multidivisional company is that it is also to some degree 'multi-technology'. It operates with a range of businesses and technologies and is unlikely to face full creative destruction across all businesses. So once threat is recognised such companies have the option of managing them by keeping their development out of the hands of departments or divisions that might not wish to develop them vigorously. If DuPont or IBM can be drawn upon, this requires a preparedness at the most senior level of the firm to survey, investigate and experimentally develop new technologies while also being able to appraise the internal political opposition and if necessary to actively bypass, suppress or outwit that opposition.

Conflict between Competition Policy and Innovation Policy

What could be more natural and right than that economic theory should be the basis for competition policy and law – where else could you find a relevant and principled basis for such policy? But as economic theory has changed over the decades, then so has the basis for antitrust law, and in the USA the practice of the law has come under pressure to change in response.

There are two, interlocking arguments that antitrust policy has harmed corporate innovation. One is based on an understanding of the evolution of economic thought and seeks to show how past theory was erroneous – therefore this theory formed an unsound basis for legal practice and the legitimacy of past cases can be suspected. The other argument is based on an examination of antitrust cases and how they evolved in practice – were the alleged anti-competitive practices proven?

There is no room here for a history of economic thought – but Weston provides a summary of this within his admitted caricature of 'Chicago' versus 'Harvard' schools of economic thought (Weston 1984). The Chicago school have been in the ascendancy from the late twentieth century and have successfully cast doubt on the idea that many 'anti-competitive' practices should be prosecuted with antitrust legislation. In this sense they can be understood to have practically elaborated Schumpeter's view that it should be a 'matter of indifference' whether markets are perfectly competitive, provided the act and capacity to innovate continues, for as a popular economics textbook comments:

> Competition policy is still typically concerned with the price abuses and restrictive practices of monopoly, and Schumpeter remains the most famous critic of the perceived need for such a government 'competition' policy. (Lipsey 1989: 429)

The paradox is that while innovation can generate transformative change through substituting competition, it may rely on firms being able to control the operation of 'normal' competition that otherwise threatens to restrict their ability to dispose of resources. Yet this control may breach certain theoretically defined standards of 'competition' and so trigger a US government antitrust case.

Weston points to unique features of the operation of US competition policy that help explain the US experience of a series of notorious 'anti-innovation' cases and why policy could change for the better from the 1980s, even with only limited new legislation. The US antitrust laws make only very general prohibitions against combination and monopoly and so

> confer enormous discretion upon our judiciary in deciding what specific practices are prohibited. ... Ultimately our Supreme Court adopted the "Rule of Reason" standard of interpretation. This requires proof that the restraints are "unreasonable" restraints upon competition after all the facts indicating the *purpose* and *effects* upon competition are shown. (Weston 1984: 270)

In addition two distinct government agencies, the Justice Department (JD) and Federal Trade Commission (FTC), have the ability to interpret antitrust law themselves and to institute antitrust suits in what they believe is the public interest. Also relevant is that in contrast to many other countries, it is possible for private interests to bring cases under the antitrust legislation – there are far more cases generated by private interests than by government (Weston 1984: 270) and so there is a danger that if antitrust practice becomes inimical to innovation, companies will be able to sue their more innovative competitors.

This brief institutional characterisation should make it plausible that fashions in economic theory can become embedded in the administration of the FTC and JD (in part through political appointments) so that their interpretation of the public interest leads to the generation of antitrust suits that must then be dealt with by the courts.

This has happened many times in US history. IBM provides some good examples, first, with that company's effective monopoly in its punch card business, with 90% of the market and 27% pre-tax profit on every dollar of revenue (Watson and Petre 1990: 216), which generated the profits that enabled it to enter computers. This same punch card monopoly also generated an antitrust suit against the company that ran from 1952 to 1956, when IBM signed a consent decree with the JD that indicated that IBM agreed to the JD's proposed remedy in this case. This involved allowing other companies to manufacture under IBM patents and to sell as well as rent IBM machines (Watson and Petre 1990: 219). This requirement to diffuse previously proprietary technology to competitors as a 'solution' is a typical feature of the resolution of such cases – with an obvious destructive effect on the incentive to research and accumulate proprietary technology.

At the heart of this type of case is the appropriate definition of the market – the JD considered the punch card business a distinct business, so IBM consequently was defined as a monopolist. Watson Snr believed it part of the wider 'business calculator market', in which IBM could not be described as a monopolist, but as a niche player. By his account, Watson Jnr simply thought the costs of fighting the case were higher than signing the relatively mild requirements of the consent decree, given that IBM was soon to exit this obsolescent business (Watson and Petre 1990: 219).

In retrospect it is easy to see that the JD had picked a monopoly that was about to be destroyed by the computer industry that IBM itself was rapidly developing. Its careful requirements to allow 'competitors' access to the punch card business were soon rendered irrelevant by the far more important process of creative destruction. Yet the truly important prospective technological competition from computers and IBM's dependence on its 'monopoly' in punch cards to finance entry into computers did not feature in the JD's case against IBM. Nor did other computer developers, the real rivals to IBM's computer ambitions. If innovation matters, then these dynamic aspects of IBM's market position should have been assessed, although this places similar burdens of technological knowledge and prescience on the antitrust authorities as fall upon firms. Had the dynamic aspects of the case been assessed as we can now assess them in retrospect, the case would surely never have been brought.

The more serious and damaging JD antitrust case against IBM began in 1969, went to trial in 1972 and was only ended in 1982 by the Reagan Administration, on the grounds that the process was simply not valid (Langenfeld and Scheffman 1989: 19).[15] This time, IBM was accused of monopoly in the computer business and the proposed solution was the break-up of the company. However, the case ran on for so long that it mutated from a monopoly case to a 'strategic abuse' case:

As the investigation and trial evolved, more emphasis was placed on strategic aspects of IBM's conduct such as model changes and product modifications and introductions, changing interfaces and creating special machines to compete with particular rivals (so-called "fighting machines"). (Langenfeld and Scheffman 1989: 20)

Langenfeld and Scheffman cite the IBM expert economist at the time on the extent of the government allegations:

> According to the government, virtually all of the System/360 computer products that IBM introduced in and after 1964 and marketed for the remainder of the 1960s were anticompetitive product introductions or "fighting machines". (Fisher 1983, cited in Langenfeld and Scheffman 1989)

In other words, the JD had come to persuade itself that the introduction of computer *models* in a response to competitors' models was 'anti-competitive':

> The basic problem in *IBM*, was that the actions challenged by the government generally represented innovations, either technical or in marketing or distribution, that customers valued. (Langenfeld and Scheffman 1989: 21)

The first IBM case illustrates the problem of applying competition law that is framed to deal with abuses of static monopoly power to a technologically dynamic sector. The second illustrates the problems with *assuming* certain business strategies are deployed to *maintain* a dominant position by deterring market entry (that is, as anti-competitive tools), rather than as the means for creating a dominant business position in the first place through innovation. The origin of this second type of case in past economic thought deserves a little more explanation, since this same thought also spawned the once-dominant, and still widely used, business and management strategy books written by Michael Porter.[16]

By the 1970s economists' attempts to explain the problem of continued concentration in the economy had come to favour the existence of 'barriers to entry', such as economies of scale, advertising and R&D expenditures, and the use of 'business strategies' to block competitors' entry into markets. Because the bias was to see existing oligopoly as a problem that needed explanation, hypothetical strategies, such as innovation suppression strategies (for example, the research, purchase and suppression of patents), were postulated as possible means by which monopoly positions were being preserved.[17] This business strategy literature ignored the issue of how these firms came to have dominant positions in the first place – and thus ignored the possibility that it was through R&D and innovation (Langenfeld and Scheffman 1989: 18). The attempt to distinguish between R&D for innovation and as an anti-competitive weapon would be left to the courts – the JD case against IBM was a typical result in that:

> It may be that some of IBM's actions were predatory (i.e. they were socially inefficient actions taken to disadvantage rivals), but almost 7 years of trial were unable to establish that fact. It was a sobering experience for antitrust authorities, lawyers, economists, and IBM which, it is estimated, incurred more than 100 million dollars in expenses defending the case. (Langenfeld and Scheffman 1989)

Damage *was* done to IBM as an innovator – not only the $100 million spent on defence at the trial, but the absorption of senior management time over a decade, distracting them from their business obligations, and possibly conditioning their future innovative behaviour to avoid the possibility of future antitrust suits (change of behaviour is, after all, one of the intentions of antitrust suits).

The failure of the US government to achieve successful prosecutions in all but one 'business strategy' case and the obvious damage done to the companies prosecuted, legitimated a policy reaction under the Reagan Administration of the 1980s and the rise to political influence of the 'Chicago' school of economic thought and its hitherto minority view that the many government-generated antitrust suits of previous decades had done nothing to aid 'competition', but had, if anything, weakened corporate incentives to innovate (and thereby the corporate ability to generate real Schumpeterian competition) (Langenfeld and Scheffman 1989).

This administration therefore attempted with some success to relax antitrust activity, through changes to the antitrust law (the passage of the National Cooperative Research Act in 1984 that legitimated R&D joint ventures) and through key appointments (for example, to the FTC).

This is not simply old history demonstrating a past victory for common sense, now firmly established. The Clinton administration returned – once more – to a more active antitrust policy, once again through political appointments to the JD and the consequent generation of the antitrust suit against Microsoft. Once again there was available a fashionable economic theory, described in Chapter 3, concerning path dependency and the theory of lock-in to suboptimal standards. This theory helped generate and legitimate the more active policy in the case of Microsoft (in particular the argument that control of the PC operating system was a monopoly) and once again, one might conclude, that theory was found not to be adequate to the task.

To Modern Times – the Microsoft Trial

Many of the same problems for antitrust action arise again in the Microsoft antitrust trial: the problem of defining a market in which monopoly control can be alleged (Microsoft's control of the PC operating system); the need to prove that this control is being used in a way that is 'anti-competitive' and harmful to the public (Microsoft was alleged to have 'leveraged' its control of the operating system to extend its monopoly to other markets, especially in respect of Internet browsers); and most important, the appearance in the real world and at the same time as the trial of a previously unconsidered form of technological competition that weakened the JD case.

In the Microsoft trial, the allegation of leveraging centred on the status and use of the Netscape Navigator browser as the major alternative browser to Microsoft's Internet Explorer. Judge Jackson found in his 1999 judgment against Microsoft that since Navigator was able to work on many types of incompatible computer (not just Windows on IBM PC clones) its continued independent existence created the possibility that applications writers would write Web-based software in the Java language, which would run on a variety of operating systems (Kehoe and Denton 1997). In other words, an independent Netscape Navigator sustained a potential competitive threat to Windows and Microsoft

(Hamilton 1999) and if Navigator were to be completely eliminated, Microsoft would be closer to being able to set future Internet technology standards in ways that advantaged Windows and itself. Therefore Microsoft's use of 'anti-competitive' practices to damage Netscape was all the more serious.

As so often in these cases, real-life events threatened to supersede the slowly and carefully established facts and judgments of the trial. When America Online bought the declining Netscape for $4.2 billion, Microsoft's defence promptly claimed that this merger would encourage applications writers to start writing software to run over the Internet, so threatening Windows (Wolffe 1999). Microsoft was now using in its defence the realisation of the scenario that the judge had accused it of trying to avoid through anti-competitive behaviour. Observers were able to conclude that 'anti-competitive' behaviour this ineffective hardly justified an antitrust trial.

The trouble with the judge's 'anti-competitive' scenario was that it was just that – a scenario of future innovation-related events that no one could be completely confident could be engineered by Microsoft, however powerful. Liebowitz and Margolis provide a persuasive general 'counter scenario', that even if Navigator or the company Netscape had been eliminated, if a genuinely innovative form of software provision via the Internet was possible, then something like Netscape Navigator would have been reinvented (see the last chapter in Liebowitz and Margolis 1999). With the benefit of hindsight the emergence of Linux as a rival open-source operating system is even more convincing evidence that Microsoft's monopoly does face competition and is not necessarily permanent.

As a result of all this one might conclude that not only has US competition policy proved highly variable through time through its over-dependence on twists and turns in fashion in economic theory, but also it has likely damaged, to an unknown extent, the capacities of some US companies to innovate, even though the cases were brought in the name of the public interest.

On the other hand these experiences with innovation and antitrust can be treated as the results of giant experiments in the laboratory of the economy. Repeated failure by the courts to establish evidence of behaviour assumed by certain economists to take place gradually eroded those very theories. Other countries have quite different competition policy procedures and institutions, but they are all influenced by the product of the great US experiments.

Despite the swings, there is also a cumulative and evolving aspect to US policy. As an example, US antitrust law was designed in such a way that cooperative R&D between companies was likely to be judged as a form of collusion. Once cooperative R&D became credited with aiding successful Japanese penetration of US high-technology markets (see for example Anchordoguy 1989) efforts were made to amend the law on this specific matter. Jorde and Teece describe their lobbying efforts for such an amendment, what became the National Cooperative Research Act of 1984, on the Schumpeterian grounds that real competition is driven by innovation, therefore cooperative activity between companies that increases their ability to innovate should *not* be judged collusive and against the public interest (Jorde and Teece 1991). Immediately after the passing of the Act there was a rapid growth in the number of US R&D joint ventures, so demonstrating the suppressive effect of previous antitrust law. Such amendments and the remembered experience of major antitrust cases such as *US* vs. *IBM* suggest that the evolution of antitrust practice has a progressive as well as a chaotic aspect.

Many of the arguments rehearsed here will prove important in the next chapter as the basis for an understanding of the operation of the patent system.

Concluding Comments – and Does the Study of Innovation Necessarily Follow a 'Chicago' or 'Schumpeterian' School of Thought?

The complicated case of the alkali industry resolved into a case where the established Leblanc producers could not risk wholesale late entry into the developed Solvay technology. Successful switch in response to substitution threat was generally associated with early recognition of said threat and early action – but very early action was bedevilled by the difficulty of building a consensus around the 'correct' perception of the nature and extent of the substitution threat within a community practised in the established technology. The problem of whether, when and how to manage the problem of a switch in technology is one of the toughest management may encounter.

This chapter began by citing Schumpeter, but ended by following with approval the 'Chicago school' reformulation of antitrust law as applied to innovation. It used Donald McCloskey's work, a former member of Chicago University's Economics Department, but the book's introduction cited Misa, Swedberg and Granovetter in criticism of the tendency of the New Institutionalists – another Chicago-derived school – to ascribe motivations of efficiency to observed firm behaviour. In the next chapter, the legal scholar Edmund W. Kitch's economics–unorthodox view of the patent system will be discussed, and any devotee of 'schools of thought' would almost certainly class Kitch as 'Chicago school' in nature.

The reason Chicago economists feature is because a long-standing strain in the Chicago school of economics has been an antipathy to quantitative *modelling* of economic processes, whether macroeconomic and Walrasian or game theoretical.[18] In other words, the Chicago 'school' gain much of their *apparent* coherence from their refusal to follow mainstream economics. Perhaps it is like this; innovation as the study of uncertain decisions that generate qualitative economic change over time has more often entered the concerns of economists in the Chicago school than in mainstream economics. However, the differences between Chicago economists over time have been considerable[19] and if an understanding of innovation is the object, there is really no alternative to a case-by-case consideration of the merits of each argument from whichever 'school' it derives.[20]

Notes

[1] Within the management literature the related idea of variable competitive pressure within a population of firms is a cornerstone of the organisational ecology literature, for example in Hannan and Freeman's book (Hannan and Freeman 1989). However, this literature assumes away innovation and the associated uncertainty over opportunities in favour of a static analysis of the

population of given industries. Many other dubious assumptions are bundled into the approach for the sake of making quantitative modelling possible.

[2] There is no better evidence for the effect in Utterback's other case, of natural ice harvesting.

[3] This was probably unavoidable because only a year separated the work of Reader and of Lindert and Trace. Lindert and Trace's work was almost certainly in press when the Reader work was published in 1970.

[4] 'Soda' is in the dictionary as shorthand for the sodium compounds of carbonate, bicarbonate and bleaching powder, but this meaning is absent from modern chemistry textbooks and probably should be considered as archaic as the Leblanc process.

[5] The full description of the Leblanc process follows: it had more wastes and byproducts than the Solvay process. It began with the treatment of salt with sulphuric acid, which produced hydrochloric acid. This was originally discarded, but by the second half of the nineteenth century it could be used as a key input for the manufacture of bleaching powder, for which there was a new and growing market. Later, chlorine could be extracted for another distinct market via the Deacon process.

The second stage of the Leblanc process involved the reduction of sodium sulphate with limestone and coal or coke to produce either the carbonate or the hydroxide (caustic soda) and 'tank waste', which was mostly calcium sulphide. So caustic soda could be made at the expense of soda production and was *more cheaply made* via the Leblanc process than the Solvay (Lindert and Trace 1971: 251). As demand for caustic grew, the Leblanc producer deliberately switched production from soda, so reducing losses and boosting profits at the same time (Haber 1958: 96).

However, whenever bleaching powder was made *either* sodium carbonate *or* caustic soda would have to be made. This meant that despite the growth of these new markets, the Leblanc producers would retain a limited interest in the state of the sodium carbonate market.

[6] The sulphur could then be recycled to the production of sulphuric acid, which itself had a growing diversity of uses through this period.

[7] In contrast to the UAC's reasonable behaviour in reaction to the rise of the Solvay process, there is good evidence that it bungled the chance to acquire the basic patents for today's electrolytic method for producing caustic soda and chlorine, thereby ensuring its final decline (see Howells 2002).

[8] This argument about the role of anti-industrial attitudes in British universities is taken up more fully in Chapter 7.

[9] See Howells (2002).

[10] An antitrust action against Bell gave it an unusual interest in disseminating the technology widely.

[11] Further information on the evolution of calculators can be found at desktop.calcmuseum.com/.

[12] For example, Hayes and Abernathy criticise Ford for buying a calliper brake supplier, investing heavily in its automation, then being reluctant to write off the investment when disc brakes were developed elsewhere (Hayes and Abernathy 1980).

[13] Wilson reviews many examples of corporate venturing in the late 1980s as established corporations sought to reap the gains of the venture capital industry (Wilson 1985). They largely failed – there are good reasons why the institution of venture capital works as it does.

[14] See for example the problem of new paints (Hounshell and Smith 1988: 143).

[15] This account draws on Langenfeld and Scheffman's brief review, but there is no need to be brief – each side wrote a book justifying their position once the trial had been ended. For the case against, see DeLamarter, a senior economist for the Justice Department; for the defence, Fisher et al. (DeLamarter 1986; Fisher et al. 1983).

[16] Porter's work evolved, but it never lost either the assumption of barriers between defined market areas or its advisory focus on building entry barriers and deterring entry to market areas. Porter's work is characterised by its *assumption* that firms can and do control these matters. The origin of this confident assumption is the economics literature assumption that firms with dominant positions must be 'doing something successfully' to maintain a position that otherwise 'ought' to disappear.

This dependence of Porter's work on an old stage in the development of economic thought should make would-be readers of his books on competitive strategy wary (for example, the widely used (Porter 1980; Porter 1985).

[17] See Chapter 5 for a discussion of the evidence for the abuse of patents to suppress innovation.

[18] A more detailed analysis of the Chicago school, internal differences and changes over time can be found on the New School's website, cepa.newschool.edu/het.

[19] See the New School website.

[20] Another reason why Chicago school economists are difficult to avoid is implied by the information on the University of Chicago website that 22 Nobel Laureates in economics have been awarded to university students, researchers or faculty who work or have worked in the university (economics.uchicago.edu/). The Chicago 'school' have long been associated with the most successful and innovative thinkers within the US economics profession of the last half-century.

5 Intellectual Property Law and Innovation

The Patent Institution as an Aid to Innovation[1]

Patents were introduced as a general means of stimulating innovation in the early modern period: England has had patent laws since 1624, while the US constitution made specific allowance for such laws at its inception (Rosenberg 1975: v).

A description of the function of the patent can be obtained from one of the many accessible national patent office websites and would usually run as follows: intellectual property law is the means society has for creating conditional property rights in several distinct intellectual fields. These rights are called patents, copyright, trademark and registered or unregistered designs. The most relevant to technological innovation is the patent, granted to an applicant for an invention, for a limited period of time during which the property rights can be bought, sold, or licensed to others to use. In return the applicant deposits a description of the invention with the patent office where it is available for anyone to view. Once the patent – and any extension – has expired, anyone can use the described technology without fear of legal infringement action.

This might appear simple enough, but the academic literature contains radically different assumptions about the role patents play in innovation.

Kitch makes a nice argument that the prevailing understanding of the function of the patent system among economists, what he calls the 'reward' theory, must be supplemented by an understanding of the institution's function as a means of managing technological prospects, prospects that are uncertain in both cost of development and potential return (Kitch 1977: 266). The significance of these distinct views will be developed here at some length, for it is the reward theory, spilling into the management and business history fields, that is largely responsible for the belief that corporations suppress useful innovation with the aid of patents. It has also been the reward theory that has supported the application of US antitrust law to restrict corporate patent positions *in the belief* that the monopoly power that these patents represent must be being abused.

This chapter explores the strengths and weaknesses of patent protection of innovation through the example of Dyson's invention of the cyclonic vacuum cleaner. The evidence for the suppression of innovation through patent abuse is then addressed, first in Reich's history of radio innovation and then in other cases: although the evidence for innovation suppression is lacking, a useful set of scenarios for the strategic exploitation of patents is developed. The larger idea is introduced that the intellectual property law has been and continues to be adapted to specific technology development needs. Finally, it is argued that the complexity of the machinery of intellectual property law does enable certain kinds of limited corporate strategic abuse.

A Contrast between the Reward and Prospect Views of the Patent Institution

In Kitch's terms, the 'reward' theory assumes that a patent represents an economic monopoly, granted as a reward for past inventive efforts. The usual economic justification for this grant is that a patent *should* allow an applicant to capture the full returns on the investment made to obtain the invention, these returns being otherwise appropriable by imitators. So the public function of the patent system is to offer a financial reward through a temporary monopoly that in general 'tends to make the amount of private investment in invention closer to the value of its social product' (Kitch 1977: 266). With this view, it then becomes a concern that this very same patent monopoly may damage the social welfare because the owner extracts monopoly rents from the sale of the invention. One's judgement of the value of the patent system now tends to depend on whether one believes the value of its posited 'invention incentive' function is outweighed by the assumed social losses of an increase in monopolistic pricing practices. The many economists who have followed this line of reasoning have tended to have equivocal judgements of the value of the patent system.[2]

This classic economic argument assumes that the development and cost of the invention is *prior to the grant of the patent*. Instead, it is common for important patented inventions, like the laser, to require further development and when they do the value of the patent's exclusive development right is eroded. This argument therefore pays no regard to the uncertainty of costs and returns in further development. It also assumes that a patent grants an economic monopoly, but this is by no means certain. Many relatively minor inventions, such as process technology inventions, may be useful without having an independent economic market; in such cases there is therefore no textbook economic monopoly (see the later section on this topic).

Kitch begins with a view of technological innovation as a series of development 'prospects', each with its own costs of development and probability of return, then argues that all patent systems in some degree work as a system of ordering 'claims' on these various prospects, akin to the way property claims were issued for mining prospects on American public lands in the nineteenth century (Kitch 1977: 266). The problem is that each technological prospect can be 'worked' by any number of firms and large amounts of resource can be spent on a prospect without the knowledge of other firms.

> This process can be undertaken efficiently only if there is a system that tends to assure efficient allocation of the resources among the prospects at an efficient rate and in an efficient amount.... The patent system achieves these ends by awarding exclusive and publicly recorded ownership of a prospect shortly after its discovery. (Kitch 1977: 266)

The prospect function of the patent serves to avoid multiple expenditures on the same prospective innovations. The public recording of patents at the patent office serves to warn rivals that work is in progress on this particular area and that competition can be avoided if they stake their claims (their R&D effort) elsewhere.

R&D proves many patents/claims are worthless, but if there is a successful 'strike' the patent system of property rights should protect the subsequent development and appropriation of the economic value of the claim.

It is probably not possible to make a general estimate of the economic significance of the problem of resources wasted on innovation prospects, but instances of such waste are common enough when a single innovative opportunity becomes suddenly and widely recognised, development ability is widely possessed and the patent system for some reason cannot be applied. This scenario is common among the innovations that are derived from semiconductor chip advances. Examples are electronic calculators, when the development of the microchip removed all barriers to entry, and disc drive innovations, developed for each smaller version of the PC. In both technologies simultaneous mass entry in pursuit of profitable opportunities ensured wide-spread profitless activity (Majumdar 1977: 156ff.; Christensen 1997). In both, venture capital played a role in generating too many companies in too short a period.

The exception that proves the rule is high-density polyethylene, where the patent system failed in its prospect role precisely because Karl Ziegler, the uni-versity academic who held the patents, had no intention of being the developer and sought only to license the patents widely to maximise royalty income. As described in Chapter 2, this triggered multiple firm entries and long-term excess capacity that depressed prices below cost for this rapidly expanding and useful product (Hounshell and Smith 1988: 493). If the patents had been in the hands of DuPont as developer, it is safe to assume there would have been an orderly expansion of capacity and higher prices that yielded a profit through the efficient use of capital.

Kitch points out that many rules of practice within the patent institution are intelligible only in terms of the prospect function. This is shown by patent claims being written not as descriptions of the patentee's specific invention but with much greater scope and in order to apply to a general class of artefact designs – to allow for future development. There are also many rules that encourage early and competitive filing of claims, such as the priority given to the first to file a claim over other claimants.[3] Such rules tend to ensure that the right to develop is secured by a single potential developer and that this right is publicly known, so that wasteful 'development races' are prevented.

Judicial Understanding of the Function of the Patent System Affects Court Judgments

We are now in a position to understand some of the spectacular reversals of court judgments on patent issues that are otherwise so inexplicable to a casual observer. Where the dominant understanding of the function of the patent system is the narrower reward theory it can influence a court to judge as invalid a patent previously granted. Kitch gives as an example a dispute over the patentability of an automated system for cleaning waste from dairy farms that reached the US Supreme Court. The lower court of appeals had upheld the patent, but the Supreme Court reversed its decision.

If one looks at the patent from the perspective of the reward function, one sees an unimaginative application of the natural forces of water, controlled by known automation devices to move cow droppings from one point to another. The Supreme Court conceived of the question to be decided as: Is this worth a monopoly? If one looks from the perspective of the prospect function, one sees all the problems of designing and marketing a reliable, durable and efficient system for automatic barn cleaning. ... The investment to achieve these objectives will be more efficiently made if the patent is held valid. (Kitch 1977: 284)

Kitch argues that the test of substantial novelty should be sufficient for the grant of a patent. Another effect of the widespread belief in the reward theory has been change in US patent law and practice to increase the amount of information disclosure in a patent, on the basis that the monopoly granted is in exchange for a public release of this otherwise secret information. This certainly raises the cost and complexity of patents, but where they function as prospect claims there is not a complete description of the innovation to be made public – further development is required (Kitch 1977: 287).

If the major economics-derived effort to produce a general justification of the patent system has faults, that does not invalidate the effort to produce a justification of the patent system as an aid to the conduct of the law. But that effort must surely be – as it is with Kitch – rooted in thoughtful abstraction from the role of patents in many diverse cases of innovation *practice*.

Individual entrepreneurs who invent and then attempt to develop that invention are very much dependent on the patent system. So entrepreneurial strategies of patent exploitation can provide an illustration of the patent both as uncertain prospect and as a highly flexible instrument for the management of development, given that uncertain prospect. This flexibility derives from the patent's status as a form of property and the ability of the holder to grant conditional licences for its exploitation to other developers. It was through exploitation of a series of conditional licences that James Dyson was eventually able to create his own company to develop his cyclonic vacuum cleaner invention for the European market.

Dyson as Heroic Entrepreneur and the Patent as Flexible Instrument of Managing Development

Heroic inventor stories require success against great odds – and this is how James Dyson presents his development of the cyclonic vacuum cleaner (Coren 1997). The cyclonic vacuum cleaner did become a great commercial success; in 1996 Dyson's company, Dyson Appliances, was the fastest growing manufacturing company in Britain and by 1997 controlled 43% of the British vacuum cleaner market, larger than the combined share of the established multinational rivals, Hoover and Electrolux (Coren 1997: 246). By the year 2000 turnover had grown to £300 million, of which approximately 18% was being reinvested in R&D (Patel 2000).

Another figure gives a clue to how he became a success: by 1997 machines built to Dyson's patents had £2 billion of global sales turnover (Coren 1997: 247). The extensive licensing of his patent allowed Dyson to reap significant royalties from these sales.

Dyson's invention is the elimination of the conventional paper bag 'filter' of the vacuum cleaner by a superior dual cyclone air filter. The cyclonic filter accelerates and spins incoming dirty air to speeds that force airborne dust and debris to settle out into a container. Dyson had observed the method of cyclonic dust removal on an industrial scale, but by his account he had the idea to adapt the technology into the domestic vacuum cleaner because he was an irritated user of the dirty and ineffective 'bag' vacuum cleaner.

In 1982, after three full years of developing and testing 5127 cyclone prototypes he was sufficiently in debt that he decided to exploit the patent through controlled licensing (Coren 1997: 121). In debt, without reputation, without any means of developing his invention himself, he was completely reliant on the patent as a bargaining chip in negotiations with established companies.

He spent the first two years negotiating with European manufacturers but without success. Dyson encountered the great problem of credibility, both of himself and of the technology. Sometimes companies gave him no time to present the technology, his interpretation being that they assumed it could not work (Coren 1997: 133), sometimes the individual chosen to make an initial judgement 'judged' the technology as useless and so ended prospective negotiation (the marketing manager from Hotpoint). Dyson met the manager of the Electrolux worldwide domestic appliance division and heard him receive engineers' reports to the effect that the cyclone technology worked better than the bag. Although this manager then said he would back acquisition, his proposal was blocked by Electrolux's production centre, who were more concerned with the imminent launch of a conventional model. Dyson's experience was that while single individuals were sometimes enthusiastic, scepticism elsewhere in the organisation killed the deal or led to an attempted tightening of the licence conditions with the same result, as in the case of Black and Decker. The achievement of an independent, positive assessment of the technology and, more difficult still, a positive internal consensus to develop it, proved beyond the ability of these companies' internal political processes. The irony of this result would be that the European market would be left to Dyson to develop, in competition with the same manufacturers that had turned down his technology.

The problems of licensing from a position of weakness are demonstrated when Dyson was finally able to sign a licensing agreement with Amway, a US company that regularly deals with individual inventors. He was to find that a deal was not a deal; Amway agreed a deal, changed the terms to its advantage at the last moment before signing, then some months after the deal was signed sued Dyson for fraudulent conduct and deception (Coren 1997: 153). Dyson arranged a settlement and the return of his patents only to find that just as his next US licensee was ready to start production, Amway had begun to market its 'own' cyclonic vacuum cleaner. Now the irony was that Amway's infringement of Dyson's patent weakened his position with respect to his new US licensee, Iona. Iona had agreed royalty payments to Dyson in exchange for an exclusive

US production licence, but this action by Amway threatened to introduce premature competition to the US market whether or not Dyson had connived at it (he had not). Dyson had no independent finances at this point so he renegotiated the deal with Iona to receive less in royalties in exchange for Iona's pursuit of a patent infringement suit against Amway. This case cost Dyson £300 000 per year, continued for three years and almost forced him into dropping the infringement suit through exhaustion (Coren 1997: 182). Fortunately, after three years, Amway decided to settle out of court and become a co-licensee in the USA.

This was not the only deal that was not what it appeared to be. Various devices were used by the licensee to gain discretion over the level of royalties, despite the terms of the licence. So in a case where Dyson had agreed a licence that gave him a royalty as a percentage of the lower, wholesale price of the cleaner, rather than the retail price, the company Rotork introduced its own selling agent between Rotork as manufacturer and the wholesaler. The effect was to allow Rotork to reduce its initial selling price and hence Dyson's royalty (Coren 1997: 166). The ability to rig prices in such ways enabled a company to strike an agreement with an inventor according to the inventor's ideas of what was his or her due and then to adjust the financial outcomes according to the company's idea of his or her due. From the inventor's side, the only response to such practices was to preserve the right to revoke the patent. Of course it helped to have extensive experience at negotiating patent licences, experience that Dyson gradually acquired.

Then there were the problems of verifying that the licensee was abiding by the deal. Dyson felt that his Japanese licensee, Apex, had deliberately marketed the cleaner in Japan as a niche product with a high final price tag of £1200 (Coren 1997: 165) in order to 'stiff me, in no uncertain terms, of a considerable amount of the money I was rightfully due' (Coren 1997: 165). This was because the licensing deal had been agreed as 10% of Apex's sale price, but Apex sold at £200 to a series of wholesalers and distributors before the product reached its final retail price of £1200. This final price limited sales volume, so that Dyson only ever received the £60 000 minimum royalty from Apex. Despite his suspicions over its reported sales volumes, he gave up trying to verify its figures. Nevertheless, this early Japanese licence was of crucial importance to Dyson's finances and credibility when negotiating other licensing deals.

A major problem hindering a negotiated licence in such circumstances is that crank inventors bombard companies with claims to have improved a company's established products. Investigation of possibly valuable inventions is dogged by the problem that the crank may later sue the company for some development that the crank believes derives from the disclosed invention. The established firm then takes steps that secure its future position but damage the prospects of current negotiations. This was why in one of his encounters with Black and Decker, the company arranged to have a neutral patent agent meet Dyson off their main premises to assess his technology. This was why Hoover demanded that before any meeting, Dyson should sign a legal document that assigned the results of their prospective meeting to Hoover – he declined to meet them.[4]

It had taken over three years from Dyson's decision to license in 1982 to obtaining the first valuable revenue stream from the Japanese from 1986. He added income streams from a licence to exploit the US domestic market and a licence to exploit the worldwide market for commercial cleaners. By 1991 he had freed himself from the Amway lawsuit. It was only from this date onward that he was in a position to raise the finance to begin production on his own account for the British and European domestic markets. He had depended throughout on the flexibility of the patent licence contract to build personal and financial credibility from near nothing to the point that he could build his own business.

For good reasons, chronic mutual lack of trust runs through Dyson's account of years of negotiation with companies over licence deals. Yet an approach to these companies was only ever *possible* because Dyson's patent gave him sufficient legal security that he could risk disclosing design secrets to strangers he had no reason to trust. And in general, the legal basis of a patent provides the basis for most know-how or trade-secret licensing (Kitch 1977: 278).

The story is a good example of a patent working as an efficient prospect claim, first because the long periods of time 'wasted' in search of licensing deals can be considered 'development' time. Second, the further development of production facilities was required to exploit the licensed patent for a specific market, something that represented a degree of risk in itself. Third, the time-limited and exclusive nature of the patent gave Dyson every incentive to vigorously pursue realistic development deals and to manage that development flexibly through the licensing function. There is no doubt, of course, that exploitation of the patent finally secured Dyson a reward for his efforts.

Patents as a Means of Suppression of Innovation?

The theoretical economics model of the patent conceptualises its abuse as the abuse of the monopoly conferred by the patent – the holder would be expected to raise prices and so reduce social welfare. But if you believe a patent confers an economic monopoly, it becomes possible to believe in a distinct and far more serious scenario of abuse of the patent system – the theoretical management strategy of using patents to *suppress* the development of new and useful innovations.

In the management literature Dunford has assembled empirical evidence that purports to show that this corporate strategy of innovation suppression is widespread (Dunford 1987). The technology historian Basalla also writes as if this form of abuse is general and explains that 'once the corporation gained control of patents, the monopoly was used to suppress any inventions that might harm its own products or enhance those of a rival' (Basalla 1988: 121). MacLeod, writing from an industrial corporate position, dismisses the idea, but refers to it as a popular prejudice amongst independent entrepreneurial inventors, convinced as they tend to be both of the value of their inventions and the nefarious motives of corporations (MacLeod 1988).

So the idea has an appeal and a degree of plausibility and these alone make it a candidate for critical review. It will be shown that for Dunford's and Basalla's material, the belief that this strategy occurs is the result of the *assumption* that a patent represents an economic monopoly. The review of the evidence they cite has another positive goal: the exploration of real scenarios for the corporate exploitation of patents.

A Patent Does Not Necessarily Confer an Economic Monopoly

This confusion of a patent with an economic monopoly is Kitch's chief 'elementary and persistent error' in the economic analysis of patents (Kitch 2000: 1727). A patent, like any other property right, is a right to exclude others from the use of something, but this right does not necessarily grant an economic monopoly;[5] an economic monopoly is defined as the possession of sole control over the price at which a developed commodity may be sold into some specified market. In other forms of property the distinction between an exclusive right and an economic monopoly is obvious: one may own a house but not have an economic monopoly in its sale; on the other hand one may be a trader who possesses a monopoly in some good for some market without owning the good (Kitch 2000).

First, as argued for patents as prospect claims, a patent cannot provide an economic monopoly until the invention that it covers has been developed for commercial sale. But as Kitch argues, even after full development, the patent will only protect an economic monopoly if the patent claims 'cover all of an economically relevant market, i.e. there is no alternative way for competitors to provide the same economic functionality to their customers without infringing the claims' (Kitch 2000: 1730). This is nothing less than a return to the issue of the nature of competition in innovation. Where the innovation is an improvement on an existing technology and there is therefore an old technology substitute, the degree of improvement and the overlap in uses matter and together they affect the rate and degree of substitutability and limit the pricing power protected by the patent.

Kitch makes the general argument that by the very process of their creation, patents that protect monopoly power must be rare. A patent claim only applies to what is new and 'non-obvious' and the patent office examiner makes a search of existing knowledge (the so-called 'prior art') to check that this is so; for example, but not exclusively, in the patent office database of published patents. All depends on the scope of the patent claim.

> A patent issued in a well-developed field of technology will inevitably contain narrow claims. There is an opportunity for broad claims in patents on inventions in new fields…[but] since the technology is new there is usually very little demand for it. (Kitch 2000)

In other words, narrow claims may be too limited in scope to confer an economic monopoly; broad claims could do so in principle but have long development lead-times that tend to exhaust the time-limited protection of the patent. Kitch's point is not that patents never confer an economic monopoly, but that this should not be the typical scenario to be used in the analysis of the patent system. His presentation of these basic issues is useful when we turn to the possibility of the abuse of patents to suppress innovation.

The Nature of the Evidence for Innovation Suppression

The plausibility of the existence of suppression strategies would appear to be greatly enhanced when it is realised that there is no US patent or antitrust law requirement that a patented invention *must be developed*; there is nothing directly illegal about the *non-use* of a patented invention or the *suppression* of a patented technology. However, 'suppression can be deemed to have occurred for anticompetitive reasons and can therefore be the focus of successful antitrust action' (Dunford 1987: 514).

No company would want to boast about its suppression of useful technology and so Dunford's method is to use documented cases of US antitrust suits brought against the strategic abuse of patents as a rich source of evidence of a practice that would otherwise be difficult to research (Dunford 1987).

However, as discussed in the previous chapter, the use of legal cases as evidence must take into account that antitrust law is itself subject to modification through economic assumptions. The reality is that in the USA, antitrust law has in the past been actively adapted to combat the *supposed* anti-competitive abuse of patent-conferred economic monopolies. Even so, most cases have not involved the validity of the patent itself, but its 'interface' with antitrust law. This interface largely concerns the type of restrictive conditions of use that may be inserted into patent licences, for example to control or influence markets unconcerned with the patent at issue.

Dunford claims to have created a list of distinct suppression strategies from 'a core of several hundred cases'[6] of such legal judgments. However, the case of radio is a particularly important illustrative case of innovation suppression for both Dunford and Basalla and deserves attention in its own right.

Reich's Account of Patents in AT&T's Strategies for Radio Technology

The role of patents in the development of radio was particularly complex, and Dunford uses AT&T's patenting strategies for radio research to illustrate three types of patent strategy. The two most important suppression strategies he calls 'patent blitzkrieg' and 'patent consolidation'. Patent blitzkrieg is the hypothetical strategy of taking out many minor patents *in order to block* any potential competitor's entry into a technological field, and patent consolidation is the practice of acquiring a few key patents that because of their broad scope and fundamental nature again confer control over an entire field (Dunford 1987: 517).

Basalla also uses radio patenting practices as a significant case of patent abuse and follows Reich in extending the argument to cast doubt on the generally assumed value of the modern industrial R&D laboratory: if patents are used to suppress innovation and if R&D is routinely used to accumulate patents, then the R&D laboratory is far less valuable a social institution than we thought. However, both Basalla and Dunford depend for their evidence of patent abuse in radio on the account by the business historian Leonard S. Reich (1977).

If the difference between an exclusive right and an economic monopoly is kept in mind, it is possible to reinterpret the claims of abuse in the accounts by Reich, Basalla and Dunford.

In Dunford's account AT&T simply acquired the key patent on the triode from its inventor, Lee De Forest, for technology suppressive purposes (Dunford 1987: 515). According to Reich, the real reason was that AT&T wanted to escape competition in the supply of local telephone services through the development and control of long-distance telephone lines in the USA. For this AT&T needed a technology of amplification. The triode, the functional forerunner of the transistor, could act both as a signal amplifier and receiver and AT&T 'launched a massive attack on its theory, its mathematics, its construction, and its behaviour' (Reich 1977: 213). The triode enabled AT&T to build the first transcontinental telephone lines by 1915.

AT&T understood that as an amplifier and receiver, the triode would also be a key component in radio broadcasting and receiving, and that by controlling the triode AT&T would be in an excellent position to control radio.[7] But given the triode's importance to telecommunications, we cannot judge the purchase of the triode rights as *primarily* a defensive act and certainly not as a *suppressive* act, since AT&T would soon vigorously enter radio research as 'the first company to launch an organised effort aimed at perfecting a functional system of transmission and reception' (Reich 1977: 214). However, with the triode, AT&T would only fully control *one* of the fundamental technologies necessary for a working radio industry.

The First World War complicated matters because to promote radio production for military purposes the US government suspended the patent system in radio technology. As might be expected, radio producer companies moved into each other's 'radio-relevant' technological areas, and by the end of the war, 'Each manufacturer, pushing its researches as far as possible, repeatedly encountered some device or circuit that was already patented by another' (Reich 1977: 216). When patents were made valid again at the war's end, all these companies were technically in infringement of each other's patents.

The solution evolved by the radio producing companies was the creation of a jointly owned development vehicle, the Radio Corporation of America (RCA). In the 1920 deal, RCA gained access to all the relevant radio technology from AT&T, GE and Westinghouse and in exchange these companies obtained RCA stock and rights to future RCA patents, but also agreed restrictions on RCA's future exploitation of certain technological fields. It would be this last element of the agreement that would appear, at last, to provide documented evidence of suppressive intent.

Now we run into one of the potential weaknesses of historical business accounts based upon management quotations. What a particular management regime says it wants to control, and acts to control, is not necessarily *controllable* within the market economy. So under the RCA agreement, AT&T retained exclusive rights in wire communications under all patents and limited rights in radio linked to wire networks (Reich 1977: 217). If RCA was restricted from developing these areas, the question becomes what AT&T intended in obtaining these rights and, more importantly, what AT&T actually did with them.

AT&T did indeed reserve the right from RCA to develop two-way radio because it was very concerned that when developed this technology would substitute for its core business of local telephone services. Reich can cite a memorandum by Frank B. Jewett, a chief engineer of AT&T, expressing the view of AT&T radio research on the RCA agreement that 'if we never derive any other benefit from our work than that which follows the safe-guarding of our wire interests we can look upon the time and money as having been returned to us many times over' (Reich 1977: 220). Reich writes that AT&T had 'stymied the competition of two-way radio with its extensive wire network' (Reich 1977: 220), and it had 'blocked the "perennial gale" of competition before the first telephone line was blown over' (Reich 1977: 220).

A little further thought about the subsequent development of two-way radio – which did take place – suggests that two-way radio was then not in fact, as it is not now, a serious threat to telephone communication.[8] As it was not a very good substitute for local telephone wire communication, there was no 'gale' of competition to be blocked. Nevertheless, if this was not understood at the time, the statement by the chief engineer can probably be taken to be an expression of suppressive *intent*. The problem is that, had two-way radio proven to be a highly profitable technology with full wire substitution capabilities, we can doubt that AT&T's success in obtaining an 'agreement' with its competitors for the control of two-way radio could ever have been made into a *genuine* suppressive strategy.

This conclusion is supported by what soon transpired in the development of radio. AT&T's *attempt* to control two-way radio development was only part of the AT&T strategy and as part of the RCA agreement the company also secured exclusive rights to manufacture and develop radio broadcasting equipment.

Within a year AT&T reneged on the 1920 agreement. This happened as the company realised that it had grossly underestimated the commercial significance of the markets for broadcasting, but especially radio *receiving* sets, the right to which it had signed away, despite the technology being 'an art to which its engineers had made significant contributions' (Reich 1977: 221). In other words, AT&T decided it had signed away too much for too little – it had made a mistake. Rather than be constrained by the voluntary technology 'agreement' with its competitors, AT&T began an aggressive attempt to enter the operation of radio broadcasting and by 1924, in possession of superior receiving technology, began selling receiving sets permanently tuned to AT&T's new broadcasting stations (Reich 1977: 223). In other words, AT&T threw away its agreement with its competitors and voluntarily began to compete in pursuit of the lucrative spoils of full radio development. The gloves really came off when RCA's attempt at independent arbitration of the 1920 agreement went against AT&T. AT&T's response was to find a legal opinion that the agreement it had once found so satisfactory was a violation of the Sherman Antitrust Act, because it blocked AT&T from selling the radio sets it was legally empowered to make under its own patents (Reich 1977: 229). In other words, in the attempt to control *future* radio technology and markets, AT&T was prepared to stoop to use the law to undermine its own illegal agreements.[9]

The significance of this story for AT&T's *expressed intention* to control two-way radio development is that, had two-way radio proved viable and lucrative, the agreement to allow AT&T to suppress the technology would probably have proved as worthless as the agreement to assign the rights of radio receiver development to RCA. Rather than technology suppression, this case is one more demonstration that the combination of desire for profit, freedom of business entry and awareness of the benefits of successful development that characterise firms in a working market economy will triumph over the natural desire of established firms to seek protection against entry into their existing businesses.[10]

The example provides a useful reference when one is confronted by contemporary expressions of suppressive intent. They appear common enough: Dyson cites the vice-president of Hoover appearing on the BBC's *Money*

Programme to regret that Hoover had not bought the Dyson patent, because then it would have been able to suppress the technology (Coren 1997: 248). After the convoluted nature of the evidence regarding radio, it is refreshing to find such bald statements of anti-innovative *intent*. Unfortunately, we can reasonably doubt that the company would have suppressed the patent had it obtained it because if Hoover had acquired the said patent and developed the cyclonic cleaner itself, it would have gained a head start over potential competitors. If it did not develop, then it would have bought itself some years of security, but when the patent expired, the company would face the same problems of development, but without the head start. The deliberate non-development of a patent *understood to represent a useful prospect* makes no kind of sense. Had Hoover really acquired the patent and understood its economic significance, once the competent elements in the company's management hierarchy had thought through their real options, they would surely have concluded that they must develop the technology.

Reich's other arguments in favour of suppression are worth brief comment. He expresses his strong position as follows:

> In a significant sense, then, science has been compromised to the extent that research funds and researchers have been sacrificed to the essentially unproductive work needed to gain or maintain monopoly position, and pursuit of patents has been at the heart of this process. (Reich 1977: 234)

'Unproductive' for Reich means that the researching company does not use its patents in its *own production*. Reich finds three 'unproductive' uses of patents: to prevent competition, to enable patent trade and 'to prevent competitors from acquiring a strong patent position through their own research' (Reich 1977: 232).

Reich's example of 'prevention of competition' is patent blitzkrieg in the form of AT&T's 'thousand and one little patents' on the district telephone exchange system. This prevented competition by 'making its product ... distinctly superior to what the others could offer' (Reich 1977: 232). This quote represents an extraordinary example of how the beneficial workings of real-world competition are obscured for Reich by his belief in the assumptions and models of the economist, as if *they* were true.

Research in order to 'trade' patents, that is to trade the right to *exploit* the patent, only aided the *development* of radio technology. The relative bargaining positions of the researching firms did depend on the value of the component technology that they controlled, so they had a powerful incentive to attempt to make useful improvements to the technology.

The last of Reich's 'unproductive uses' for patents refers to RCA's requirement that its equipment licensees should give it an option on any radio-related patents they might develop. This is the better candidate of the three for an anti-competitive abuse, but it did not prevent the development of radio, rather it sought to lever RCA's existing radio technology dominance to maintain future dominance. It relied on RCA continuing to deliver state-of-the-art radio equipment that businesses would want to lease and so RCA retained a strong incentive to exploit all useful patents and technology. Whatever the rights and wrongs of the attempt to privilege RCA's development opportunities over other companies, there is no question here of the technology itself being suppressed. We can conclude that all three of Reich's 'unproductive' uses for patents were in fact productive uses.

The role of patents in the development of radio was complicated, but it is not a story of technology suppression. On the contrary, what it clearly demonstrates is the effective coordination of radio development through patent acquisition and licensing despite the extraordinary suspension of radio patent validity during the First World War. The result of that suspension was the extraordinary agreement between radio producers to create RCA. The written evidence of that agreement does record the *desire* of a chief engineer of AT&T to suppress one *potential* element of radio development, but in retrospect, there is no evidence of *actual* suppression of any element of radio technology.

Improvement Patents in Combination with the Use – or Abuse – of the Patent Pool

If radio failed to provide illustrations of practical patent abuse, it did provide some illustration of the problem of dispersed ownership of related patents and in RCA one of the solutions: the creation of a jointly owned company that itself had unified control over the relevant patents. If this was something of an extraordinary solution to the problem generated by the war, the more normal solution to dispersed ownership of related patents is the 'patent pool' (Dunford 1987: 518). The patent pool is the use of the patent licensing function to organise the exchange of patent rights between owners in order to enable technology development. It offers the interesting sight of competing companies cooperating to ensure technology development can proceed.

The content of the cross-licensing agreements varies greatly and once again suppression may be suspected, as when Dunford writes:

> Rarely is the pool aimed purely at removing barriers to the development of technology ... it is also a basis for attempting to control the members' common markets by eliminating competition from outside the group. (Dunford 1987: 518)

This 'elimination of competition from outside the group' is again better termed the controlled exercise of the exclusive property rights of the patents, and we should expect those excluded to be the same as those who have nothing to contribute to future technology development.

Nevertheless patent pools do provide some of the more complicated scenarios of technology development. Two particular features have generated much heat: when the original patents that form the basis of the pool expire, it has often proved possible for companies to continue exclusive technology development through the control of many improvement patents. The other complicating feature has been the frequent involvement of the US antitrust authorities in disputes over the contents of the patent pool licences.

A classic case that demonstrates both of these features is the US Supreme Court patent pool judgment *Hartford-Empire Co.* vs. *US.*[11] The original legal charge against this patent pool of glass machinery and glass manufacturers was that they

> agreed, conspired and combined to monopolise, and did monopolise and restrain interstate and foreign commerce by acquiring patents covering the manufacture of glass-making machinery, and by excluding others from a fair opportunity freely to engage in commerce in such machinery and in the manufacture and distribution of glass products. (*Hartford-Empire Co.* vs. *US* 1945, para 2)

This patent pool developed in a typical manner. As glass manufacture had become automated several developer companies found they possessed patents on parts of several processes, so they created the pool as a means of exchanging the various rights to process technology development. When the original basic patents had expired the developer companies continued their control over the process technology through the acquisition by research and purchase of over 800 'improvement patents' (*Hartford-Empire Co.* vs. *US* 1945, para 21).

The Supreme Court rejected many of the charges against the companies upheld by lower courts, but did find them to have engaged in two forms of abuse of patents. They had abused the patent licence to 'allocate fields of manufacture and to maintain prices of unpatented glassware' (*Hartford-Empire Co.* vs. *US* 1945, para 28). This 'leveraging' of the licence to control behaviour outside the direct scope of the patent is a typical focus of attack in antitrust cases and will be commented on below.[12]

But the members were also found guilty of combining to maintain a dominant patent position in their own technology by applying for many improvement patents for the sole purpose of denying them to firms outside the pool. It was this alleged prevention of outside interference in their basic patent licensing policies that was judged anti-competitive. They were also found to have applied for patents on potentially competing inventions simply to block their development. The Supreme Court upheld the decree restraining *agreements* and *combination* with these objectives of technology suppression through patent acquisition (*Hartford-Empire Co.* vs. *US* 1945, para 58) and so legitimated the idea that corporations engage in deliberate technology suppression.

The edited transcript of the Supreme Court judgment does not contain the evidence on which this judgment was based, or evidence on the economic achievements of the patent pool, so it is difficult retrospectively to judge the companies' overruled defence, that the patents they bought were often trifling and of nuisance value only. The scenario is the by now familiar one, where the independent inventor sells patent rights to a corporation but is dissatisfied with the lack of exploitation of the inventor's former patent property. In this case the Supreme Court judgment is to an extent in the inventor's favour, and the real issue is – why did the glass manufacturers buy patents that they then did not exploit?

If resort is made to cases with better documentation and with the simplification of single-company ownership of the patents, it can be seen that it is reasonable for improvement patents to provide a period of continued control, albeit more tenuous, following the expiry of the basic patents for development.

The role of a portfolio of improvement patents as a means of control of technology is clear in General Electric's development of the incandescent lamp market (Reich 1992: 9). A small number of basic patents such as the tungsten metal filament patent would provide a period of control running from the 1920s until patent expiry in the early 1930s. The improvement patent period would run from then until 1945 and although some improvements concerned product upgrading, such as the introduction of frosted lamp glass, they consisted mostly of manufacturing patents. Throughout this period General Electric (GE) would maintain super profits of between 20 and 30% return on investment (Bright 1949: 256–405; Reich 1992: 4).

In the period of validity of the tungsten filament and other patents, GE had been able to exercise strict control over the number of licensees and their market share, but several features of the improvement patent period of control show that it was a weaker form of control.

First, unlicensed competitors, mostly Japanese, became important. These could legally manufacture lamps using GE's patent-expired technology in uncontrolled competition with GE's patent-protected improved lamps. The Japanese manufacturers relied on cheap labour and prices to compensate for the technological 'gap'. Their lamps had 'low efficiency, short life, and uneven performance' but they were half the cost of GE's lamps and had taken 9% of the US market by 1932 (Reich 1992: 9). The minor process improvements from continuing innovation collectively increased GE lamp output per worker-hour from 30 in 1926 to 95 in 1942 (Reich 1992: 9). This provided GE with its main competitive defence – an ability to reduce prices while quality improved.

Second, the improvement patents were individually less robust and unlicensed producers (especially the Japanese) regularly bypassed them with alternative designs, or simply infringed them. GE responded by fighting more trademark and patent infringement suits, 'the situation resulted in extensive patent litigation and a continual testing of the GE patents' (Bright 1949: 273). Through a combination of infringement suits and process innovation GE succeeded in forcing Japanese market share back to 2% by 1940 (Reich 1992: 9).

Third, GE's largest licensees such as Westinghouse only remained as licensees in the patent improvement period in return for greater market share and lower royalty rates (Bright 1949: 257). So GE's dominance within the continuing, controlled industry was reduced by agreement and in accordance with its weakening control of the technology. Because GE knew its control of the technology was becoming more tenuous it had every incentive through the improvement period to research, acquire and apply useful innovations to its lighting technology. That it did so and that it was forced to pass on the benefits to consumers are evidenced by the regularly falling price of GE lamps through *both* periods of patent control.

The weaker control of the improvement patent period is a common innovation phenomenon. With the expiry of Dyson's basic cyclonic vacuum cleaner patent Dyson's company has entered such a period, with the company claiming to have

> accumulated over 100 different patents on the machine, many of them cyclonic improvement patents, which will lengthen indefinitely our period of exclusivity. The only way to keep possession of your invention is to keep strengthening it. (Coren 1997: 206)

As for the Hartford-Empire case, the same competitive pressures evident in the GE case can be expected to have borne down upon the patent pool members in their period of 'improvement patent control'. In such circumstances, it is entirely possible to find the Supreme Court's decision to uphold the decree restraining 'agreements and combination with the objectives of technology suppression through patent acquisition' as neither necessary, nor harmful in itself – because these firms had every incentive to pursue *and develop* useful invention. The exotic strategy of 'patent blitzkrieg', the *deliberate* purchase and development of patents in order to suppress invention, can be considered unlikely even in this case.

The Evolution of US Antitrust Policy to 'Contain' the Patent System

As Kitch describes the situation, in the past antitrust law has been used to 'confine the operation of the patent system to its "proper sphere"' (Kitch 1977: 267). In a now familiar way, Kitch argues that belief that the patent system *should be* so confined derived essentially from the conventional economic assumption that the patent system was a monopoly-based reward system. In his view this manner of attack probably damaged the prospect function of the patent system.

The 'interface' between antitrust law and the patent system appears to have been largely concerned with the alleged abusive extension of the patent licence to areas outside the direct scope of the patent. In the 1970s US courts' intolerance of such practices reached an extreme in the informal policy towards licensing that came to be called the 'nine no-nos' (Langenfeld and Scheffman 1989: 11). These were nine patent licensing practices held to constitute anti-competitive action and to warrant antitrust action, for example the so-called 'tie-in' when a licensee was required to buy non-patented products for use with the licensed product. The reform with regard to tie-ins has been to drop the 'rule' that such practice is *necessarily* abuse, and to require case-by-case proof of this abuse.

Whatever one thinks of this form of alleged anti-competitive abuse of the patent licence, it is quite distinct from technology suppression, as it continues to imply the active development of the patented technology by the licensor. Dunford's reference to hundreds of antitrust cases involving the patent system is likely of this type, concerned with the 'abuse of extension'. If one agrees with the Reagan Administration's view[13] that the 'nine no-nos' were 'containing more error than accuracy' (Weston 1984: 281) then, with Kitch, one would see many of these historic patent–antitrust interface cases as simply wrong judgments.[14]

Uncertainty in the intellectual basis for antitrust law breeds variation in its application, as we have seen in this chapter with respect to patents and in the previous one, and more generally, with respect to innovation and competition.

Concluding Comments on the Misinterpretation that Patents Are the Basis for Innovation Suppression Strategies

In current cases it is rarely possible to obtain definitive information on actual company motivation and so we are often in the position of having to construct credible interpretations of expressed company motivations and actions.

Unfortunately the tendency to find suppression strategies is reinforced by a strong combination of circumstances: the widespread confusion between a patent as an exclusive right and an economic monopoly; the accumulation of cases of US antitrust prosecution of supposed patent abuse; and last but not least the occasional encounter with managerial expressions of suppressive intent and the suspicion that that these are likely the 'tip of an iceberg'.

Once it is understood that a patent does not necessarily represent an economic monopoly, explanations of company behaviour could be found that did not require the attribution of suppressive behaviour. It is reasonable to adopt a sceptical stance towards most claims that useful technology is being deliberately suppressed, above all when patents are involved.

That does not mean suppression never occurs: one must consider the evidence on a case-by-case basis. One of the more persuasive cases is Clark's example of the suppression of magnetic recording technology after 1940 by – once again – AT&T Bell Laboratories (Clark 1993). Clark can show that Bell Labs patented and advanced the technology to the point of developing a prototype magnetic recording answering machine by 1934, but 'delayed in offering an answering machine to its customers until the early 1950s, almost twenty years after the production of a successful prototype' (Clark 1993: 530). Clark is able to cite internal memoranda which reveal that some senior managers 'sought to suppress the commercial exploitation of magnetic recording for ideological reasons stemming from the corporate culture of the Bell system' – and they did so successfully (Clark 1993: 533). Apparently senior managers thought that if magnetic recording devices became widely available and attached to the telephone system, they would lead to a lower level of privacy and therefore of trust in the telephone service – and a consequent lower level of usage. However, to complicate matters, Clark states that it was not the economic reasons that were stressed in the internal memoranda, but the issue of public trust in the telephone service and in AT&T. AT&T already had 'constant public relations problems, largely a result of antitrust investigations' (Clark 1993: 534). This is not an 'ideal case' of the 'free choice' of technology suppression by a private sector monopolist – something that would have tended to strengthen our suspicion that the practice was prevalent. It shows antitrust action intended to protect the public against monopoly-abuse damaging the incentive to innovate, and so damaging the public interest. It should also be said that Bell's suppression of its own technology may have damaged its own economic interests, but if it delayed the general commercialisation of magnetic recording devices it was only by some number of years, not decades, since by the 1940s other corporations were vigorously marketing their own, technologically much advanced, versions of the technology (Clark 1993: 538).

Technology suppression claims are not likely to disappear in the near future and the probability is that we shall have to exercise judgement with insufficient evidence as is illustrated by one last variant of the genre. Dunford cites several authors who during the oil crisis years of the 1970s accused the oil companies of researching solar energy solely in order to acquire patents and block the development of that technology (Dunford 1987: 516). Paranoid suspicion of corporate motivation might sometimes be justified, and it is surely encouraged by the necessary ignorance of contemporary observers. But in this case we have the benefit of hindsight and with any patents from that time expired, we know that solar energy research in the 1970s simply failed to produce the economic development opportunities that its advocates claimed it represented. And with the example of the US valve researchers before us, we could doubt the ability of the oil companies to suppress solar energy, even had they wanted to and had it proved profitable: their researchers would have left to create start-up companies. As the valve example suggests, corporate ineptitude is a more likely explanation for the failure to develop good technological prospects than a secretive strategy of suppression.

On the positive side, the examples that have undermined the credibility of suppression strategies have given us a better understanding of the nature of uncertainty in the anticipation and development of new technology and a better appreciation of the role of patents in managing the development of innovation.

None of this is to say that the patent system and its operation are free of problems, such as other, milder forms of corporate abuse. One such provides a direct contrast to the idea of deliberate innovation suppression: the idea of patent claims as a form of competitive *deception*, facilitated by the operation of the *machinery* of the patent institution. Once again, in the evidence for this behaviour, there is the opportunity to observe the working of the patent institution, this time for application of the rule that a patent should be 'non-obvious'.

The Rule of Non-obviousness Applied to Revoke Pfizer's Viagra Patent

The basic rules for determining the patentability of an invention appear clear enough in patent office literature:[15] an invention must be novel, capable of some industrial application, non-obvious, and *not* specifically 'excluded' from eligibility. To demand that an invention be novel and useful is to demand that it really *is* an invention, while exclusion limits the scope of the patenting system (as 'life' was once excluded, until the advent of biotechnology). The value of the non-obviousness rule is perhaps the most obscure of these standard rules, because it introduces the problem that, although the patent system must work in terms of discrete inventive steps, there is no inherent standard for what constitutes such an 'inventive step'.

Its application to an important case can be examined in the published version of Justice Laddie's finding that Pfizer's commercially valuable Viagra patent was invalid for reasons of obviousness.[16] This case also demonstrates the danger of uncritical reading of newspaper stories of corporate innovation.

The Non-obviousness Rule

The courts administer the rule of non-obviousness by attempting to judge evidence through the eyes of a 'skilled but non-inventive man in the art'. In the words of the British Mr Justice Laddie:

> This is not a real person. He is a legal creation. He is supposed to offer an objective test of whether a particular development can be protected by a

patent. He is deemed to have looked at and read publicly available documents and to know of public uses in the prior art. He understands all languages and dialects. He never misses the obvious nor stumbles on the inventive. He has no private idiosyncratic preferences or dislikes. He never thinks laterally. ... Anything which is obvious over [i.e. can be easily deduced from] what is available to the public cannot subsequently be the subject of valid patent protection even if, in practice, few would have bothered looking through the prior art or would have found the particular items relied on.[17]

At first this reference to a fictional person as part of a process of making a judgment probably appears quite bizarre. The definition and use of this fictitious person has been evolved through decades of experience of court cases as a useful guide in the sifting of evidence that bears upon whether or not an invention represents a genuine inventive step. Rosenberg describes how, in the USA, 10 'negative' rules of invention emerged 'from the myriad of court decisions grappling with the requirement of invention [in] a number of recurring situations in which the baffling quality of invention was rather consistently found wanting' (Rosenberg 1975: 118). In other words, much effort has been expended in the attempt to patent variations of the known and these negative rules are a reaction to certain patterns in these attempts. Rosenberg describes the 'standard' of the 'skilled man in the art' entering the 1952 US Patent Act 'in an attempt to foster the establishment of at least a modicum of certainty and consistency' in the application of the law (Rosenberg 1975: 119). The requirement that an invention be non-obvious also allows patent examiners to disallow trivial variations to existing technology, such as variations in colour or style, and can be understood as an attempt to restrict the patent system to socially significant inventions.

How Justice Laddie Concluded that the Inventive Step for the Critical Viagra Patent Was Obvious

An illustration of this fictitious person 'in action' is the successful Eli Lilly – ICOS challenge to Pfizer's patent position on its blockbuster drug Viagra on the grounds that the drug represented an obvious step at the time when the patent was filed, in 1993. The challenge succeeded: in November 2001 Justice Laddie revoked the Pfizer patent as an exclusive right to exploit the entire class of compounds, the phosphodiesterase (PDE) inhibitors, of which Viagra (sildenafil citrate) is one, and although Pfizer did retain rights to sildenafil citrate, would-be rivals such as Eli Lilly and ICOS are now free to develop closely related competing compounds.

A key piece of evidence became a scientific paper, referred to in the judgment as 'Rajfer',[18] that was published in January 1992, 18 months before the 1993 filing of the relevant Pfizer patent. This 'prior art' described for the first time the entire biochemical pathway that lies behind the operation of smooth muscle tissue, such as that which controls the penis. It described the experimental use of a PDE inhibitor, Zaprinast, on strips of penile material to demonstrate the relaxant effect of such drugs on smooth muscle (*Pfizer* vs. *Lilly ICOS* LLC 2000, para 73) and even suggested 'that interference with this pathway might cause some forms of impotence and that it could be treatable' (*Pfizer* vs. *Lilly ICOS LLC* 2000, para 116). This paper appeared to point directly to such a treatment as Viagra.

To defend the patent counsel for Pfizer concentrated on arguments that it was a *non-obvious* step from Rajfer and other prior art[19] to the idea of Viagra. These arguments included, for example, that other development routes than Pfizer's would have been *more* obvious for the 'skilled man in the field', that specific features of Pfizer's development such as the oral form of the medicine were non-obvious, and perhaps most surprising of all, that because others, *including Pfizer*, did not think of the idea immediately, it must be non-obvious (*Pfizer* vs. *Lilly ICOS LLC* 2000, para 114).

Such a failure on the part of skilled individuals to think of the 'obvious' is an example of what Rosenberg calls 'circumstantial' evidence that can weigh on the judgment of obviousness. Evidence of similar circumstantial status are commercial success and the satisfaction of a long-felt need, both present in the Viagra case. However, Rosenberg comments that the law has learnt to put the technical evidence of obviousness above such 'circumstantial' evidence (at least in the USA (Rosenberg 1975: 127)). In this way, the law acknowledges that when attempting to disentangle the evidence bearing on the novelty of an invention, real expert workers sometimes *miss* the obvious (as in the Viagra case) and sometimes treat an inventive step *as* the obvious (*Pfizer* vs. *Lilly ICOS LLC* 2000, para 63). As Justice Laddie expresses it, real experts often use in the expression of their views special knowledge acquired in the course of their research work that the court must count as *not knowable* by that construct, the 'skilled worker in the field' (*Pfizer* vs. *Lilly ICOS LLC* 2000, para 64).

The potential burden of investigation this places on patent lawyers is evidently large. It does imply that the published judicial findings in such cases are likely to yield wonderful source material for the analysis of invention and innovation. This was certainly so for Pfizer's 'discovery' of Viagra.

Pfizer's Research Discovery Story Derived from the Patent Trial

Pfizer had experimented with sildenafil citrate from 1990 as a treatment for many conditions, including angina and hypertension, and had *scheduled* an evaluation of the drug as a treatment for erectile dysfunction by August 1991, well before the publication of the Rajfer paper (*Pfizer* vs. *Lilly ICOS LLC* 2000, para 86). However, when the experiments took place in early 1992, the drug was injected into the penises of anaesthetised monkeys and evidence of any erection noted. Pfizer's Dr Ellis commented on the experimental monkey results in court, that sildenafil citrate

> did not work in this model, inducing only a very transient partial erection. This indicated that the basal nitric oxide drive was insufficient for sildenafil to be effective in this model. We were disappointed at this result and did not have the conviction to continue exploring the utility of [sildenafil citrate] in MED (male erectile dysfunction) in the absence of other supportive data.

Indeed, I do not recall seeing any formal report of this study. (*Pfizer* vs. *Lilly ICOS* LLC 2000, para 88)

These trials were designed to test compounds that would directly generate an erection. The reason they failed to indicate sildenafil citrate's promise was because it is not a 'direct acting compound', but one that inhibits the enzymatic destruction of nitric oxide, the chemical generated by arousal and essential in the maintenance of erection. Anaesthetised animals are not experiencing natural sexual arousal and so there is no nitric oxide to be preserved by sildenafil citrate.

A further twist to the discovery story is that a Pfizer scientist had read and circulated a copy of Rajfer *one week* before the trials were to begin with a note in the margin 'Should we not try out [sildenafil citrate] in impotence? Have we seen any beneficial S[ide]/e[effect]s?' (*Pfizer* vs. *Lilly ICOS* LLC 2000, para 75).

With this evidence, Justice Laddie accepted as fact that Pfizer's scientists had not understood sildenafil citrate's method of action despite the distribution of Rajfer, but could nevertheless conclude, by appeal to the legal principles outlined above, that 'these are matters which a skilled man in the art would have understood from Rajfer' (*Pfizer* vs. *Lilly ICOS* LLC 2000, para 90).

Justice Laddie's acceptance of the Pfizer scientists' immediate failure to understand had potential significance, for this was presented as evidence of the non-obviousness of the inventive step. However, the Justice commented that because the gap between the publication of Rajfer and the filing of the Pfizer patent was only 18 months, it was too short a period to be *good* evidence of the non-obviousness of the step to Viagra (*Pfizer* vs. *Lilly ICOS* LLC 2000, para 115). Justice Laddie found the 1993 Pfizer patent invalid for obviousness based on a consideration of the significance of Rajfer alone (*Pfizer* vs. *Lilly ICOS* LLC 2000, para 118).

Necessary Representations? Comparison of Trial Evidence with Viagra Invention Stories Pre-trial Finding

The November 2000 judicial finding and review of evidence give us a rare perspective on the newspaper 'stories' told about the discovery of Viagra after its commercial launch in 1998, a time when the 1993 patent was still valid.

A standard 'discovery story' derived from Pfizer[20] can be found in the *Wall Street Journal*. After describing the discontinued effort to explore the effect of sildenafil citrate on angina, the *Wall Street Journal* continued:

The program was about to be shelved permanently in 1993, when Pfizer researchers noticed something quite unexpected: several men who had received higher than usual doses in a small study told doctors they had achieved improved and more frequent erections than before. At the time, it seemed like a side effect rather than a remedy. But the Pfizer scientists, trying to salvage a drug they had worked on for years, believed the erection effect might represent a significant advance. (Langreth 1998: B1)

The essential feature of the story is that *the company* made the inventive step linking sildenafil citrate to erectile dysfunction, following a classic 'lucky observation'.

If we take into account the evidence Pfizer presented in the patent trial we may examine one possible view of how events occurred.

In the complete picture we must believe that although Pfizer had been experimenting with sildenafil from 1990, its scientists only noticed 'accidentally' and then understood 'independently' the significance of the erectile effect in human beings in 1993, a date that by bad luck was at least a year *after* the publication of Rajfer in 1992. We are also asked to believe that a copy of the paper that described the effect of PDE inhibitors on erectile tissue was circulated shortly after its publication, marked with the explicit suggestion that Pfizer might look for the effect of Viagra on impotence. We are then asked to believe that despite the circulation of the Rajfer paper, Pfizer scientists were *not* able to infer that sildenafil citrate might be used to treat MED. Instead, Pfizer scientists continued to conduct human trials with sildenafil citrate and it was from these trials, with no aid from the circulated paper, that they made their 'lucky observation' and therefore 'independent' inventive step that justified the award of the patent.

This remains a possible sequence of events and we are in no position to prove it right or wrong. Nevertheless, we can invoke the 'cynical individual' who might find it difficult to believe that Pfizer's scientists were so incompetent as either not to have read or not to have understood a paper recognised as key to their field, that directly suggested a commercially significant path of research and that was circulated in their firm. Such cynics would observe that both the newspaper story and the trial evidence were compatible with the commercial need to preserve the patent. They might go on to build a different construction of events than that related here, despite there being no definite *proof* for their position being the correct one. Such a construction would be that, very late in the day, Pfizer researchers understood *from Rajfer* the true mode of action of sildenafil, perhaps hurriedly set up some confirming experiments, then filed for what would be the valuable *additional* patent to cover MED, and last but not least constructed a discovery story that would satisfy the public and support the claimed patent position.

In support of such an interpretation, the cynical individual might note that it took five years, until 1998, for Viagra to pass through the various mandatory tests before market launch and until 2000 for the patent to be revoked. During this period, Pfizer's patent was apparently valid and if it acted as a deterrent to research by imitators, it would have been a valuable aid to Pfizer's achievement of a head start in the development of this novel and lucrative market. The cynic would observe that such were the commercial stakes that even if the company understood the grounds by which the additional patent would prove unsound if challenged, it was worth filing and fighting for.[21]

The cynical individual might also ask whether it would be reasonable to expect a drug corporation, dependent for its continued existence on patent-protected drug innovations, to admit voluntarily that it had no defensible patent position on a blockbusting, record revenue-earning drug. The cynic would also point out that the Viagra case has obvious implications for the value of innovation stories that derive from company sources and that involve the viability of a patent position. Historical studies based on company archives make a better basis for the understanding of corporate patent strategies.

Academic Science Versus Industrial R&D for New Pharmaceuticals

At the heart of this case is uncertainty over the relative contribution to innovation of corporate as opposed to academic research. This issue has strong implications for public policy on prescription drug prices, a hotly debated issue in the USA, where prescription drugs currently account for $170 billion of the $1.4 trillion annual health expenditure (Relman and Angell 2002: 27). If you judge that the pharmaceutical industry is the principal engine of drug discovery and development then it makes sense to allow high prescription drug prices to support the vital private sector R&D activity. If, however, the Viagra case represents a more widespread reality of drug discovery and development, you might be inclined to believe that because the public are already paying for much drug discovery work through publicly supported academic research, they deserve some controls on prescription drug prices.

An award-winning article by two eminent US social medicine professors argues the latter position[22] (Relman and Angell 2002). They begin by pointing out that the large US pharmaceutical companies make more in profits, and spend more than twice as much on marketing and advertising, than the $30 billion they spend on R&D. They also point to the large sums spent on political lobbying, such as the $60 million annual budget of the industry's collective association (Relman and Angell 2002: 27) and the accounting manoeuvres used to generate the industry's preferred, and very high, public figure of $800 million development cost for the most novel drugs: Relman and Angell estimate the true cost of development for such drugs to be lower than $266 million (Relman and Angell 2002: 30). In any case, Relman and Angell argue that most 'new' patented drugs are minor variations on older drugs and so are not the novel breakthroughs the industry likes to associate itself with (Relman and Angell 2002: 28).

In other words, the industry is good at developing and politically promoting information and stories about its R&D expenditure in order to win support for its preferred policy of price-setting freedom.

The other type of evidence cited by Relman and Angell is represented by an unpublished internal National Institutes of Health (NIH) study that found that 16 of the 17 key scientific papers that led to the discovery and development of the five top-selling drugs of 1995[23] came from outside the pharmaceutical industry. Relman and Angell argue that even allowing for the greater incentives of academic researchers to publish, the pharmaceutical industry is clearly dependent on much basic work performed in the academic research sector. This they support with summaries of the evolution of understanding necessary before commercialisation of other recent top-selling drugs became possible.

Given the threat of regulated drug prices and the drug industry's resort to political lobbying, Relman and Angell are concerned to redress the balance of opinion on the relative importance of academic versus private sector innovative activity. The picture they draw is by now familiar from other industries: pharmaceutical companies may excel at the development and marketing of drugs, but in large part they remain dependent on academic research for the improved understanding of biological processes that may generate breakthrough innovations.

Mutual Adaptation of Intellectual Property Law and Technology

It is already clear from the discussion of the patent–antitrust law interface that there is a degree of variation in the practice of intellectual property law. The following sections introduce examples of how significant changes in technology create the possibility of a significant adaptation of intellectual property law; adaptations that may or may not aid future innovation.

Adaptation of the German Patent Law to Enhance the Patent Prospect Function in Azo-dyes

The patent system depends on being able to define an inventive step, but this is particularly difficult to establish when a technology is characterised by a continuum of design variations, all with the same use-value. According to the industrialist MacLeod, electronic microcircuit design became such a technology and

> it has become the practice to take a patent on every variation, however trivial, in circuit design. Patent examiners have long since given up the hopeless task of identifying genuine novelty and have adopted the safe line of accepting anything that is not absolutely identical with a previous specification. (MacLeod 1988: 258)

In such a circumstance, the patents granted could offer little commercial reward or protection. If protection were wanted, a sensible strategy would be to save time, secrets and money by not filing for patents.

For other technologies, an adaptation of patent law practice may enable the definition of commercially valuable inventive steps and thus patents. This issue of how to define the appropriate inventive step arises forcefully in the case of azo-dyes and the amendment of German patent law.[24]

A German patent law was passed in 1877, but it had to be made meaningful in the technological context of the dye industry. The basic issue was *what constituted novelty* and so what could be patented and protected, in the new industry. The general law had the following characteristics:

> The law excluded the protection of chemical substances; a chemical invention would be patentable only in so far as it concerned "a particular process" for the manufacture of such substances. But the law did not define … "a particular process". More generally, even the term "invention" was left undefined. (van den Belt and Rip 1987: 151)

These features of the law would prove problematic for the dye industry because azo-dyes are all produced by the same basic chemical reaction. The 'coupling' reaction links two carbon rings through a double nitrogen bond and the number of possible combinations was estimated to be over 100 million (van den Belt and Rip 1987: 151). The first German R&D departments were created to sort systematically through these combinations in a search for useful colours.

The large number of possibly useful dyes combined with the indeterminacy of the German patent law left open a number of opportunities for defining novelty in the new dye industry.

> At one extreme the coupling reaction in its generality could be interpreted as a "particular process", leaving no room for the patentability of the more specific "processes" within its compass. At the other extreme one could argue that only a process leading to one and only one azo dye could count as a "particular" process in the sense of the law. (van den Belt and Rip 1987: 153)

If one of the extreme interpretations governed the award of azo-dye patents then patent law would have proven useless in its prospect function. Control of a patent on the coupling reaction would have given one firm the exclusive right to sort through the 100 million dyes, probably to the detriment of the social good that the field should be rapidly exploited. But if every possible combination were in principle patentable, there would be 100 million possible patents, many for dyes with the same effective colour. In this latter case, despite enormous patenting effort, a company could not be sure of a secure 'prospect' in its patents, because of the possibility of the existence of a patentable, distinct chemical formula with the same colour and commercial value. The same situation would have arisen as in MacLeod's electronic circuit example. Competitors would be able to target any particularly valuable dye for imitation and the exploitation of the innovative potential of the coupling reaction could be retarded.

So from the social and commercial point of view, either of these extreme definitions of 'a particular process' would create a defective patent system, considered either as an efficient reward or prospect claim system. In the face of this legal uncertainty, firms claimed patents for 'the widest possible territory' (van den Belt and Rip 1987: 153) typically for 200 compounds in each application.

Eventually the Congo Red case arose, a test case for the fundamental issue of the inventiveness, and so patentability, of a particular azo-dye. Some eminent expert witnesses called to the court demonstrated the complete absence of inventive effort in the generation of the Congo Red dye. Van den Belt and Rip quote from the memoirs of Carl Duisberg, another participating expert:

> I tried to make clear to my opponent... that through his point of view he actually delivered the death-blow to the whole chemistry of azo-dyes and thereby also brought most of the patents of his own firm to the brink of the abyss ... he did not withdraw one iota of what he had said earlier. (Duisberg quoted in van den Belt and Rip 1987: 154)

Duisberg decided to appeal to the court to consider that the fate of the synthetic dye industry hung on its decision. According to van den Belt and Rip, this may have been decisive in persuading the court to uphold the Congo Red patent and establish the principle of the 'new technical effect' for the award of dye patents. This allowed that if a dye had new commercial value, the inventiveness in its generation could be acknowledged and a patent granted (van den Belt and Rip 1987: 154). This was to make legal a concept of inventiveness intermediate between the two extremes defined above and the doctrine of the new technical effect represented an adaptation of the patent law to the characteristics of the azo-dye technology. Now the patent system could function as it was intended in principle to function, as a means of coordinating and rewarding research.

Change in Internet Technology and Copyright Law

Not all the pressures to adapt intellectual property law are necessarily benign. The intellectual law of copyright covers the expression of an intellectual work and when media technologies change, the control of the expression of intellectual work threatens to change also. Today it is the relationship between copyright law and the new medium of the Internet that is at issue.

The challenge that new media technology can pose to existing copyright became clear when a small start-up software company called Napster created 'one of the most frequently downloaded software applications in the history of the internet' (Editors 2002). This enabled its users to exchange music freely in the form of MP3 compressed data files.

Most of the music exchanged was in violation of copyright law. The US music industry in the form of the Recording Industry Association of America (RIAA) successfully sued Napster's owner, Shawn Fanning, for facilitating illegal copying. This 'success' would appear redundant, as the technology for exchanging the files is widely diffused and many 'Napster clone' sites now exist, some in other countries than the USA. It appears that the pursuit of site owners under current copyright law in order to close or limit each site would be an expensive and ultimately hopeless strategy. This case has added significance because with a few years' delay for the perfection of compression and transfer technology for digital film files, a similar systematic violation of film copyright can be expected.

Such cases might be understood to support the idea that the technology of the Internet has destroyed the relevance of copyright law and with it the attempt to enforce legally a balance between the rights of producers and the rights of users of intellectual and artistic works. It is difficult to resist the temptation to speculate about a future unregulated world, where the large music corporations of today have withered as they are unable to secure revenues from their legal ownership of copyright, and where the financial reward incentives for intellectual and artistic innovation are drastically reduced.

Such speculation is probably misguided because it forgets that copyright law has never existed in isolation from technologies of control and transmission of intellectual material, but works with them and in acknowledgement of them to achieve its objects. The Harvard lawyer Lawrence Lessig makes a powerful argument that if existing copyright law fails to create enforceable ownership rights on the Internet, there is a strong incentive for commercial interests to use technological controls for this purpose instead of legal controls. As a result, argues Lessig, far from it being not possible to regulate the Internet, it is likely that it will become regulated and a tool for *control* of its users (Lessig 1999). Lessig argues that this will begin as commerce pursues its natural interest in establishing a robust means of *identifying* users. Indeed, since Lessig wrote his book one of the unanticipated ways this has happened is through so-called 'spyware' and 'adware'. One of Lessig's 'technological futures' for the Internet offers insights into the possible interaction between law and new technology. Lessig considers the proposed technology of 'trusted systems', which promises to 'achieve what copyright law achieves … *without the law doing the restricting*' (Lessig 1999: 129).

Trusted systems work by tracking and controlling all copies made of a work and so they depend on that work being limited to exchange between 'trusted system' networks. If this can be achieved, a fine degree of authorial control over user access to the work becomes possible. In principle, different access contracts could exist dependent on whether the user wants to read the work once, 20 times, or use just a section of it (for example, of an electronic newspaper). This is in contrast to the sale of intellectual property in the form of books, where the original seller has no control over the number of times it is read, or if it is resold, or copied (there being no effective means to prevent copying). In other words, if 'trusted systems' technology becomes the solution to the current problem of the erosion of the value of copyright, it will work by giving owners near absolute control over their material.

Lessig argues that what is wrong with such a technology is that it ignores the fundamental purpose of intellectual property law. The purpose of granting property rights to authors is to strengthen their incentive to produce, in exchange for eventual public access and free use of their product. Authors' rights are intended to be limited in many ways, for example through the legal concept of 'fair use', which is the 'right to use copyrighted material, regardless of the wishes of the owner of that material' (Lessig 1999: 134). Fair use supports the critical review of books in public by allowing a degree of quotation – copying – as of right. Authors' rights are not absolute and 'the law of copyright is filled with such rules' (Lessig 1999: 134).

Yet the trusted systems technology threatens to make authors' rights absolute, and to abolish concepts like 'fair use' as use becomes conditional on each author's individual wishes. Lessig argues that such a technology of control ignores 'copy-*duty*' (Lessig 1999: 127).

> Whether you consider it a problem or not depends on your view of the value of fair use. If you consider it a public value that should exist regardless of the technological regime, then the emergence of this [technological] perfection should trouble you. From your perspective, there was a value latent in the imperfection of the old system that has now been erased. (Lessig 1999: 138)

If one values the existing balance between authors' rights and limitations on those rights, one would want the law to limit the new technology so as to preserve that balance (Lessig 1999: 139). Lessig stresses that we have a choice that we have to make. An obvious danger is that *inactivity* – on the part of both citizens and legal institutions – may allow a choice to be effectively made by the promoters of the technology and their allies. The ability of corporate interests to triumph over the notion of the public good is apparent in the successful attempts to lengthen the period of copyright, for example in the European Union Directive of 1993 extending copyright to the lifetime of the author plus 70 years (Lessig 2002: 17). Lessig's thought experiment does remind us that exactly how the law is adapted to these future technologies will depend on the values of the society that the law is supposed to serve.

New Research Streams, Changes to Patent Practice and Reform of the Patent Law

The appearance of new streams of research has the potential to 'wrong-foot' patenting practice. The flood of patent applications for gene sequences, especially those associated with the successful effort to sequence the human genome, is likely an example of this. For many of these patents it is not clear what the prospective use of the gene sequence would be, while there is nothing novel about the process of sequencing. Strictly speaking, there is no inventive step. Rather the establishment of a gene sequence resembles a 'discovery' with no immediate or obvious application – and discoveries are not normally patentable. According to the British Society for Human Genetics, once the first patent applications were granted a culture of 'defensive patenting' took hold where researchers applied for patents on gene sequences only because everyone else did (British Society for Human Genetics 1997). Although the 1998 European Directive on the legal protection of biotechnological inventions sought to *strengthen* patent protection for biotechnological research, by 2002 the European Commission had announced that it would investigate the scope of patents on human gene sequences (Cordis News 2002). This gives some idea of the time it takes to achieve the political consensus necessary to reform the law to correct practice gone astray – too long to save the enormous waste represented by the defensive patenting activity.

In a similar way the growth in the importance of software and the Internet led to two changes in US patent practice: from copyright to patent protection for certain classes of software and the granting of 'business process' patents. It is difficult to judge these changes as necessary and useful to the promotion of innovation. The change to patents on software hands the principal owners of old software companies, like IBM and Microsoft, a stronger form of protection than copyright can provide.

However, the introduction of patents to software creates a possible problem for open source operating system software, because the coordination costs of arranging patent licence exchange between different 'owner–users' may make the future cost of development and use prohibitive (Lessig 2000: 17). As IBM and Microsoft seek to have their software patents validated in European courts, Lessig argues that Europe is in danger of adopting the US approach without having first analysed the costs and benefits of such a change.

If such changes have uncertain value, others that offer clear *economic* aid to innovative activity are nevertheless politically problematic. There is a European patent office, but at present it coordinates an applicant's claim in each of the European Union countries and produces a bundle of national patents. The cost of obtaining patent coverage in eight European states remains five times higher than the cost of a single patent within the USA.[25] Language translation makes the major contribution to the higher European costs. The same report also estimated that for a sample of claims it took 46 months to process a claim in Europe, twice as long as in the USA (Eaglesham 2000: 16).

In November 2000 the European Commission made a proposal for a single European patent.[26] The idea is straightforward enough and greatly simplifying: a single Community patent publishable only in English, French or German and administered post-grant through a new section of the European Court of Justice (currently national courts arbitrate in disputes, so enforcement of rights requires involvement in up to 15 member states' law courts). If it is simple in conception, it is not novel; the Commission has been working towards this end for three decades with limited success (Eaglesham 2000: 16). Member states have been reluctant to cede legal decision-making powers to the European Court of Justice and reluctant to see law made and expressed in another language than their own, but which is nevertheless enforceable in their sovereign territories (Eaglesham 2000: 16).

At the time of writing (2004) member states had once again baulked at the point of implementation of the Commission's proposals and there is still no 'single European patent' nor any sure prospect of it.

Concluding Comments on Innovation and the Patent Institution

As for competition law, economic theory of a certain kind has also proved to have been an important influence on the application of US intellectual property law. The derivative idea that patents could be abused to suppress innovation proved to be without empirical support, and yet the idea was a useful foil in the examination of the function of the patent system. The patent proved to be an adaptable instrument for the coordination of development and the documented cases of its exploitation make a rich source for the study of innovation and its management.

So what was truly significant about the case of radio was how, despite extraordinary circumstances, with the ownership of technological prospects dispersed and developers having every reason not to trust their rivals, the cross-licensing of patents nevertheless allowed the coordination of development. Dyson also provided an example of the role of patents in allowing negotiation over development possibilities despite the negotiating parties' natural distrust of one another. With the pace of development high in biotechnology and IT, patents should be expected to serve similar purposes.

Nevertheless, the contrast between trial and pre-trial corporate statements on the discovery of Viagra should make us wary of contemporary corporate stories of R&D and invention if patents are involved. The viagra example raised the issue that the machinery of the patent institution itself provides some scope for strategic manipulation.

The ability of that machinery to respond well to new streams of research was also discussed. In the instance of azo-dyes, the adaptation of patent law that was the 'new technical effect' can be judged to have had both social and private benefit by making dye patents effective prospects worth developing. In the more recent example of patents granted for human gene sequences it has often not been clear that there is novelty or prospect of use and the European Commission has not proved able to reform the law in time to save the wasted effort that such patents represent.

Notes

1 Simple descriptions of intellectual property law can be found on websites run by most country's patent offices, for example in the UK: http://www.patent.gov.uk/patent/definition.htm.

2 The long-standing tendency for economists to doubt the value of the patent system has been noted by a diverse range of authors writing on patents, for example by Taylor and Silberston (1973), Basalla (1988) and Schiff (1971).

3 'First to file' is used by most countries in the world with the important exception of the USA, which uses a 'first to invent' rule.

4 These and other grim experiences with established companies lead Dyson to end the book in a classic 'inventor's rant' against 'bean counters' (accountants are villains throughout), the degeneracy of 'marketing' (when divorced from technological understanding or any idea that this is necessary) and advertising (lumped in with marketing, but receiving special attention because of its self-identification with 'creativity'). It makes for a stimulating read, but should be contrasted with the equally justified 'mature-corporate' cynicism about the value of the many self-styled inventor–entrepreneurs that one can find in the work of industrial writers on innovation, for example MacLeod (1988).

5 The same point is stressed in Rosenberg's *Patent Fundamentals* (1975: 275).

6 Dunford states that these cases were selected from standard legal texts recommended by four independent legal experts.

7 Both Reich and the historian of AT&T, Noorbar R. Danielian, use a selective quote from a memo by John J. Carty, AT&T's chief engineer at the time, that outlines the arguments in favour of research into a 'telephone repeater' or amplifier (Danielian 1939: 104). The evidence for a 'suppressive' intention with regard to radio comes when Carty gives, in his words, 'One additional argument' for working on the repeater, that of its obvious pivotal role in the development of radio (Danielian 1939: 104). So Carty did not say that the technology, if controlled, *should be* suppressed, only that control of the technology was desirable. As described in the text, subsequent AT&T behaviour shows that the company used its control of amplifier technology *in order to reap benefit from the successful joint development* of radio.

8 Broadcasting bandwidth limitations restricts the utility of two-way radio systems. They tend to use high frequencies that attenuate over reasonable distances, so that other distant users can use the same frequency. Users are forced to be close to each other and to be limited in number.

9 The later story of AT&T's foray into radio development involves an adverse patent judgment and commercial failure in the broadcasting business, so that by 1926 the company settled with RCA and essentially withdrew from the markets of radio receivers and broadcasting (Reich 1977: 229).

10 The same argument applies to other statements by AT&T managers that Reich and his source, Danielian, choose to present as evidence of suppression, rather than merely indicative of suppressive *intent*. So Danielian is able to cite a desire by AT&T managers to 'protect' existing investments through the control of radio (Danielian 1939: 107). If this were an expression of suppressive intent (itself a matter for interpretation) *in practice* it would not be realised.

[11] The Supreme Court judgment has been made available in an edited form on the Internet at http://www.ripon.edu/faculty/bowenj/antitrust/hart-emp.htm.

[12] Rosenberg in his *Patent Fundamentals* argues that even price-fixing stipulations in such contracts *may* serve a pro-competitive purpose – it all depends on the case in question. So a small developer company may at once wish to license its technology to a larger rival, but also want to fix that rival's selling price to guarantee that the developer itself is able to develop the technology (Rosenberg 1975: 276).

[13] Enunciated by an administration appointee, Deputy Assistant Attorney-General William Baxter, a former professor of antitrust law at Stanford University (Weston 1984).

[14] However, the administration failed to have its preferred rules governing patent licensing codified into law (Langenfeld and Scheffman 1989: 62).

[15] These criteria are from the UK Patent Office home page at www.patent.gov.uk/patent/defin.

[16] Dates and events referred to in this section can be found in Justice Laddie's published findings (2000).

[17] Mr Justice Laddie's findings can be found at http://news.findlaw.com/hdocs/docs/viagra/viagrapatent.html.

[18] The full reference is Rajfer et al. (1992).

[19] Other prior art included, for instance, including Pfizer's own, earlier patents on PDE inhibitors that did not specify their potential use in erectile dysfunction (2000, para 23).

[20] The short version can be found in *Business Week*, but the claim that the company independently made the inventive step remains the essential feature (Barrett 1998).

[21] The commercial advantage of such a practice is enhanced if the 'machinery' of filing and trying cases is slow.

[22] The policy impact of the article won it the George Polk Award for Magazine Reporting, 2002. Arnold S. Relman is professor emeritus of medicine and social medicine at Harvard Medical School, Marcia Angell a senior lecturer in social medicine at Harvard Medical School.

[23] Zantac, Zovirax, Capoten, Vasotec and Prozac.

[24] The following account is derived from van den Belt and Rip (1987).

[25] 49 900 euros in Europe compared with 10 330 euros in the USA (Eaglesham 2000: 16).

[26] The proposal can be downloaded from http://www.eurunion.org/legislat/iiprop/patents.htm.

6   Finance – Techniques, Institutions and Innovation

Innovation and development are expensive activities and no wonder that in the private sector – but not necessarily the public – they are undertaken for the prospect of financial return. In a fantasy world without uncertainty, projects could be ranked and developed by the relative scale of their profitability; in reality uncertainty dogs the evaluation of development costs and future returns. In these circumstances the strict application of textbook financial evaluation may generate dysfunctional patterns of investment. This chapter aims to review some of the diversity of means that organisations and society have developed to evaluate and finance innovation. With this departure point, both internal techniques of financial appraisal and external institutions such as industrial investment banks and venture capital are included. The emphasis is on how financing practices have been adapted through experience to aid innovation and how and when practice may sometimes obstruct it.

Development and Internal Financial Evaluation

Investment Decisions and Problems with Calculated and Projected Rates of Return

Historical Role of Return on Investment in the DuPont Management Hierarchy

It was within the new multidivisional organisations of General Motors and DuPont that return on investment (ROI) techniques first acquired a role as a management tool of control. As described by the latter company's research historians in Chapter 3, the technique was first used to make meaningful comparisons between corporate divisions' achieved returns, *not* to evaluate prospective returns on individual projects. In DuPont's post-war free-spending R&D heyday, the company backed the most credible projects generated by its scientists. Return on investment calculations usefully demonstrated the reality of unforeseen, but unfortunately experienced, development difficulties in innovations like Kevlar. In turn, the carefully accumulated record of declining real returns sapped the company's faith in its strategy of funding all 'promising' innovation projects and a retreat from that policy began.

Kevlar also reveals the significance of the element of choice in financial evaluation. DuPont used evaluation criteria based on its historic ability to achieve high returns and its desire to maintain those returns. Product development costs were charged against profits for only the first 10 years of a product's market life, discounted at some appropriate interest rate. As described in Chapter 3, only three of the products introduced between 1952 and 1967 had positive net worth at an interest rate of 10% (Hounshell and Smith 1988: 533). Kevlar, discovered in 1964, took a record-breaking 15 years of technology development before commercialisation and had cost $500 million by 1982.

Kevlar 'failed' when judged against DuPont's demanding standards, but with a longer timescale, reflecting the product's actual life, and a lower imposed interest rate, it may well have been profitable. Hounshell and Smith suggest this was true of many of the expensive products DuPont developed in the 1970s, because these continued to bring in income, like Kevlar, during the 1980s and for longer than any 10-year evaluation period (Hounshell and Smith 1988: 8). So while return on investment calculations reliably showed a macro-trend for decreasing returns, single product calculations were not reliable guides to the total returns on products still in the market. And it is clear in cases like Kevlar why the product was difficult to evaluate. Development costs were high because there were many distinct product markets (bullet-proof vests, fibre optic cable, and so on). These were not all identified and developed at once and both development costs and returns were distributed in lumps over time. A 10-year assessment period would capture much of the development cost, but not the returns on the late developed markets.

The history and use of the technique in DuPont provides good grounds for doubting the reliability of financial evaluation as the sole guide for the selection of *prospective* innovation projects. Financial evaluation of past returns influenced the company's confidence in its judgement, but 'judgement' remained, in DuPont's case, devolved to its scientists.

The Effort to Adapt Discounted Cash Flow (DCF) Techniques to CIM

The evaluation of the adoption of new technology presents an intermediate case between the uncertainty of novel innovation projects and the 'known risk' of fully established technologies. An example is the IT-related automation technologies, CNC tools, FMS (Flexible Manufacturing Systems) and robotics, collectively referred to as computer-integrated manufacturing (CIM). These technologies began to be promoted vigorously in the engineering industry in the 1980s, in response to Japanese advances in their production and use. The danger understood by many (Hayes and Garvin 1982; Primrose and Leonard 1984; Kaplan 1985; Primrose and Leonard 1988) was that in this early period of diffusion many firms would make a routine DCF analysis and wrongly reject a profitable investment opportunity. If they recognised the drawbacks of routine evaluation and then ignored DCF, they risked investing optimistically and blindly and so unprofitably. The problems with DCF are interesting in themselves, but so are the solutions that were evolved, representing the adaptation of the financial techniques to the characteristics of the technology-in-use.

For CIM the problems with existing evaluation practice were that, first, firms tended to use unrealistically high discount rates – of 15–20% or more according to Kaplan (1985: 4) or 25–40% according to Hayes and Garvin (1982: 268). Second, textbook application of these techniques required that firms compare the returns on the prospective investment with *the returns available elsewhere* and 'in most of the capital expenditure requests I have seen, any new investment is evaluated against an alternative of no new investment' (Kaplan 1985: 9). The problem here is that the scenario of 'no new investment' typically makes the easy assumption that the future market position and revenues of 'firm A' will be the same as today – an assumption of technological stasis. But if a competing 'firm B' invests blind and happens to make the good but low-return investments that firm A does not make, then the relevant future scenario for firm A is likely to be a decline in sales and revenues as customers switch to the improved products of its competitor. Hayes and Garvin describe this as a 'logic of disinvestment', because if in a second round of technological improvement firm A continues to assume technological stasis and to impose a high discount rate for its DCF calculations, the result is more likely to be negative within its now contracted business. The result would be that firm A will gradually and quite 'logically' decide to exit from what should have been a profitable business (Hayes and Garvin 1982: 271).

Another scenario of 'dysfunctional' investment described by these authors occurs when a series of small-scale investments are proposed to improve the operation of an *existing plant* (Kaplan 1985: 12). It is possible for the series of small improvement projects to have positive net present values, because they make a local improvement to some small section of plant operations, but also assuming no change in operations elsewhere in the plant. The 'correct' alternative investment comparison is with the scrap and rebuild of the plant – this may prove to have a higher net present value than the sum of the small-scale improvement projects. However, Kaplan suggests that 'the company seldom brings itself to scrap the old facility and replace it with an entirely modern, integrated, rationalised plant because, at any point in time, there are many annual, incremental investments scattered about whose investment has yet to be recovered' (Kaplan 1985: 13). In other words, an initial failure to make the correct alternative investment comparison, reasonable if the plant is relatively new and small-scale improvements are at first appropriate, can lead to an entrenched practice that obscures the profitability of scrap and rebuild.

A third type of problem is directly concerned with the contingent value and genuine uncertainty in some of the revenue streams from new, complex technologies like CIM. An example is the release of large areas of floor space that should follow the reduction in inventory that follows successful implementation of FMS. The value of this physical change is contingent on the firm's circumstances: if the firm is growing it could now make direct use of the freed production floor space; if the space is owned outright, value may be realised through its sale or rental. On the other hand, value may not be realisable if the space is leased and dependent on the contract terms of the lease.

Primrose and Leonard recognised the numerous contingent dependencies of future financial returns to CIM with an elaborate software program into which these contingencies, such as the existing equipment, could be entered. This software enabled the calculation of the returns on different piecemeal CIM equipment purchases, so that of the many paths of equipment adoption, there was more chance that the more profitable for a particular firm context would be adopted (Primrose and Leonard 1983).

With successful adoption of CIM an increase of sales would be reasonable to expect, derived from increased part quality and reduction of throughput time that increases the ability to deliver to order. The difficulty for the adopter of accurately assessing the value of such changes to its customers is clear: increased part quality may enable not only existing but new users to make not only the present products better, but also to do the same for entirely new products. In the face of such unknowns, a typically conservative evaluation approach is to ignore them – along with the equivalent, hard-to-estimate costs, such as the disruption and productivity dip associated with the implementation of the technology on the shop floor. Yet the conservative approach is hardly satisfactory because there is the problem, as some researchers have found, that increased sales may be the most significant benefit of this new technology (Primrose and Leonard 1987; Primrose and Leonard 1988). If uncertain, but likely increased, sales are ignored the formal financial evaluation is rendered highly misleading and hardly useful.

Kaplan's advice on this issue is to include some estimate of increased sales only if the initial analysis of the more precisely quantifiable revenue streams proves negative, and so the danger of a foregone but profitable investment arises (Kaplan 1985). It is also possible to make conservative, but positive estimates of the change in sales and to construct upper and lower DCF scenarios. The firm may still make a wrong decision to invest, given uncertainty, but it will not have allowed itself to be misled by a false sense of certainty derived from misleadingly precise evaluations.

What the solutions of Primrose and Leonard and Kaplan have in common is that the financial evaluation process itself is imaginatively adapted to characteristics of the technology in question.

The Evolution of Managerial Accounting and the Quest for Exact Product Cost Knowledge

The last subject most people would consider to be related to innovation would be accountancy, but there are a number of interfaces between forms of accounting and innovation. When a firm grows and becomes dependent on internal accounting procedures rather than market prices as a means of understanding the relative costs of producing products, in principle it becomes possible for faults in the design of the cost accounting system to deform investment decisions. A highly influential book by Johnson and Kaplan reviews the historical development of management accounting to argue that the subject took a 'wrong turning' in its early development which destroyed its value as a guide to investment throughout the late twentieth century (Johnson and Kaplan 1991). If this argument is correct, and as far as firms rely on their internal cost calculations to make their investments, then we have another source of distorted investment decisions.

Johnson and Kaplan succeed in showing that the various engineers attempting to develop cost accountancy method at the beginning of the twentieth century all understood 'product cost' as a sum of different portions of the various forms of overhead cost, where overhead may include energy consumption, lighting, marketing and R&D – anything other than actual (direct) production costs (Johnson and Kaplan 1991: 56). The great problem was how to allocate different proportions of this fixed overhead between multiple product lines. According to these authors this early understanding would become lost to both management accounting practice and even theory during the course of the twentieth century, (Johnson and Kaplan 1991: 129), hence the title of Johnson and Kaplan's book in reference to management accounting, *Relevance Lost*.

If we ask why this happened, Johnson and Kaplan's explanation is that, first, firms proved unable to operate any of the proposed new cost accountancy methods, apparently because the information collection costs were too high (Johnson and Kaplan 1991: 128–9). By default, then, firms adopted the simple procedure of aggregating all overhead costs and dividing them by the quantity of product to produce an 'average' cost per product – a procedure that necessarily loses any information on internal differential product costs (Johnson and Kaplan 1991: 132).

Second, the rise of public accountancy would consolidate the practice of aggregating overhead. A growing need by firms to tap capital markets in the early twentieth century led to a demand by investors for a standard form of public financial reporting (Johnson and Kaplan 1991: 130). The professional accountants that developed these standards served the needs of investors, not the internal cost control needs of the firm (Johnson and Kaplan 1991: 130). They sought to make public accounts *auditable* – so, for example, they demanded that public accounts should be based on original recorded transaction data that could be checked, rather than actual market prices that might aid cost control. They also imposed on firms the associated procedure of 'inventory costing', where the firm divides the total manufacturing cost for an accounting period by total output in this period, so it can then use this 'average' cost per unit of product to give a value to both product sales and inventory.[1] For these external public accounting purposes, all that was required was the aggregate cost overhead (Johnson and Kaplan 1991: 130).

So Johnson and Kaplan explain the lost relevance of internal cost accounting procedures by a combination of the information costs of implementing good procedures and the development of public accounting with its lack of concern for internal cost control. Hopper and Armstrong supplement this explanation by pointing out that the wave of horizontal mergers between US firms at the end of the nineteenth century gave the merged firms great control over prices in their product markets (Hopper and Armstrong 1991). From this time it therefore became less imperative to understand and drive down internal costs; on the contrary, rises in internal costs could be relatively easily passed on to the consumer through the firm's monopoly-derived power to set prices. We have met this argument before – this increased degree of monopoly was the basis for the foundation of new forms of overhead function, such as the R&D department. Firms also used their new-found pricing power to extend security and other benefits towards their primary workforce. So the period when internal costing procedures were inadequate was a period when their inadequacy did not really matter, because US firms did not need the precise control of costs as a tool for vigorous rationalisation of their workforce and other costs. It is not surprising, then, that the calls to reinvigorate internal systems of cost control have coincided with the rise of effective Japanese and overseas competition that has inevitably undermined US companies' price-setting abilities.

To return to Johnson and Kaplan's account of the development of cost controls, professional public accountants would eventually turn to the problem of designing cost allocation systems to aid management decisions, but by this time the original engineering cost accountant contributions were largely forgotten. The 'new' approach to cost accounting attempted to use the information prepared for public accounts *based on aggregated overheads* and so were intrinsically flawed – they could never address the needs of the complex multiproduct firm. Yet according to Johnson and Kaplan, these were the kinds of cost control systems that became established in practice[2] and that were then automated without improvement in the USA from the 1950s onwards (Johnson and Kaplan 1991: 184).

Whatever the variations between these cost accounting systems:

> Virtually all companies ... allocated ... costs to products based on direct labor ... typically this fully burdened cost center labor rate was at least four times the actual direct labor rate paid to workers. In some highly automated cost centres it was not unusual for the rate to be ten or even fifteen and twenty times the hourly labor rate. (Johnson and Kaplan 1991: 184)

The practice of allocating overhead costs to direct labour was reasonable in times when overheads were small as a proportion of labour cost. The danger in today's more capital-intensive manufacturing environment is that if management pay any attention to such cost systems, they will tend to an excessive focus on the elimination of 'high-cost' direct labour. This may result in expensive over-automation or, when planning new products, management will tend to subcontract rather than manufacture in-house because outsourcing *appears* cheaper.

Another result was the unaccountability and greater security of those employed in staff functions that counted as overhead compared with those employed in production. Whereas overhead staff were subject only to the scrutiny of a management committee and treated as a fixed cost, production staff were treated as a variable cost and so were subject to the greater scrutiny and major cost-reduction drives (Armstrong 2002: 100). One of the results is that where management paid attention to the numbers produced by such inherited accounting systems, one could expect those who plan their career paths to avoid a production route.

Because of overhead aggregation and the allocation of overhead to labour, such systems can neither *precisely* relate costs to products, nor provide useful cost control information.

> The rationalisation for their production and existence seems only for the periodic, usually monthly, financial reports prepared for senior management. (Johnson and Kaplan 1991: 195)

Even if the figures did represent a meaningful ROI for a particular cost centre, they could not be a tool for management to *change* the material and organisational sources of internal costs, or to cost prospective new products – the only ways return on capital can truly be raised. In these circumstances, senior management pressure for improved performance in ROI terms (see below on the conglomerate movement for why such pressure should be applied) leads to middle management manipulation of figures to create an *appearance* of higher ROI – and so Johnson and Kaplan join the many authors who have criticised the abuse of ROI and devolved profit centres as a tool for management control (Dearden 1969; Dearden 1987; Johnson and Kaplan 1991: 195–205).

Whatever the reasons for the many decades of neglect of internal cost accounting controls, Johnson and Kaplan's analysis of the inadequacy of the inherited practices is widely recognised to be valid. Johnson and Kaplan today advocate the creation of a new management accounting system designed as a tool to represent and control internal costs and separated from the system used for public reporting purposes. In other words, a return to the principles that were developed and understood a hundred years ago, when managerial capitalism was first developed by technologists. In recognition of the role public accountancy has played in the generation of the current situation, they add:

> Accountants should not have the exclusive franchise to design management accounting systems. To paraphrase an old saying, the task is simply too important to be left to accountants. The active involvement of engineers and operating managers will be essential when designing new management accounting systems. (Johnson and Kaplan 1991: 262)

Activity-based Management as the Solution?

Johnson and Kaplan's contribution was to provide a plausible historical explanation for management accountancy's failure to become a guide for investment decisions. Their book invigorated attempts at change, but the widely touted cost control technique of 'Activity-Based Costing' (ABC) represents not a new approach, but a return to the early engineer cost accountant approach. It follows that the success of ABC today depends on whether the high implementation and processing costs that Johnson and Kaplan suggest hindered its development in the past can be reduced today by innovative IT solutions.

It is too early to judge whether ABC will transform industries such as general engineering, from which many of Johnson and Kaplan's examples are drawn. In Armstrong's review of the growing number of implementations of ABC, the enthusiasm of ABC advocates is drawn from ABC's promoted ability to cut staff overhead costs. Yet for the technique to deliver benefits, overhead staff activities must be able to be allocated to specific production activities. Armstrong points out that there are staff activities that simply cannot be attached to specific production activities, while there are forms of value added that derive even from activities such as purchasing (through its potential to manage the supplier relationship) that will be lost if ABC is implemented as if its assumptions are reality (Armstrong 2002).

This discussion of cost evaluation and control techniques has shown the experienced limits of the promise that such techniques can reduce investment and management approval decisions to uncontroversial routine.[3] In contrast, the striking feature of both decisions is that effective appraisal requires technological knowledge applicable to the project in question. In its absence, simple-minded management by the numbers is likely to lead to dysfunctional decisions.

A last point would be that Johnson and Kaplan's contribution is to show that it was during the unprecedented period of post-war economic growth and technological advance that cost control information had decayed into irrelevance as a guide to management. This gives some ground to doubt the urgency of their remedies. We can certainly doubt that innovation opportunities in the economy were foregone; it is only that, as far as their argument applies, certain firms that relied on such figures might miss opportunities that others exploited through other means, such as luck, intelligent investment gambles and deep knowledge of the prospective technologies. Although numbers matter, innovation decisions have not and cannot be reduced entirely to routine appraisal of numbers.

Are Accountants the Enemy of British Innovation?

Quite apart from accounting techniques there is the issue of whether accountants, as managers, are predisposed to be inimical to innovative projects, as alleged in the 'inventor's rant', an example of which was cited from Dyson's book.

Two forms of quantitative evidence on British accountants at first sight appear to support the inventor's case against the accountant. First, by 1991 a fifth of British company chairmen and more than a fifth of all directors had professional accountancy backgrounds (Matthews et al. 1997: 411). By this date too, there were twice as many professional accountants represented on British boards of directors as lawyers and engineers together (Matthews et al. 1997: 412). Second, a stark contrast between countries is revealed by a measure of the total labour force as compared to the number of professional accountants. In 1990 this measured 198 in Britain, 427 for the USA, 5066 in Germany and 5800 in Japan (Matthews et al. 1997: 410). With this data alone, it becomes rather tempting for technologists to assume that if Germany and Japan have developed without professional accountants, it is the overrepresentation of professional British accountants in British management that explains why British management fail to support their ideas.

Matthews et al.'s review of the development of the British accounting profession suggests a more plausible explanation. First, it is not the case that accounting was neglected by Britain's industrial rivals, rather 'it is probable that the US, Germany and Japan led the British in the development of management accounting' (Matthews et al. 1997: 421). These countries had other sources of accountancy expertise than an accountancy 'profession'. In Germany, the subject of 'business economics' was accountancy dominated and the graduates from these schools became the bookkeepers and internal auditors of German business (Matthews et al. 1997: 421).

It is therefore possible (though it cannot be quantified with our present state of knowledge) that there were as many employees performing accounting functions in companies in other countries as there were in Britain. What distinguished the British companies was the source of their 'bean counters', and their reputation for employing many accountants may be something of an illusion based on the fact that these are easily identified as such by their professional qualification. (Matthews et al. 1997: 422)

There were very few British university courses in accountancy until after 1946, and so the apparatus for regulating the training and supply of accountancy expertise in Britain became that of a profession (Matthews et al. 1997: 422). The professional apparatus developed early in Britain because Britain possessed the most developed capital markets in the world by the nineteenth century. Large-scale frauds of public companies had created a demand by shareholders for the regulated financial audit of such companies. The consequently greater volume of audit work in Britain provided the basis for the early development of the accountancy profession, both compared with other countries and compared with other potential sources of managerial talent within Britain.

If the differences between these countries in absolute numbers of accountants are a mathematical illusion, what remains to be explained is the much greater success of British accountants at reaching the highest levels of management and the boardroom. Matthews et al.'s tentative explanation again relies on the early and rapid development of the British accounting profession. These authors suggest that both the quality of recruits to the accountancy profession and the examination standards set by the professional bodies were high. These features 'brought early aspects of a meritocracy to the profession, unusual in British business life' (Matthews et al. 1997: 425). In addition, these authors suggest that although the accountancy curriculum was then, as it is now, narrow compared with the needs of companies, the audit experience of British accountants gave them what amounted to the best and most systematic management-relevant training available in Britain. This would remain true for most of the twentieth century, in part because, while graduates of various sorts were the feedstock for the development of other countries' corporate hierarchies, Britain was long distinguished by the absence of large-scale graduate provision by British universities. There was also a dearth of management training by British companies – in 1985 over half of all British companies provided no formal training for their managers (Matthews et al. 1997: 425). In these circumstances the British accountancy profession became the preferred source of managerial talent and training and understood to be the premier route to senior positions by those seeking a career in business.

This analysis gives a subtle picture of the role of accountancy in British business life. It is still possible to suspect that the overrepresentation of accountants at the highest levels of British management does nothing to aid the assessment and support of innovative projects. Such a suspicion is strengthened by the results of research that show that the 'collective mobility project' of the British Chartered Institute of Management Accountants has led to 'a long term tendency to reduce the emphasis on process knowledge' within that body's professional exams (Armstrong and Jones 1992). Professional accountants have been busy distancing themselves from knowledge of particular technological processes, when we know that such knowledge is one of the requirements for successful involvement in the assessment of innovative investment projects.

In conclusion, one can have reservations about the content of accountancy training and about the tendency for accountants to be overrepresented at senior management levels. However, there are good reasons for this overrepresentation in Britain and despite the many anecdotes told about accountants and technology, one should be careful about making the generalisation that these often able people are systematically unable to cooperate with technologists at and below their level. Nevertheless, this analysis does raise the question of why the engineering professions were not able to develop equivalent routes to the top for able technologists.[4]

Development and the Market for Corporate Control

The Conglomerate Movement – the Origin of the Abuse of ROI Measures in Corporate Control

Johnson and Kaplan's preferred explanation for 'management by the numbers' was the exposure of post-war senior managers to the content of pre-war university public accounting courses. This asks us to believe, like the criticism of the over-representation of accountants in senior British management, that accountancy education and training permanently limit the abilities of those who suffer them.

One might suspect there was something more durable behind the spread of management by the numbers. Lazonick provides such an explanation in his critique of the operation of the 'market for corporate control': the operation of the market in company ownership has promoted organisational forms such as the conglomerate that are inimical to sustained innovative performance (Lazonick 1992). Because the conglomerate form has endured, and because it is implicated in alleged cases of technological 'asset stripping', it is an interesting study in its own right.

At the peak of the US conglomerate merger movement in 1969, there were 6000 mergers of mostly unrelated businesses into conglomerates (Lazonick 1992: 176). The movement was based on the idea that the conglomerate could manage its constituent divisions through their ROI results, an 'innovation' that only *appeared* to be an extension of the organisational methods pioneered by DuPont and General Motors. At the risk of repetition, DuPont used ROI to indicate past divisional performance and to create a meaningful overall corporate result for the historic rate of return on capital. ROI reflected the results of its policy of backing research scientists' development ideas. Conglomerates could never do this because of their characteristic senior management structure, as described by Chandler:

> The only staff not devoted to purely legal and financial matters was for corporate planning (that is, for the formulation of the strategy to be used in investment decisions). As a result the conglomerates could concentrate more single-mindedly on making investments in new industries and new markets and withdraw more easily from existing ones than could the older, large, diversified companies. On the other hand, the conglomerates were far less

effective in monitoring and evaluating their divisions and in taking action to improve divisional operating performance. They had neither the manpower nor the skills to nurse sick divisions back to health. Moreover, because conglomerates did not possess centralised research and development facilities or staff expertise concerning complex technology, they were unable to introduce new processes and products regularly and systematically into the economy. (Chandler 1977: 481)

With this structure, and in the classic scenario, if a division began to underperform because a competitor had successfully introduced new technology, the conglomerate management *tended* not to be capable of assessing the detailed reasons for decline or providing the development and implementation of an investment-led recovery plan (Lazonick 1992). Because increased investment would also have the immediate effect of depressing ROI 'numbers' further, it would be easier to sell the declining division. Similarly, when conglomerates tried to impose 'ROI targets' on their divisions, they lacked the resources or technological ability to understand or distinguish between the different means by which a target was reached or missed. The insistence on meeting targets combined with the inherent inability of senior management to understand the detail of divisional activity to give conglomerate middle managers an incentive to manipulate operations to create the short-term *appearance* of value (Hayes and Abernathy 1980; Johnson and Kaplan 1991: 200).

The proliferation of this organisational form goes some way to explain the spread of 'management by the numbers'. Yet it seems clear from Chandler's description of the conglomerate that it was never intended as a means to manage innovation and this conclusion is reinforced when Chandler describes the typical industries that were targeted for acquisitions, namely textiles, shipping and machine tool industries, but not the capital-intensive, mass production and research-intensive industries (Chandler 1977: 481).

This modifies somewhat the significance of the critics' case for the danger of this organisational form to the economy-wide process of innovation, notwithstanding individual examples. It was possible, as described in Chapter 3, for the new leadership of RCA, a company that had historically relied on innovation for its continued existence, to fall foul of the 'fashion' for making technologically unrelated acquisitions and to make the fatal mistake of allowing its R&D activity to drift. But more significant, even the 'low-tech' industries that conglomerates targeted like machine tools and textiles did experience innovative challenge over time, especially from overseas. Whenever this occurred, it could be expected that the conglomerate structure would tend to prove inadequate.

Lazonick cites research by Ravenscraft and Scherer into the financial fortunes of a selection of the conglomerate-driven acquisitions of this period (Ravenscraft and Scherer 1987). By the 1970s one-third of the conglomerate acquisitions had been resold, usually in conditions of financial distress, and the evidence strongly supports the idea that loss of operational control was worse with merger than under independent control (Ravenscraft and Scherer 1987: 193). The actual problems encountered were diverse, but it was the inability of conglomerate management to react quickly or with knowledge of context that proved most damaging to their subsidiaries finances. Although data on the performance of 'surviving' merged entities are difficult to isolate, Ravenscraft and Scherer estimate that compared to pre-merger profitability, there was a decline in average post-merger profitability (Ravenscraft and Scherer 1987: 194).

This generally poor long-term performance raises the question of what financial logic drives the conglomerate acquisition process. The conglomerates sought to acquire low price–earnings (P/E) ratio companies, that is companies whose shares were cheap compared with company earnings. The consolidation of the conglomerate and acquired company accounts allowed a revaluation of the acquired company's assets that generated a once-off boost to the conglomerate's own earnings per share, 'which in turn generated a higher P/E for the conglomerate shares, which in turn permitted the conglomerate to use a given number of shares to make more acquisitions' (Lazonick 1992: 178). As long as there were undervalued companies to acquire, the conglomerates were able to create the upward trend in their financial indicators that investors desired. When they ran out of acquisition targets, or when innovation and inadequate control of the industries of the acquired companies created the problems outlined above, growth stalled, and they were faced with real problems – paying the interest on the debt they had issued and managing their acquired firms to create real value, but without a senior management structure that could achieve this. In these circumstances senior management could attempt to 'manage by the numbers' and abuse ROI. However, after a series of disposals, the ultimate fate of many conglomerates was to become the basis for a restored expert management hierarchy in some residual set of businesses[5] (Ravenscraft and Scherer 1987).

In this way the creation and growth of the conglomerates was just another example of the 'fictitious' creation of value that drives bubbles in financial markets, the greater and more recent example being the Internet/dot.com boom and bust. As Ravenscraft and Scherer, and Lazonick, point out, this financial logic is not confined to the 1970s, but is inherent in capital markets that allow hostile takeover. So the conglomerate as an organisational form may rise and fall in fashion, but it persists because 'In America ... the most common fast path to great wealth has entailed buying and accumulating assets of uncertain value. Many fail at the game, but a few succeed spectacularly' (Ravenscraft and Scherer 1987: 215).

Lazonick believes the real legacy of the conglomerate movement is the relative rise in number through the 1960s and 1970s of specialist marketing and finance managers employed in senior US executive positions once occupied by engineers (Hayes and Abernathy 1980; Lazonick 1992: 179). This rise helps explain the continuing denunciations of 'management through financial technique' long after the conglomeration movement declined: for example see Hayes and Abernathy in 1980;[6] Johnson and Kaplan in 1987.

This second proposed long-term change could be considered less enduring. Financiers may have become 'entrenched' (Lazonick 1992: 179) in many US corporations' senior management, but they, and their management values, could be 'dug out' in a crisis when it became understood that they were themselves harming long-term value. In Pascale's analysis of organisational change in Ford, it was because the company was a financial mess by 1950 that Henry Ford II began to promote financial experts as a management 'solution'. Pascale describes Ford's legacy of over determined elites.... From Henry Ford 1's elite cadre of designers and engineers, to the operations dominated environment of the forties, giving way to a finance driven company from the fifties onwards – Ford always took things to extremes. Each emergent elite had its "answer" for what would make Ford successful. Each demonstrated initial value. But ... each took its "solution" too far. (Pascale 1990: 148)

It is to the point that none of the professionally specialised elites showed an ability to retain permanent power. Although the organisational grip of 'finance'-specialised managers helps explain the slow response of Ford and especially General Motors to Japanese competition, by Pascale's account, at least Ford gradually fashioned an appropriate response that involved raising the prestige and discretion of operations and technology functions.

Such stories of change encourage the reasonable view that there are no management specialities permanently ranged against 'innovation'. Different management specialities have a tendency to vie with one another for dominance. Where one succeeds in dominating the others its limitations are likely to be revealed in time as external circumstances and the company's problems change.

The Critique of the Market for Corporate Control

This leaves us with the conglomerate as a persistent organisational form implicated in 'collateral damage' to the management of innovation. The grand thesis has it that the Anglo-Saxon financial markets have evolved so that they now operate to damage the creation and sustenance of innovative capability. This has been popularised by Hutton in Britain (Hutton 1995), developed by Lazonick for the USA and supported to a degree by Chandler (Lazonick 1992; Chandler 1990). Rather like the case against financial specialists, it is possible to bring a critical eye to some of the marshalled evidence, in particular the case evidence that often appears so striking. One can have reservations about the general thesis. Yet an important truth about the relationship between finance and innovation does emerge from the analysis.

A succinct expression of the critical position by Lazonick is that:

The belief in the efficacy of the market for corporate control is inconsistent with the history of successful capitalist development in the United States and abroad over the past century. The history of successful capitalist development ... shows that value-creating investment strategies increasingly require that business organisations exercise control over, rather than be controlled by, the market for corporate control. (Lazonick 1992: 153)

The historical basis for the critique gives invaluable perspective on the changing role of financial markets with respect to innovative enterprise and it can be briefly summarised here.

Most of the large US corporations of today had been created by the early decades of the twentieth century. The process that separated their ownership from control was the desire by the original owner–entrepreneurs to exit their businesses upon retirement. The issuance of shares financed their exit and transferred ownership to a broad class of wealth owners who acted as portfolio investors; that is, they invested passively, not actively, and attempted to min- imise the risk of investing by distributing their investment among a broad range of stocks (Lazonick 1992: 159).

While ownership remained dispersed and the financial track record of these firms was good, management were left to finance investment through retained earnings and bond issues – stock issues never being an historically important source of funds for expansion. But post-war, ownership became increasingly concentrated amongst pension and mutual funds, whose managers were rated by how well their portfolios performed against general stock market perform- ance. These investment managers were therefore *required* to 'manage by the numbers' and by the 1960s their aggressive buying and selling had created something without historical precedent, a genuine market in corporate owner- ship and control: 'For the first time, individuals, groups, and companies could obtain control of well-established companies in industries in which the buyers had no previous connections simply by purchasing their shares on the stock exchange' (Chandler 1990: 139). So there is no doubt that the historical record shows that the periods of *growth* and consolidation of US enterprises from Ford to General Electric and DuPont owe little or nothing to the stock market, and this tends to shift the onus of proof to those who would advocate that the oper- ation of the stock market is at the centre of capitalism, understood as a system that encourages successful innovation.

Lazonick argues that the novel financial practices enabled by the market for corporate control pressure the corporate manager to raise short-term financial performance to the loss of any long-term investment strategy. These include direct institutional fund manager pressure, the creation of the conglomerate form, or in the 1980s the corporate raider's use of leveraged buy-outs to fund acquisition; the argument is that these all contribute to a pervasive *fear* of acqui- sition that encourages behaviour which seeks to deter takeover. This might take the form of reduced capital expenditure for higher dividend payments, and a higher share price, or increased debt and service obligations to make takeover unattractive (Lazonick 1992: 165).

In Lazonick's view, this behaviour is only made more likely by the increasing proportions of top-level executive remuneration that come from stock owner- ship, rather than salary; by the early 1960s ownership compensation for the top five executives in a sample of 50 Fortune 500 companies was over four times salary compensation (Lazonick 1992: 173). As Lazonick points out, ownership in a new venture must be a great motivator, since unless that venture generates an income there can *be* no compensation. The motivating effect of ownership in a going concern is a more equivocal matter and Lazonick is forthright: 'My view is that the access of top managers to substantial amounts of ownership income weakened the innovative response by providing them with individualistic alter- natives to personal success that could best be achieved by choosing adaptive strategies' (Lazonick 1992: 175). By 'adaptive strategies' is meant behaviour such as raising the dividend to create a higher personal income stream, as an alternative to investment-led strategies of company growth.

Problems with the Critique of the Market for Corporate Control

That, in outline, is the 'strong argument' that the market for corporate control is destructive of innovative capacity. But it has problems. First, within each element of the 'case against' can be found examples of the market for corporate control being used to *recreate* managerial hierarchies with incentives to plan long-term investment in innovation, and this is acknowledged by Chandler and Lazonick. For example, although leveraged buy-outs (LBOs) were involved in several notorious assaults on integrated management hierarchies in the 1980s, between 1980 and 1987 the majority of LBOs were used to fund divisional buy-outs from the failing conglomerates of the 1960s and 1970s. These buy-outs reintegrated operational competence with financial control of investment (Lazonick 1992) and repaired damage done by the conglomerate movement.

Second, the problems of over-investment in which an 'owner-independent' management hierarchy can become involved can be considered to be at least as bad as alleged problems of under-investment generated by the market for corporate control. If this is so, the market for corporate control can be understood as a means of controlling a management tendency to over-investment. Some insight into the market conditions that promote over-investing behaviour is provided by Chandler's classic and spectacular example of how the relatively independent management hierarchies of US railroads came to waste investors' capital on an unprecedented scale in the nineteenth century.[7]

Railroads – Management Hierarchy Out of Control

The problem arose once a basic nation-wide railroad network had been completed which consisted of many companies that each owned several hundred miles of railroad, necessarily interconnected with their rivals (Chandler 1977: 134). At first companies followed a 'territorial strategy' (Chandler 1977: 134) of making informal alliances to immediately connecting and competing railroads to set rates and cooperatively manage the flow of traffic across their section of the network. But with a growing volume of through traffic over local traffic, the need to cover the unprecedented high fixed costs of the railroads provided an irresistible temptation to undercut agreed rates and 'steal' through traffic from rival routes.

Investors and managers understood very early on that tit-for-tat rate cuts for long-distance rail traffic could undermine the profitability of the entire industry. They made great efforts to establish nation-wide cartels to control rate-setting, but these had failed by the mid-1880s[8] and an uncontrolled period of 'system-building' followed where managers paid little attention to real demand but reacted to any strategic moves by their rivals with acquisition and route-building plans of their own (Chandler 1977: 170).

> The systems were not built to reduce costs or increase current profits. The strategies of growth were not undertaken to fill any need for or to exploit the opportunities resulting from improved administrative coordination.... The basic motive of system-building was, therefore, defensive: to assure a

continuing flow of freight and passengers across the roads' facilities by fully controlling connections with major sources of traffic. (Chandler 1977: 147)

The most obvious form of capital waste occurred when, if neighbouring companies could not be bought, parallel lines were built in an effort to secure a particular organisation's control over the sources of traffic (Chandler 1977: 159). Lead railway financiers like J P Morgan sought without success to deter investment in profitless parallel and extension lines by threatening to cut off malefactor companies' access to funds (Chandler 1977: 172). Investors in individual companies proved ineffective at controlling their management, and:

> During the 1880s, 75 000 miles of track had been laid down in the United States, by far the greatest amount of railroad mileage ever built in any decade in any part of the world. And between 1894 and 1898 foreclosure sales alone aggregated over 40 000 miles of track, with a capitalization of over 2.5 billion dollars, the most massive set of receiverships in American history. (Chandler 1977: 171)

J P Morgan took a lead role in the refinancing and restructuring of many bankrupt companies. The new stable structure would consist of 32 railroad systems controlling 80% of US railroads and with a degree of cross-shareholding between neighbours to ensure cooperation on rates (Chandler 1977: 172). Most significant, the new typical form of corporate control would consist of a professional railroad manager as president of the company, but answerable to a board of directors containing a majority of financiers and whose role was to monitor capital spending. Chandler calls this final form of railroad company organisation a 'variant' of managerial capitalism, for its outstanding characteristic was that finance was the ultimate source of control (Chandler 1977: 187). Yet this 'form' was part of the reaction to the waste sanctioned by relatively independent operational management that *had* existed and it can be understood as insurance that such waste would never occur again.

This not only establishes that capital waste occurred at a particular stage of railroad development, but strongly suggests that the danger of its occurrence is when relative independence of management from ownership control combines with a relative exhaustion of investment opportunities for a set of organisations that have previously grown through capital investment. It is also significant that these were managers expert in the operations and technology of the railroads: to caricature the contrast, if under-investment is the accountant's vice, over-investment is the engineer's vice. If beneficial investment opportunities recede, but leave the technologists in control, there arises a temptation to turn the organisation's once socially useful investment abilities to the socially wasteful task of preserving the organisation itself.

The purpose of this argument is not to say this scenario is common, but only that it should be balanced against the scenario of financiers unable to understand or sanction investment in technological and operational change. It would seem relevant to an analysis of the Japanese economy of the 1990s, where industrial companies originally organised for sustained technological development have been extremely reluctant to cut capital investment in the face of years of financial losses (see the discussion of Japanese banking below).

Hostile Takeover and the Destruction of Innovative Companies

Hostile takeover certainly varies in frequency between countries: between 1945 and 1995, Germany experienced *four* contested takeovers, while in Britain there is an average of *forty* hostile takeovers per year (Hutton 1995: 156).

But does the prevalence of hostile takeover necessarily imply the destruction of innovative companies? Hutton chooses to illustrate the negative effects of hostile takeover on innovation with the apparently compelling example of the predatory conglomerate Hanson's hostile takeover of Berec, the one-time leading British manufacturer of Ever Ready batteries.

So Hanson acquired Berec by the start of 1982 and in that year, in classic predatory–acquirer fashion, sold the head office, various loss-making overseas operations, one of the R&D units, and recouped 40% of the purchase price by selling the European operations to the great rival, Duracell (Brummer and Cowe 1994: 131). Hanson then sought to make its profit by 'sweating' the remaining assets to meet its 20% internal rates of return. This meant that battery prices were raised at the cost of continually falling market share. When Hanson sold Berec again in 1992, the company was no longer market leader but had retreated, 'a mere shell; milked for profits and starved of investment' (Hutton 1995: 164).

Hanson effectively destroyed this *company*, but no damage was done to battery innovation and additional context explains why Hanson was able to buy Berec cheap – it had mismanaged its business and innovation opportunities.

The worst problem was the sadly familiar one of Berec having specialised in the 'wrong' technology, here zinc–carbon battery technology. Alkaline batteries were smaller and longer lasting, but 'Berec insisted that this was a small, specialist niche market which was not worth pursuing' (Brummer and Cowe 1994: 128). Yet this was not solely management's fault – Berec had owned a share in Mallory, the manufacturer of Duracell alkaline batteries, since just after the war, but in 1977 the British Monopolies Commission forced its sale. By the late 1970s 'Duracell was making huge inroads into its UK base, attacking new distribution channels such as supermarkets as well as traditional outlets' (Brummer and Cowe 1994: 128).

So we can reinterpret Hanson's destructive action as a *positive* one for the development of battery technology in Europe. First, the European division of Berec was sold to the great rival Duracell, today owned by Gillette. This would evidently aid Duracell's rise to its current position as European battery market leader, based on alkaline battery technology. Second, although Berec launched a programme of new investment from 1977, at the time of Hanson's bid its profits were in decline, there was a general recession and the outcome of this investment was unclear – it could properly be feared that it might only destroy the ability for either firm to establish price control in the market. These were the circumstances that drove the share price well below asset value so that hostile takeover became possible.

This case was supposed to exemplify the damage that hostile takeover can do to innovation. It is not only *unconvincing* evidence of damage, it can be understood as the market for corporate control consolidating the control of the technologically successful management of Duracell and possibly also preventing capital waste by a management team with a solid record of serious error.

A perusal of Brummer and Cowe's book on Hanson's growth shows that Hanson would take *any* sort of opportunity for profit through hostile acquisition and certainly did *not* make a 'habit' of attacking R&D-intensive innovators, simply because they were innovators, successful or otherwise. This is the key to understanding the one other intervention that Hanson made into the world of British R&D-intensive innovation – its assault on the British chemical company ICI.

Hanson acquired a 2.82% stake in ICI in 1991, a typical first step towards attempted takeover, but a formal takeover bid would never come. The best explanation for this is that, first, ICI moved to secure the attractive £500 million surplus in its pension fund by placing its management with a trust – if takeover proceeded, the pension fund would not become the property of the acquiring company. Second, ICI launched an investigation of Hanson's corporate structure and tax filings. These and the newspaper investigations that followed revealed a complex set of Panamanian and Bermudan subsidiaries that 'senior Hanson officials' later acknowledged were part of their 'sophisticated tax strategies' (Brummer and Cowe 1994: 232). In other words, Hanson could make hostile acquisition profitable, not *only* because of poor management performance, but also because of flaws in British corporate tax law and the law covering the ownership of pension fund surpluses.

In such circumstances there is little point attempting a definitive judgement in the debate about the value of hostile takeover as a discipline on management, for the institutional framework to support 'discipline' as the major function of hostile takeover has not yet been properly built.[9] Franks and Mayer in a survey of hostile UK takeovers confirm that the acquired companies were *not*, in general, poorly performing (Franks and Mayer 1996). But if hostile takeover does not have the value normally ascribed to it and loudly proclaimed by the predator companies and if Hanson's record is regarded as evidence, neither is hostile takeover practice *systematically* at the expense of innovative companies. This does, however, leave the *fear* of takeover as a possible deterrent to the risk taking that accompanies innovation, but the remedy would appear to be better maintenance of the legal framework for corporate governance.

Conclusions on the Market for Corporate Control

Selected spectacular examples of hostile takeover, of the obstructive influence of entrenched financial interests and of conglomerate activity – none of these have provided unequivocal evidence of the *market for corporate control itself* as destructive of innovative capability. When the operation of that market is shown sometimes to have restored technically competent and committed management and when the problems of abuse by uncontrolled management are admitted, the difficulty of a decisive judgement can be recognised as at least highly complex and perhaps impossible.

Writers such as Goldstein also seek to explain the many periods of financial excess of the sort reviewed by Lazonick, but they do not primarily see a process destructive of innovation. Rather the process is driven by competition within the financial markets that acts to change the behaviour of financial intermediaries in a similar way to industrial companies. First it erodes margins in the safest lending markets so that banks are prompted to experiment with higher risk forms of lending. Then it promotes a scramble for new market share and the relaxation of standards of lending. This behaviour tends to generate financial bubbles of varying severity and in the aftermath the lending institutions' survival is endangered and so is the financial health of borrowing organisations. In Goldstein's analysis the true problem is the absence of sufficient volume of real investment opportunities and returns that could provide the financial institutions with a secure future (Goldstein 1995). This description of the operation of the financial markets gives them a role in causing 'collateral' rather than 'systemic' damage to innovation capabilities, and with the promotion of over-investment as likely as the promotion of under-investment.

Rather than search for general innovation-damaging characteristics within the Anglo-American financial systems, it may be more fruitful to search for the presence or absence of close and informed bank–corporate relationships that could aid innovative strategies at critical moments.

Development and Financial Institutions External to the Firm

The Bank–Corporate Lending Relationship

Davis points out that in 1985 the majority of bank loans to medium-sized firms in Germany were long term and with fixed rates, of the type that should be expected to encourage long-term investment. In Britain medium-sized companies made more use of the equity market and the majority of their bank debt was in the form of short-term bank loans typically with variable interest rates[10] (Davis 1994: 99). Associated with the greater use of the equity market was a tendency for British firms to pay a larger proportion of earnings as dividends, a tendency that Davis, like Hutton, attributes to the fear of hostile takeover. The implication is straightforward – publicly owned British firms pay higher dividends to make takeover expensive, but at the cost of foregone investment opportunities.

The differences in cost and structure of loans in Britain and other industrial countries are typically and plausibly associated with institutionalised differences in the relationship between banks and the corporate sector. Both Germany and Britain have institutions for the management of corporate restructuring, with German corporate restructuring activity managed by the banks and British corporate restructuring occurring through takeover, buy-outs and buy-ins within the financial markets.

A major difference is the degree of access banks are able to obtain to inside corporate information, and the *variety of devices* that enable this. Davis describes the German situation:

> For the largest public companies, which have supervisory boards on which stakeholders are represented, banks will tend to have representatives at board level. For all public companies, banks may hold equity stakes and

exercise voting rights on behalf of custodial holdings. Banks may also have non-marketable equity holdings in private companies. Meanwhile, for small and medium firms, banks will be more likely to have exclusive relations (the housebank). Banks are legally obliged to obtain accounting and balance sheet data before lending. (Davis 1994: 100)

Davis broadly characterises the German corporate–banking relationship as one of 'commitment', for greater access to information on corporate conditions generates a greater willingness to lend long term, when expansion is possible. Mayer (1997) casts some doubt on this general characterisation, by pointing to the high degree of collateralisation of German bank loans, the typical low frequency with which supervisory boards meet and the high degree of turnover of share stakes in German companies,[11] albeit through the banks selling and acquiring those stakes. Mayer's comments apply to the German banking relationship in general, but it remains clear that the German model is at least capable of providing committed banking support where it is most needed, as in cases of innovative investment.

It is worth noting that there is a significant and misleading difference in the *visibility* of corporate restructuring between these countries. Whereas takeover for the purposes of restructuring is on public display in the Anglo-American financial market for corporate control, similar behaviour by a German bank would more likely be kept secret or masked by anodyne public press releases. When one takes into account that the great bulk of firms by number have a relationship with a bank but not to the stock market, it becomes plausible that intelligent corporate restructuring for 'efficiency' as well as for innovation is more a German than an Anglo-Saxon characteristic. This position contradicts the conventional wisdom frequently peddled in, for example, the *Financial Times*, as criticism of German institutional practice.

It now makes sense to examine further how a close bank–corporate relationship operates in practice to support innovative risk taking. A classic example is Reich's study of the rescue of the Japanese manufacturer of Mazda cars, Toyo Kogyo, by Sumitomo Bank (Reich 1985).[12] This case is also given great weight by Hutton as support for his denunciation of the British banks.[13]

A Comparative Account of Two Corporate Bailouts – Toyo Kogyo and British Leyland

Toyo Kogyo had become Japan's third largest car manufacturer through its radical development of the low-pollution rotary car engine (Reich 1985: 175), but this engine's high fuel consumption led to falling sales and a financial crisis during the 1973 oil crisis (Reich 1985: 175). Sumitomo Bank's early advice to cut production and halt the capacity expansion programme was ignored, so in response and over two years, Sumitomo Bank installed 11 of its own managers in key internal functions. This new management enforced cuts in production and costs. Most impressive is that despite the debt to equity ratio having risen to four to one (Reich 1985: 176) Sumitomo Bank's management developed a plan for a series of new piston engine models to secure the company's future and continued to increase lending to ensure that this plan could be realised. Sumitomo Bank also acted to restrain minor lending banks from withdrawing their loans and persuaded other companies within the Sumitomo 'keiretsu' (the grouping of industrial companies and a trading company around a significant bank, see following section) to make further loans and purchases of Toyo Kogyo 'assets'. Lastly, Sumitomo Bank arranged for Ford to buy a quarter of Toyo Kogyo's equity for cash (Reich 1985: 178). By 1980 Toyo Kogyo was profitable once more, debt to equity had returned to two to one and production had reached a million cars, well above the pre-crisis level (Reich 1985: 179).

The Toyo Kogyo bailout is a spectacular instance of relationship banking when it works, but we may ask what the appropriate lessons are from this example. The idea that this is a better way to run a financial system appears to be supported by two characteristics: a bank's ability to acquire early, detailed information on problems and its ability to judge and then intervene in management to plan the detail of a return to financial health through change in organisation and technology.

On these dimensions, Reich makes a contrast in favour of the Japanese with the role of the British government and banks in a similar crisis for the British car manufacturer, British Leyland. In 1974 when British Leyland ran into losses, its banks refused to make any more loans despite a debt to equity ratio of one to one. They had no information about the nature of its internal problems and no capacity to develop a route out of the crisis. The electoral liability of a British Leyland bankruptcy meant the British government was forced to manage the company's bailout, but it also lacked the information and ability for precise intervention within the management hierarchy. When it developed a plan for British Leyland's future it was essentially dependent on the views of the existing, compromised management and so failed to address the problems with that management (Reich 1985: 205). In retrospect, what the British government achieved was a managed, rather than a precipitous, decline for British Leyland.

Yet it would be hasty to judge that the British banks were wrong to refuse loans to British Leyland at *this* crisis point. Toyo Kogyo was an astonishing success that had been caught with the 'wrong' technology – its past growth, production and engineering excellence and the growth of the market for Japanese cars demonstrated a potential future. British Leyland inherited chronic problems from the past: too many minor plants without economies of scale, industrial relations problems and poor quality in production (Reich 1985: 170). The point about a close bank–corporate relationship is that *if* there is a scenario for restoring the company, the interested bank is more likely to be able to evaluate and enact it. Rather than the lesson being that the British banks should have intervened in 1974, it is more plausible that a lack of long-term *and critically informed* finance over decades was *one* of the causes of British Leyland's problems.

Two principal questions now arise – what are the origins of the systematic and persistent differences in bank–company relationships and is it possible to judge the 'closer' relationship as demonstrated by Sumitomo and Toyo Kogyo as an unalloyed blessing for innovation? The answer to the first properly sets the scene for the second.

Origin of the Institutional Differences in Bank Lending Behaviour in Britain and Japan

The evidence that current British banks now have a 'distant', low financial risk relationship to corporate customers is convincing – but it begs the question of how this came about, and why, if this 'distant' relationship leads to foregone mutually (bank–industrial) profitable investment opportunities, new financial institutions do not arise to exploit the opportunity.

The origin of today's difference between capital markets with regard to industrial investment can be found in the nineteenth century. Local British banks had *begun* to develop local industrial lending, in response to a general rise in the fixed capital requirements of many industries:

> By the middle decades of the nineteenth century, the banking system at a local [British] level was playing a role strikingly similar to that played at the end of the nineteenth century by the German Great Banks on a national level. (Kennedy 1987: 121)

But in the second half of the nineteenth century these small-scale, local banks could no longer safely make the loans that their industrial clients demanded. The great danger of industrial lending for fixed capital investments is that it is highly *illiquid* – it cannot be turned into cash on demand, and during a recession, perhaps not at all. Yet the source of these loans was the accumulated deposits of many small depositors who naturally put a high value on the bank's liquidity: its ability to repay their deposits immediately on demand. If all went well for such a bank, new short-term deposits would balance withdrawals. The great danger was a recession-induced liquidity crisis, when net demand from short-term depositors would be for repayment and the bank's ability to service these claims would depend on the depth of its reserves and the quality of its industrial loans. In such a situation it could be quite possible for an industrial lending bank to be *illiquid* but *solvent*, that is its long-term book assets could more than equal its short-term liabilities. Many local British banks found themselves in exactly this situation in the recession of the mid-1870s.

> The rash of mid Victorian bank failures can be attributed directly to banks becoming too closely linked with local firms and over-lending as these firms attempted to expand. ... These failures reached a crescendo with the failure of the City of Glasgow Bank in 1878. ... A point had been reached where the entire system had either to be reorganised to withstand the greater risks of steadily enlarging industrial requirements or the system had to withdraw from long-term industrial involvement. The system withdrew. (Kennedy 1987: 122)

The Bank of England was the most obvious candidate bank to advance cash to solvent, but illiquid local industrial banks, like the City of Glasgow Bank. But the British central bank had evolved a concept of its role that emphasised the maintenance of its reserves and the stability of the overall financial system (Kennedy 1987: 121) and felt no obligation to help these local banks. Nor did the British state believe it should intervene in such circumstances. With the Bank of England's refusal to extend credit, banks like the City of Glasgow Bank were dissolved, so destroying the wealth of a swathe of middle-class investors and providing banks and future investors with a terrible lesson in the dangers of industrial lending.

From this point onwards the apparent paradox of British capital markets grew more marked; through merger, the British banks became 'larger, stronger, and more diversified and therefore, in principle better able to take bigger risks with less danger' (Kennedy 1987: 122). But they had learnt not to take such risks, now classified as unsound banking practice, and continued to place a premium on liquidity and disengagement from industry.

The story of the evolution of the British capital markets marks out the state and the central bank as critical actors in the shaping of the financial system. Japan provides a great contrasting example of the potential of these actors to shape the capital market system at a similar historical turning point.

As in the British case, a critical economic event prompted institutional change, but this time the change would be managed by the active intervention of the state. A general capital shortage arose in the early 1950s as firms attempted to expand capacity rapidly to meet a flood of Korean-War-related US military procurement orders. The Japanese government was determined to promote reindustrialisation, and so to make the most of this one-off demand bonanza, and after intense internal debate, emergency measures were enacted that[14]

> had a profound significance for the economy of later years and for governmental economic policy. It led to the two-tiered structure of government-guaranteed "city-bank"[15] over-loaning and newly created government-owned "banks of last resort".[16] (Johnson 1982: 200)

This was the application of industrial investment bank lending practices to an entire economy. The availability of state-guaranteed loans meant the widespread abandonment of prudential banking criteria and within a few years had generated one of the Japanese industrial system's 'most distinctive characteristics – the pattern of dependencies in which a group of enterprises borrows from a bank well beyond the individual companies' capacity to repay, or often beyond their net worth, and the bank in turn over borrows from the Bank of Japan' (Johnson 1982: 203). The result was that, whereas before the war equity finance had been the major source of funds for industrial expansion, post-war, between 70 and 80% of industrial finance would derive from bank loans, ultimately provided by the Bank of Japan (Johnson 1982: 203). The over-loaning *itself* became the reason why banks developed relationships with their industrial clients – they rapidly became dependent upon them (Johnson 1982: 205). And with the disappearance of normal risk criteria for industrial lending, banks came to compete especially vigorously to *expand* their share of loans in government-designated growth industries, 'regardless of whether it made business sense to do so' (Johnson 1982). This bank competition to make loans played havoc with the government's attempts to plan expansion; when the government wanted to foster four petrochemical companies in the 1960s, it found five more companies had been acquired by banks in 'excluded' keiretsu. These were intent on developing capacity on the same scale as the chosen leaders and 'over-production was inevitable' (Johnson 1982: 207). It is important that this feature of the system appeared early on; it can thereby be understood as intrinsic. It was relatively harmless only as long as general demand rose strongly, for this ensured that overcapacity would soon be brought into use. For the same 'relationship' reasons that the system discouraged the liquidation of a troubled company like Mazda in good times, it encouraged the construction and main-tenance of capacity even when macroeconomic conditions became poor, as in the period of low growth in the 1990s.[17]

This system, so distinct in its *extent* from the financial structures and prac-tices in other western countries, including Germany, has essentially endured. As regards innovation, the choice facing the British and Japanese states and central banks can be seen to have been the same: the degree to which priority would be accorded financial system stability in contrast to the risks of expan-sion of industrial lending. Dependent on historical circumstance that has been bypassed here, they understandably acted in contrary ways. Once they had acted to tilt the balance in favour of conservative or aggressive lending criteria, capital markets consolidated around their distinctive orientations towards industrial lending. If the British system would hinder the rise of *new* capital-intensive technologies, the Japanese carried heavier financial system risks once technology 'catch-up' with other western countries had been achieved.

The British Electrical Engineering Industry and Uninformed Finance

Perhaps the classic case of an industry that required patient, long-term finance for its effective development is electrical engineering, and in William P. Kennedy's account, the British industry serves to demonstrate the effects on management of an absence of long-term and 'patient' capital.

In 1882 an access of public enthusiasm for the future of electricity allowed a record investment sum to be raised through the stock market, equivalent to 0.1% of British GNP, but the result was to

> transfer money from eager investors to a wide variety of promoters, lawyers, and owners of dubious, fraudulent and useless patents. In this manner the financial assets of the electrical engineering industry were firmly established as "lemons". (Kennedy 1987: 135)

This bubble of ignorant technology investment enthusiasm generated investor losses and consequently an equally ignorant period of investment withdrawal from any electricity stock. Of course, such gushes of enthusiasm and with-drawal remain a feature of stock market finance (see the section on venture capital and the bubble economy). The aftermath of the British electricity enthusiasm might not have mattered had there existed, or been created, an industrial investment bank willing to search out intelligently the better companies in order to establish a long-term financing relationship. In the absence of intelligent finance, ordinary investors remained deterred from buying equity or bonds in what became the best British firms of the time, Brush and Crompton (Kennedy 1987).

These firms needed long-term capital investment, adapted to the internal demand cycle of the growing electrical equipment industry. In Chapter 8 of this book it will become clear that the British structure of local government with its tradition of utility investment became a major obstacle to the conversion of obsolete DC electricity supply systems to the efficient AC supply system. Whereas US and German utilities managed an orderly acquisition of recently developed AC generating equipment, the British home market was the scene of a decades-long political struggle between would-be developers and local political authorities to legitimate the rationalisation of electricity supply. Electrical engineering developers in Britain did not therefore have reliable and expanding demand in their home market to the extent that their overseas competitors had.[18]

Kennedy writes that in the absence of industrial finance and tariff barriers to keep out the German and US firms that were rapidly achieving the decisive advantage of economies of scale, the British firms became 'exhausted from thwarted anticipation', illiquid and 'by 1895 in the hands of cautious men who used the affluence which the boom brought to restore their companies rather than expand them' (Kennedy 1987: 137). Given that Britain was then the richest country in the world, this example forcefully represents that for technologies like electrical engineering, it is not the crude volume of capital available in an economy that matters, so much as the organisational ability to offer it on terms that can enable competent management to plan for specific development opportunities.

The Exception that Proves the Rule – the Origin and Successful Development of a British Industrial Investment Bank

The exception that proves the rule to the consolidation of the British financial system around arms-length, property-secured and high-cost lending is the British industrial lending bank 3i, now a FTSE 100 company itself and the leading British venture capital company.[19] The success of 3i suggests that there were in the past foregone, profitable investment opportunities, even as 3i grew through their profitable exploitation.

This financial institution owes its existence to wartime fears of bank nationalisation on the part of the Treasury and Bank of England. Such fears prompted these institutions to act in response to a widespread belief that the small-company sector was poorly served by the clearing banks (Coopey and Clarke 1995: 49). The clearing banks themselves were reluctantly persuaded to become joint owners with the Bank of England and to contribute to the £45 million capital of the colourfully named ICFC (Industrial and Commercial Finance Corporation). This was the political cost of forestalling direct government intervention in the financial system (Coopey and Clarke 1995: 47).

Since it was owned by the clearing banks the ICFC had every incentive to avoid direct competition and to find profitable business in the high-risk loans that they would not touch – another example of competition working in its creative sense. The ICFC developed the characteristic industrial bank approach to loans: it was prepared to make a case-by-case assessment of risk that:

> Entailed a closer technical investigation than traditional bankers were equipped to carry out, which in practice meant building up an in-house capability. (Coopey and Clarke 1995: 35)

As early as 1947 the novel practice was adopted of sending accountants to assess prospective accounts *on-site*. And when the clearers began to make centralised lending decisions and reduced their lending to small regional firms in the 1960s, the ICFC moved into the gap by building a regional branch network with devolved decision-making powers (Coopey and Clarke 1995: 83).

Some measure of its success can be gained from the fact that by the mid-1990s it had invested in over 11 000 firms, that it has been estimated that 10% of all manufacturing workers in Britain were employed by companies financed by 3i (Coopey and Clarke 1995: 1) and that although it had never received the tax privileges or government guarantees and subsidies of industrial investing banks in other countries, its net worth had reached £1.3 billion in March 1993 (Coopey and Clarke 1995: 376). The 1971 British Bolton report into small firms found that ICFC was 'by far the most important institutional provider of long term capital to small firms' (Coopey and Clarke 1995: 95).

If its success is evidence that a small-firm funding gap existed, like other industrial investment banks the scale of 3i's lending operations was limited by the degree to which it could obtain secure, long-term funding itself. It borrowed very long term through the issue of 25-year debentures and would reinvest that capital in firms two or three times over during this period (Coopey and Clarke 1995). According to Donald Clarke, 3i's finance director until the mid-1990s:

> The fact that 3i was always able to raise money to meet demand…cannot disguise the relatively minor impact which it had on the general level of activity as the only institutional source of long term and permanent capital. It would have been possible with the benefit of a government guarantee, to raise very large sums on the capital market without interfering in the government's own fundraising programme. (Coopey and Clarke 1995: 383)

In the absence of such a guarantee, the scale of 3i's operations would remain dwarfed by the equivalent, government-guaranteed German and Japanese banks. On the other hand, because it never received subsidies or external guarantees, it is a more compelling demonstration of the potential to add value purely through a combination of in-house appraisal expertise and long-term stable lending practices, and therefore an exemplary confirmation of the inadequacy of the other types of loan available in the British financial markets.

3i also provides a demonstration of the dangers of short-term pressure in a market for corporate control. Because of the long-term nature of its investments, it was itself potentially subject to short-term pressures to 'realise value', but had been protected from such pressures for as long as the clearing banks remained (forcibly) committed, long-term owners. However, it appears the clearing banks were not fully aware of the growing value of 3i, because 3i placed conservative accounting values on its long-term investments. True to their character, when the clearing banks understood 3i's likely value in the mid-1980s, they became interested in realising that value, and the 3i board prepared a proposal for the company's flotation. 3i understood the *potentially* destructive dangers of flotation, or of ownership by a single shareholder intent on early returns:

> We could not easily accept a situation in which market pressures forced us to abandon or severely modify these long-term policies in order to meet short-term income demands. It would be possible greatly to increase the short-term returns but only at the expense of cutting out new investments in high-risk areas where short-term income is not available and by forcing the realisation of immature investments. (Coopey and Clarke 1995: 196)

Whether these pressures materialised after flotation eventually took place in 1994 is beyond the scope of Coopey and Clarke's book, but the unique status of 3i within the British financial world was further confirmed by the bank becoming the lead British venture capital provider by the time of the millennium boom.[20]

Conclusions on Innovation and the Bank–Corporate Relationship

As for the debate about financial systems and innovation, it is safe to conclude that however it is achieved, a close bank–corporate relationship is an aid to investment in innovation, where innovative opportunities are available, because it enables the bank to have a more detailed understanding of the risks and opportunities of prospective innovation. Replace 'bank' with 'owner' and the argument does not change – Mayer (1997) shows that for the largest listed companies in France and Germany share ownership is highly concentrated in other companies, and in Germany, also in families.[21]

And when innovative opportunities become exhausted it is the owners' fear for their capital that motivates them to restrain the investing machines that they have created. In contrast, the scenarios of *both* under and over investment were possible when there was uncommitted arms-length ownership ignorant of, and therefore unable to judge, the detailed prospective investment opportunities of their firms. It is the quality of the relationship between financing and investing agents that matters.

The Institution of Venture Capital

Venture Capital as the Marketisation of R&D Management

Venture capital is another form of long-term investment involving a managed relationship between financiers and firm, but with the focus on the selection and development of ideas through to early-stage companies. The target has always been 'leading-edge' technology, with the first funds organised in Boston in an attempt to exploit ideas coming out of MIT and venture activity growing strongly with Silicon Valley electronics activity. By the end of the 1990s the 'industry' of venture capital consisted of several thousand professionals working at some 500 funds concentrated primarily in California, the state that between 1990 and 1996 received three times more investment than the next most active state, Massachusetts (Gompers and Lerner 2000: 5,14).

Since it is also a major object of R&D management to select ideas and develop them into viable commercial entities, it makes sense to think of venture capital activity as the 'marketisation' of this aspect of R&D management. Once again, on superficial inspection this is not obvious because of the dramatic difference in the *visibility* of the operations of venture capital as opposed to the R&D department. The part-developed, but market-unready R&D project is typically hidden by the secrecy of the corporate hierarchy, but the venture equivalent is that highly visible and peculiar object, the publicly quoted but profitless company.[22] The dependence of the venture capital industry on public flotation effectively exhibits an earlier stage of the 'innards' of the R&D development process to the public.

If in technological development terms they are comparable, the management of venture capital is quite distinct from that of corporate R&D. Since the object is the development of early-stage ideas there is a typical lack of tangible assets in the portfolio of a venture capital fund. The concomitant risks of mismanagement and misrepresentation of value between the investors, fund managers and supported entrepreneurs help account for the rise of the limited partnership as the typical organisational form of a venture capital fund – 81% of funds had the limited partnership form by 1992 (Gompers and Lerner 2000: 10). This serves to align the interests of investors as limited partners, and fund managers as general partners through a contract for each fund, specifying the duration, scope and reward structure on liquidation[23] (Gompers and Lerner 2000: 8). Likewise, a variety of mechanisms are used to bring the interests of entrepreneurs into line with those of investors, including stock options, the 'staging' of investments of capital and time-intensive, on-site monitoring of the entrepreneur's activities (Gompers and Lerner 2000: 130).

Tight management control ends with the liquidation of the fund and the return of capital to the investors, typically after 10 years of development (Gompers and Lerner 2000: 206). Perhaps contrary to popular perception, only 20–35% of portfolio firms are floated, but these *tend* to be the greatest successes and are certainly the major source of profit (Gompers and Lerner 2000: 6). This dependence on the health of the IPO (Initial Public Offering) market is confirmed by surveys of returns on funds: returns were high in the early 1980s, then fell to single figures from the mid-1980s to the early 1990s as venture capitalists became dependent on established firms to buy their start-up companies. They began to rise again as the IPO market recovered from 1993 (Gompers and Lerner 2000: 22). This dependence on the stock market matters, because high venture capital fund returns become an index of the willingness of the public to buy the shares. This is not, of course, the same thing as evidence of the success of venture capital in developing useful innovation.

It should not be surprising that unscrupulous behaviour around the time of an IPO sometimes occurs, such as the reduction of R&D simply to create an apparent earnings increase and boost the offer price, or that 'many private equity investors appear to exploit their inside knowledge when dissolving their stakes in investments' (Gompers and Lerner 2000: 206). The response to these problems has included 'lock-in' provisions that prevent venture capitalists disposing of their stakes immediately after flotation – and of course the desire for a good reputation on the part of venture capitalists who wish to build a long-term business (Gompers and Lerner 2000: 207). These kinds of problem do not mean that venture capitalists are systematically selling investors lemons. Gompers and Lerner investigated the long-run market performance of 4000 venture-backed and non-venture IPOs between 1972 and 1992 and found that for this period at least, the venture-backed companies 'appear to earn positive risk-adjusted returns' after going public (Gompers and Lerner 2000: 210).

Attempting to Evaluate Venture Capital as an Institution

It might seem reasonable to 'believe' in the superior performance of venture capital over R&D management hierarchy on *a priori* grounds, such as the clearly stronger alignment of ownership interests in the partnership form, or based on the documented and poor record of management hierarchy at choosing future 'disruptive' technologies. However, this has to be set against the dependence of returns to venture capital on the stock market and the changes in venture management behaviour this can generate.

One of the earliest US venture capital funds to be established, by American Research and Development (ARD) Corporation, founded in 1946, had trouble raising its modest target of $3 million (Wilson 1985: 19) for want of such a record. This early fund also demonstrated other typical venture capital features. In ARD's case, there was one outstanding success: in 1957 it acquired 70% of Digital Equipment Corporation for $70 000, an investment that returned $350 million dollars, no less than 80% of the value of the fund, when it was distributed in 1972 (Wilson 1985: 20).

What is striking in the achievement of this return is, first, the long timescale before success could be demonstrated, and that when successful, a high *average* rate of return was dependent on a few spectacular successes among many failures and mediocre performers. This dependence on portfolio investment to ensure a decent return, despite the care taken in the selection and management of projects, warns us to be careful not to jump to conclusions about venture capital's superior ability to 'pick winners' compared with established management hierarchy. A large amount of waste characterises venture activity *and* hierarchical R&D management – it is in the nature of the search for successful innovation.

The way that venture capital interacts with other institutions complicates attempts to evaluate its impact on innovation. The rate of return to a fund is not a good indicator, because that rate is dependent on the vagaries of the market for IPOs. Then venture capital obviously coexists with, and to an extent competes with, established firms for development projects. So, for example, in the upswing of the early 1980s' lending cycle, between 1980 and 1982, 10 new companies were created by former Hewlett Packard employees alone (Wilson 1985: 190). At first glance venture capital appears to be subtracting people, and projects, from established companies, but in some degree the venture-backed activity may not only be more efficiently managed, but also may include projects that this and other established firms would not have funded, for a variety of reasons. So Wilson observes that in the 1980s, before European countries had acted to stimulate their own venture capital activity, total R&D spend was sometimes higher in European firms but the 'portfolio' of technologies they developed was more limited than in the USA (Wilson 1985: 220).

So if there is some tendency for venture capital to subtract from or duplicate existing corporate activity, at the same time it is likely to improve the effectiveness of R&D project selection and development. At this point the prospects for an answer to the question '*to what degree* does venture capital make a net positive contribution to innovative activity?' would seem remote. When it is considered that established activity is likely to become modified to some degree in reaction to venture activity, for example by an increase in the tempo of internal promotion and recruitment, or by hikes in internal remuneration rates and perhaps more efficient project selection and management, the ultimate result might be expected to be a rise in the innovative performance of the 'system' of interacting venture capital and established firms.

Nevertheless, a pioneering effort to prise apart the relative contributions of corporate R&D and venture capital through regression analysis yields the appropriately conditional result that 'the estimated parameter varies according to the techniques we employ, but focusing on a conservative middle ground, a dollar of venture capital appears to be about three times more potent in stimulating patenting than a dollar of traditional corporate R&D' (Kortum and Lerner 2000: 675). Some of the important cautions one could place against this result are that whereas venture funding is likely to be nearly entirely focused on new developments, corporate R&D is not. So corporate R&D should be less likely to result in patenting because it can include basic research, because it includes much established process improvement and because the option of secrecy in development is more available when development occurs within the financing organisation. In contrast, venture capital projects may have an increased propensity to patent because they begin with no fixed assets and have an increased need for 'symbols' of value such as patents to justify public offer prices. Kortum and Lerner do show that among a sample of venture-backed firms the patents were more often cited and aggressively litigated – they interpret this as evidence that these patents were no less valuable than others, but other effects may have occurred, for example copycat venture funding of a narrow area leading to crowding and an excess of patent disputes.

Despite such reservations over the significance and precision of this kind of result, it is difficult not to believe, if only from *a priori* argument, in the net benefits for innovation of venture capital activity. However, the dependence of venture capital on stock market IPOs for its most profitable returns introduces a fascinating instability into the conduct of the institution.

Venture Capital and the Bubble Economy

Since the rise in venture capital supply from 1979[24] there have been two periods of stock market euphoria characterised by the flotation of high-technology start-ups with dubious prospects – the biotechnology boom of the early 1980s and, of course, the Internet bubble of the late 1990s. Galbraith documented the complicit behaviour of financial intermediaries in the great 1929 stock market boom (Galbraith 1975) and what interests us is both the process that drives venture capital behaviour to become short term and exploitative in these periods and the result of such exploitative behaviour.

Stock market booms and busts are such well-documented phenomena that they are properly seen as a natural feature of the capital markets. They often centre on some poorly understood 'novelty' that promises riches, and the nature of this novelty can vary widely, from tulips in the seventeenth-century Netherlands to the fabulous prospects for the South Sea Company in eighteenth-century England. These old manias demonstrate similar behavioural patterns to recent ones and it was common for the quality financial press to make reference to them throughout the Internet bubble. It was one of the few means of making sense of the otherwise inexplicable behaviour of investors.[25]

The Syndrome of Venture Capital Oversupply

The *potential* for short-term behaviour in venture capital activity is clear enough – venture funds make their money by selling their projects either to established firms or to the public through an IPO. If the public lose their collective financial sense and become willing to buy undeveloped projects with a solid record of losing money, venture funds can profit by providing what the market wants. On the other hand, if they wish to remain in the market long term there should be some desire to behave in an ethical manner to cultivate a reputation with investors.

The 1980s boom had as its origin the sudden access of pension fund money into venture funds, but the results were similar to the Internet boom. In the 1980s oversupply of venture capital and competition between funds for limited opportunities soon produced a variety of higher risk investment strategies. One such competition-induced strategy was what Wilson calls 'vulture capitalism', the practice of waiting for established firms to finish the expensive process of developing an idea then funding the defection of key people to start 'new' firms. Vulture capitalism was denounced at the time by some of the leading managers of established high-technology companies, such as Intel's Robert Noyce (Wilson 1985: 191). However, established companies do have the defence that they can sue this kind of activity under trade secrets law.

Another feature of the syndrome of oversupply of venture capital was 'copycat' funding, as venture capital funds competed to have their own start-up in the most obvious areas of opportunity. This resulted in as many as 100 start-up companies manufacturing Winchester disc drives for PCs so that despite the high growth in demand 'only a few' of the start-ups were profitable by 1984 (Wilson 1985: 196).[26]

The successful flotation of Genentech in the 1980s, one of the few profitable and expanding biotechnology firms, stimulated much copycat funding. The temptation to depart from strict business assessment criteria is captured by the comment of a venture fund manager whose firm had avoided biotechnology start-ups on the grounds that such projects were run by researchers, not business managers, and appeared to have no credible market for their products:

> We were 90% right. Where we were embarrassingly wrong, it never occurred to us that the public would buy these companies and finance them, that you could raise money publicly for companies that had infinite losses and little damn prospect of sales. (Wilson 1985: 65)

But the public *were* willing to buy these companies – for a short time – and so venture firms stepped in to offer them; by mid-1983 the combined public value of biotechnology companies reached $2.5 billion (Wilson 1985: 65).[27]

For duration and magnitude the Internet bubble is in a class of its own. The foundations were laid from the mid-1990s with a series of increasingly successful flotations of loss-making companies. A significant early example at the end of 1994 was when Netcom On-line Communications Services, an Internet service provider (ISP), sold 1.85 million shares at $13, valuing the profitless company at $85 million (Cassidy 2002: 79). Repeated successful flotation of profitless companies began to make new stock evaluation criteria credible, such as the number and rate of growth in subscribers. Netcom had 41 500 subscribers and:

> At 13 dollars a share, each one of them was valued at about 2 100 dollars – more than twice the value that the stock market was attributing to cable television subscribers. This was despite the fact that Internet service providers charged about 20 dollars a month, while cable companies charged upwards of 30 dollars a month. There was no obvious reason why Internet users should be valued so highly, but even at this early stage Internet stock valuation wasn't based on reason. It was based on hope and hype. The new valuation formulas were primarily an attempt to rationalise the fervor of investors. (Cassidy 2002: 80)

What appears as common sense after the event was not easy to recognise as common sense at the time. As entrepreneurs and venture capital firms understood what was possible they rushed to 'supply' the market with what it wanted – shares in start-ups. The motivation given by Cassidy for entrepreneur Jim Clark's decision to float Netscape in May 1995 is significant. Besides the usual desire for personal wealth, there was the knowledge that flotation was now possible despite the fact that Netscape gave its browser product away for nothing – and he apparently also anticipated that Microsoft would soon provide severe competition to Netscape in the browser market (Cassidy 2002: 82).[28]

At the height of the boom between October 1998 and April 2000 there were more than 300 Internet IPOs (Cassidy 2002: 192). This was the period when copycat funding became endemic and took on a serial form. A wave of Internet IPOs in 'business-to-consumer' applications generated companies that were only superficially distinct by business area and possessed no barriers to entry since anybody could set up a website. They dare not charge users and many depended on an untested advertising revenue model to justify their existence. Another popular business justification was a 'winner takes all' argument: the first firm to achieve brand recognition through advertising and marketing would supposedly dominate its 'sector'. This was somewhat beside the point if the underlying business idea was poor. No sooner had doubt about the viability of these entities begun to harden than 'business-to-business' or 'B2B' companies were touted as the next goldmine. Healtheon is a good example of what became possible, not only in terms of saleable business ideas but because Healtheon was Jim Clark's third IPO. The founder of Netscape had become a 'serial entrepreneur', another phenomenon of the Internet boom. Cassidy tells the story of how the idea for the firm was generated:

> Healtheon started life as a blank piece of paper with a diamond on it. At the corners of the diamond, Clark wrote four words: Payers, Doctors, Providers, and Consumers. In the middle he placed an asterisk. The asterisk represented the new company, which would somehow link the corners, representing a sixth of the American economy, in an online network. Armed with his Magic Diamond, Clark went to see the VCs [Venture Capitalists] on Sand Hill Road and announced that he planned to "fix the US health care system" … the only question was which VCs he would allow to back him. (Cassidy 2002: 204)

Clark's reputation for *generating floatable companies* was enough to attract venture capital support. By the time of flotation in February 1999, Healtheon had spent nearly $100 million and still had no clear prospect of a financial return. Nevertheless, five million Healtheon shares were floated at an issue price of $8. On the day of issue the share price had risen to over $31, a price that gave a value to the entire stock of the company of more than $2 billion (Cassidy 2002: 205).

The sustained and large profits from such flotations attracted large inflows of capital to the venture capital industry itself – on a more massive scale than had happened in the early 1980s. In 1996 there were 458 firms with $52 billion of funds under management, by 1999 there were 779 firms with $164 billion of funds (Cassidy 2002: 236). The pace of investment was forced upwards: in 1996 over $11 billion were invested in 2123 new ventures; in 1999 just under $60 billion was invested in 3957 companies (Cassidy 2002: 237).

The increased number of funded ventures emphatically did not represent a proportional increase in real innovative opportunities, but the efficient exploitation of gullible investors' willingness to buy shares whatever the quality of the business plan.

The ease of flotation led to a progressive relaxation of venture capital practices. So venture 'development horizons' shrank to 12 months or less. The common restriction that entrepreneurs invest a third of their net worth into their start-ups was relaxed (Cassidy 2002: 237). Even old established firms like the Mayfield Fund were eventually lured into e-commerce investments by the scale of the returns others were obtaining (Cassidy 2002: 238).

In the downturn the process of restoring normal behaviour became visible. Dot.com companies became a joke; the *New York Post* began a column titled 'Dead Dotcom of the Day' (Cassidy 2002: 303) and the website FuckedCompany.com provides a useful archive of commentary on the hype and death of dot.coms. Now that it was possible to compare the reality of failure with the recent hype accompanying the public offer, the reputations of venture capitalists and investment banks were once again visibly at stake in the public offers that they chose to support.

Venture capital paradoxically combines two features: the tight alignment of investor and management interests in the limited partnership form of organisation and a dependence on the stock market IPO for its greatest returns. The first tempts one to make comparisons with orthodox R&D management that are favourable to venture capital, but the second introduces the swings in sentiment characteristic of the stock market as a driver of venture activity. Although, an institution with complex and often wasteful economic and managerial effects, its existence can nevertheless be judged to ensure that the hunt for genuine innovation will be conducted more effectively, and the development of that innovation conducted more efficiently than in its absence.

Conclusion

Throughout this chapter financial institutions were discussed as adaptations of financial supply to various circumstances of technology development. The range of institutions was varied, and so were the problems of abuse, from which no institution was free.

For the financial lending institutions, the stability and extent of financing activity depended on how those institutions related to other parts of the financial system. Venture capital activity and returns gain their stability from their limited partnership form of organisation and volatility from their dependence on the stock market. Industrial investment banks depend on finance from elsewhere and so may themselves be subject to short-term financing pressures, pressures that can be mitigated by a variety of devices such as central bank guarantees or loans. Comparisons between Germany, Japan and Britain (3i) reveal some variety in the devices used to promote close relationships between banks and companies, but the essential common feature is that a close relationship makes possible assessment that is adapted to the characteristics and circumstances of the proposed development project. The discussion of the use of financial evaluation techniques and cost accounting also pointed to the necessity of adapting these forms of assessment to the characteristics of the new technology – when sources of finance are internal to the developing firm they cannot be *assumed* to be matched well to the prospective development projects that are available to that firm.

From this point of view the crude volume of capital available for loan in an economy, or for investment within a firm, is less important than the organisational ability to tailor the conditions of the loan or investment to the development context. It follows that the kind of financial innovation that matters for technological innovation is one that generates a more effective way of matching capital to technology development prospects.

Another common feature of these institutional forms is that the history of their founding and growth *mattered*, sometimes for whether the institution would exist at all (industrial investment banks, keiretsu) and certainly for the precise form in which they operate today. Despite the clear differences in financial institutional endowment between nations, close inspection found the differences in terms of function to be more a matter of degree than absolute. Nevertheless, in the operation of the British market for corporate control and in the example of 3i there were grounds for viewing that country as having been relatively undersupplied with the kind of 'relationship banking' that is the ideal support for innovation.

Notes

[1] As a result the value of inventory appearing in public company accounts is always likely to be adrift from its current market value. This matters most notoriously in a market downturn, when falling sales for a given rate of production will first generate higher levels of finished good inventory – a clear sign of trouble. But the public accounting procedures described here will simply assign greater value to the increased level of inventory – if the accounts are read 'at face value', they will be understood to show the company to have an increasing asset value. The reality is probably a falling asset value.

[2] Johnson and Kaplan favour university accounting education as an explanation for this. Because accounting *education* from 1900 was shaped by the need to train *public* accountants, the generation in the US educated before the Second World War had no training in management accounting. Successful post-war managers from this generation accepted and introduced managerial decision-making systems based on data gathered for public accounting purposes (Johnson and Kaplan 1991: 134).

[3] Miller and Napier describe the promotion of DCF in Britain in the 1960s in these terms (Miller and Napier 1993).

[4] For which see Chapter 7.

[5] Hanson and Tomkins are two recent British examples. Hanson was the foremost predatory acquirer in Britain during the 1980s, but as it ran out of acquisition targets it spent 10 years underperforming on the stock market until in 1996 Lord Hanson broke it up (Farrelly 2000). It is somewhat ironic that the search for break-up value in Hanson should yield a focused building materials company, and Tomkins an automotive and engineering parts company; at the end of their brief lives these conglomerates may have generated true value-adding companies in industries formerly typified by their fragmentation and lack of innovation.

[6] They went so far as to state baldly that this management style was responsible for the perceived US economic ills of the 1970s and 1980s; in particular, relatively low rates of productivity advance. This at least made their argument provocative.

[7] All the railway material is drawn from Chandler's chapter 'System-building' in *The Visible Hand* (Chandler 1977).

[8] Companies found it difficult to agree appropriate traffic shares, speculators like Jay Gould bought and operated railroads and violated rate-setting agreements and by 1887 Congress finally decided not to sanction cooperative rate setting (Chandler 1977: 144).

[9] As if to demonstrate the difficulty of ever judging the operation of the market for corporate control, ICI decided during the Hanson 'siege' and *probably* as a result of it (Brummer and Cowe 1994: 236) to solicit a City view on its restructuring. The company adopted the proposal by a single corporate financier to split the company into two, a pharmaceutical and biotechnology company Zeneca, and a chemicals business, ICI. As the market for corporate control at work, it appears very dependent on context and the individuals involved and not at all determinate in outcome at any particular time (other restructuring outcomes were possible).

[10] Davis' review draws on Mayer and Alexander and a Cranfield University survey of British firms' access to finance – but no sample size is available in Davis (Mayer and Alexander 1990).

[11] In a sample of 131 firms 30% of large share stakes were turned over in four years (Mayer 1997: 296).

[12] Of course one can follow the accounts of many such corporate rescues in the pages of the *Financial Times*; MAN and Holzmann are examples from recent years.

[13] No less than six pages (Hutton 1995: 136–42).

¹⁴ If there were an element of improvisation in this system's origins, the Japanese bureaucracy learnt to preserve and develop the system for the power it gave them to direct limited post-war resources (Johnson 1982: 204).

¹⁵ The 'city banks' receive loan privileges from the central Bank of Japan (Johnson 1982: 202).

¹⁶ These banks, such as the Japan Development Bank, were able to favour certain technologies and industrial sectors by announcing their willingness to make supportive loans in that area (Johnson 1982: 201). Despite modifications in practice, these institutions have also endured; the Japan Development Bank continued until 1999 when it was succeeded by the 'Development Bank of Japan'. Its continuing distinct nature compared with Anglo-Saxon financial institutions is demonstrated by its governor's website message: 'DBJ supplies long-term, low-interest funds to projects that are important to the long-range national interest but would not be feasible with private financing alone. ... Our basic commitments are to supplement private sector financing and ensure the redeemability of our funds' (www.dbj.go.jp/english/).

¹⁷ Throughout the 1990s the Japanese banking system has been in crisis and the system that encourages over-loaning has remained intact. A feature of this extended crisis has been the unreliability of the reported figures and despite repeated write-offs, new bad loans continued to 'appear'. At the time of writing the Japanese Financial Services Agency has announced that total Japanese bad bank loans are one-third higher than previously thought, at Y13 000 billion (Ibison 2002: 3). On the other hand Japanese corporations are now vigorously paying back their debt, at the rate of 5% of Japanese GDP per annum for the four years up to 2003 (Morgan 2003: 14).

¹⁸ A final piece of the underdevelopment jigsaw was probably the British practice of training engineers through 'pupillage', a practice that persisted through the crucial late years of the nineteenth century when electrical engineering was being established (see Chapter 7).

¹⁹ This was true in 2002, see webpage http://www.3i.com.

²⁰ According to the 3i website, 3i has a leading position in Britain (http://www.3i.com/pressoffice/deliveringthemessage/).

²¹ Mayer's sample is the largest 170 listed companies with single shareholdings over 25% in France, Germany and Britain: 80% of these firms have single shareholders in possession of more than 25% of the shares. Only 16% of the largest British firms have such concentrated ownership (Mayer 1997: 298). Mayer supports the idea that the closer relationship, or 'insider system' as he terms it, improves monitoring and control, but paradoxically suggests that 'they may be inflexible in responding to technological changes' (Mayer 1997: 299). On the contrary, one can expect insider systems to have more confidence in their ability to recognise and exploit opportunities and to act more aggressively as innovators. Mayer perhaps thinks of technological opportunities of a certain kind, the entirely new opportunity without competitive implications for existing business. But these can be developed by other financial–corporate relationships, such as those developed by venture capital.

²² The comparison reminds us that our judgement of the value of the part-developed project within an established corporation would ordinarily depend on the degree of financial commitment that company gives the project (if we could know it) – the company has other uses for its money. Once a venture becomes publicly quoted there is no 'body' whose degree of commitment can signal the *relative* value of that project compared with others. The usual stock-related inducements will ensure that the management of the publicly quoted project are committed to its success – but that does not tell us anything of the chances of its success. The only guide is the dubious one that 'investors' with little or no technological knowledge were willing to buy and hold the shares.

²³ The reward for the fund managers is typically a fixed fee plus a percentage of fund profit (Gompers and Lerner 2000).

²⁴ The boom in US venture capital supply dates from 1979 with the revocation of a US rule that prohibited pension fund investment in venture capital on the grounds of prudence (Wilson 1985: 26).

²⁵ Kindleberger's *Manias, Panics and Crashes* (1996) analyses a series of great historical manias.

²⁶ This throws light on the inability of so many disc drive companies to manage transitions between disc drive size standards, something that Christensen treats as a management-solvable problem (Christensen 1997).

²⁷ On the other hand one must be wary of the numbers often presented in newspapers in support of the idea that 'biotechnology' has not yielded profits. So in 2001, after a second gush of money into biotechnology start-ups, one US broker's estimate was that only 25 of over 400 US

biopharmaceutical companies would make a profit in the year (Abrahams and Griffith 2001). Many biopharmaceutical companies have been acquired by established pharmaceutical companies, so that the number of 400 independents is a residual of the recently created companies, within which the less desirable companies are overrepresented.

[28] This tends to make one more dismissive of the alleged 'anti-competitive' behaviour of Microsoft in bundling its browser into Windows for free (see Chapter 3).

7   Innovation and The Organisation of Technical Expertise and Work

Much popular and guru-management writing is built around the claim to possess knowledge of best practice and in addition assumes that this knowledge can be passed on through written text to the reader. The empirical evidence that a practice is 'best' is often no more than the coincidence between a firm's good financial performance and its possession of unusual practice. Nevertheless, this kind of analysis may be all that is possible when information is limited – as it was in the years of emerging awareness that Japanese quality practices were distinctive.

So this chapter begins by taking the process by which one set of techniques become established as superior to another as interesting in its own right. In the two examples, of Japanese quality management techniques and Anglo-German matched manufacturing plants, this leads into the analysis of the organisational and institutional context that supported the generation of such techniques. In these examples, institutions of apprenticeship and industrial organisation prove also to have an important bearing on the possibility of transfer of techniques between countries. The argument is that in their absence, or where, as in Britain, apprenticeship has suffered decay, the ability of management to control work is weakened and thus the ability to implement and adapt new technologies is weakened.

By demonstrating the importance for technology transfer of institutions of apprenticeship and industrial organisation, this chapter justifies the analysis of the origin and reform of such institutions, a matter that inevitably involves the state. How and why the state becomes involved in the reform of such institutions is developed more fully in the last chapter.

Explaining Spectacular Achievements in Japanese Car Production

Development in the Explanatory Power of Accounts of the Japanese Achievement

As Japanese firms gained US and European car market share from the 1960s – to reach over 30% of the US market (Womack et al. 1990) – so did the necessity of finding an acceptable explanation of that success on the part of western observers. A kind of market in explanations developed with an excess of suppliers, so that Schonberger, the writer who did much to diffuse the notion of 'just-in-time' (JIT) manufacture as a bundle of freely transferable shop floor techniques, could write 'much of what is professed in current readings consists of half-truths and misconceptions that stand in the way of rapid progress in catching up with the Japanese' (Schonberger 1982: vii). Schonberger made only passing reference to the role of Japanese industrial organisation; in contrast, Monden's account of the development of the Toyota Production System for shop floor management not only described Toyota's pioneering ingenuity in *generating* an evolving set of techniques, but allowed that this ability depended on differences in industrial organisation and social structure (Monden 1983).

Four years after Monden and five after Schonberger, the authors Clark, Chew and Fujimoto published research which suggested that nearly half of the leading Japanese car manufacturers' advantage in quality and productivity derived from their management of the new model design process (Clark et al. 1987). So in retrospect both Monden and Schonberger were misleading by omission. That was understandable, because their method had been personal observation of highly visible shop floor practice, supplemented by published accounts and, in retrospect, very clearly limited company access.

By 1990 a popularising book *The Machine that Changed the World*, by Womack et al.,[1] usefully combined descriptions of the shop floor and new model design techniques with comparative quality and productivity data for global car manufacturers (Womack et al. 1990). These tended to show European car manufacturers at the bottom of the quality and productivity league tables. Womack et al. did not hesitate to make dire warnings of the consequences for European manufacturers if they failed to adopt these bundles of best practice techniques. Yet 10 years after that book was published the dire consequences have not yet materialised for most European manufacturers, despite the Japanese retaining a lead in product quality and productivity. A reasonable speculation would be that the aesthetic appeal of certain new European models – Peugeot–Citroen's range, or BMW's new Mini – had begun to offset significantly the persistent advantage in productivity, quality and price of rival models that remained aesthetically dull.

Aesthetic advantage need not mean beauty. By the end of the 1990s US car manufacturer profitability had been temporarily restored by the rise in popularity of the gas-guzzling, monstrously heavy SUV (Sports Utility Vehicle) and the misleadingly named 'light' pickup trucks that together now account for more than half of all US new-vehicle sales (Easterbrook 2003: 27). Easterbrook cites survey evidence that such vehicles are bought for the – bogus – psychological sense of safety that they convey to their users.

When reviewing the history of popular management and economic analysis of the car industry it is difficult to avoid the conclusion that even when it has access to good data, at best it represents what has already happened and rarely does it predict future competitive circumstances well. Sometimes it merely sensationalises: in 1994 a best-seller was titled *Comeback*, in reference to the apparent success of the big three US producers in fighting off the Japanese challenge. Now comes Maynard's book titled *The End of Detroit* that once again points to the persistent underlying problem of the US car manufacturers as an 'inability to provide good-quality vehicles that Americans keep wanting to buy, and at premium prices' (Grant 2003: 10).

As regards quality in Japanese production, an impressive feature of this account was that if one were an academic and dependent on western publications to identify and describe the technical source of Japanese advantage, for much of the time one would be in possession of a partial and outdated picture of practice. In contrast, western firms were in a position to gain a direct understanding of Japanese practice by buying into Japanese firms (for example, Ford acquired a stake in Mazda) and setting up joint manufacturing ventures. But as we saw in Chapter 4, even such close organisational relationships did not guarantee an adequate organisational response.

With this overview of changes in analytical fashion in mind, we are in a position to reassess the value of the stories of Japanese unique shop floor and model design practices, for the clues they provided that the Japanese car manufacturers were doing something different with their organisations. By this we will be examining an important example of a very common process: how firms identify and attempt to transfer the proprietary know-how and techniques of successful rivals. Today, this is sometimes referred to as 'benchmarking', but the way that term is used often implies that the relevant information is available if you can be bothered to look. In practice, as to an extent we have already seen, the problem is how to establish the value of the available, but probably incomplete, 'information' when there are competing explanations for your rivals' success. The rivals have no interest in helping you. It is a problem of developing knowledge, or understanding, not simply of comparative metrics – although metrics remain a useful way of organising data to aid the detection of patterns in practice.

Innovative Shop Floor Practice as the Basis of Competitive Advantage

A listing and description of the full set of Japanese manufacturing techniques after two decades would be besides the point; what they collectively demonstrated was a sustained and determined creativity on the shop floor by engineers and workers. Some, like kanban,[2] were original attempts to control inventory and others, like preventive maintenance, were already well known but not widely used in the car industry. Some more clearly demonstrated engineering ingenuity on the shop floor, such as the much cited reduction of set-up time in the Toyota sheet steel press room; this involved the creation of a revolving table of adjustable height so that press dies could be slid on to the table, the table revolved and a new die slid from the table into position on the press (Monden 1983: 76–9). The numbers attached to the effort to reduce set-up time also gave some sense of the sustained effort involved in generating the achievements. According to Monden, the Toyota press shop set-up time was reduced from 2 or 3 hours in 1954 to 15 minutes by 1964, and after 1970 this was further reduced to 3 minutes (Monden 1983: 8). These achievements were not the product of an overnight flash of genius, but of a long period of trial and error experimentation by both engineers and workers on the organisation of work and the design of shop floor artefacts.

Another long-lived insight came when comparisons between established western and Japanese practice revealed that some of the academic theorising about mass production operations had inadvertently helped *block* the development of better practice, rather than aid it. This was the case with the idea of an 'economic order quantity' (EOQ), a theoretically ideal number for the size of a batch of manufactured parts. The derivation of an EOQ was once a staple of operations management textbooks and was calculated as a trade-off between the overhead cost of machine set-up time, which reduced per unit part as batch size was increased, and the variable 'carrying cost' of stored parts that could not immediately be used. The problem was that to model the problem in this way, the EOQ *assumed* that set-up time was fixed, rather than a form of cost that could be driven down through creative shop floor thinking.[3]

A last instructive contrast that was made at the time was between a 'western' presumption that quality was the final responsibility of inspectors at the end of a production line,[4] and the 'Japanese' method of keeping responsibility for quality with the workforce. If the quality of delivered products depended on final inspection, then it was natural to associate the achievement of higher customer-delivered quality with the greater costs of more thorough final inspection. The rule of thumb became to assume that higher delivered product quality cost more. In contrast, in the Toyota Production System, the reduction of inventory and costs was closely associated with the solution of a host of quality-related problems in cooperation with the workforce, so that rising production quality was associated with *decreasing* costs.

A lesson here was that the mathematical modelling of existing production practices had proved to be a diversion from the creative transformation of that practice. Another was that the immediate objects of new practice were savings in simple physical terms, for example less floor space for storage, less time for machine set-up, reductions in numbers of breakdowns, reductions in inventory. Success in the reduction of these physical parameters in the working production line could be relied upon to translate into economic gain.

Innovation in New Model Design and Competitive Advantage

The design of new models of car is an extremely costly activity, as shown by General Motors GM-10 project for four new models in the 1980s, budgeted to cost $7 billion and take 5 years, but running to 7 years (Womack et al. 1990: 106). Such high design costs obviously limit the frequency of redesign and the variety of models mass producers are willing to offer their customers. So Clark et al.'s evidence of a great gap in the efficiency of the model design process between Japanese and western producers (Table 7.1) implied that an enormous competitive advantage lay with the Japanese.

Table 7.1 New model car design compared in USA, Japan and Europe, abstracted from Clark et al. (1987: 741)

	Japan	USA	Europe
Number of projects evaluated	12	6	11
Millions of hours engineering time	2.7	4.9	6.4
Months from first design to new deliveries	43	62	63
Size of average product development team	485	903	904
Delay of introduction date (as fraction of project lifetime)	1/6	1/2	1/3

Note: Number of engineering hours has been standardised by an attempt to take account of variation in project scope (Clark et al. 1987: 744). Size of team was estimated from a different selection of data (Clark et al. 1987: 755).

Not only was the Japanese *design process* cheaper and faster, but Womack, et al. also document some evidence that Japanese design was a significant source of quality and productivity advantage on the shop floor. This occurred because the Japanese were good at designing cars that were easy to manufacture; in practice this means that cars were designed to be built with fewer parts, and parts were designed to be easier to assemble[5] (Womack et al. 1990: 96).

As discussed above, this new evidence reduced the explanatory burden previously placed on Japanese shop floor technique. Altogether, the evidence was overwhelming that the Japanese were able to manufacture cheaper, higher quality (lower number of defects in production) cars because of their distinctive achievements in the reorganisation of both the design process and the shop floor.

An obvious question is whether their western rivals would be able to copy their achievement once they understood it. The answer to this question is not strictly separable from the answer to another: what enabled these Japanese companies to move from a technologically backward position to a leading position? The answer to both questions involves the characteristic Japanese industrial organisation of the keiretsu and the associated institution that it serves to maintain: the 'Japanese employment system'. The next section teases out the connection between work practices and industrial organisation, using the example of the organisational practices most clearly linked to the achievement of efficiency in new car model design.

From the Organisational Practices that Support Superior Design to Distinctive Institutions

At least the Japanese shop floor practices were visible to informed plant visitors such as Monden and Schonberger. Clark et al. had to identify the less visible project management practices behind the achievement of more efficient and effective new model design. They explain that in the best of the Japanese projects,

> a heavyweight project manager leads a multifunctional team, in which problem-solving cycles are overlapped and closely linked through intensive dialogue. (Clark et al. 1987: 766)

This requires some explanation. The design of a new model range in the car industry historically took place within the functional departments of the companies and was coordinated by the normal management hierarchy. From the 1960s a trend became established to manage new product development as a distinct project, a trend most pronounced in Japan. Yet the degree of project management organisation varied considerably and for convenience Clark's researchers distinguished between 'lightweight' and 'heavyweight' project management: lightweight managers had *responsibility* for the coordination of development, but little control over the content of the project. The development work continued to take place in the functional departments. In contrast the heavyweight project manager is

> not only a coordinator but a concept champion with direct responsibility for all aspects of the project. He or she has strong influence outside the

development group, works directly with the engineers (creates a project team) and has high status within the organization (equivalent to a major functional department ...). (Clark et al. 1987: 752)

This relative power and prestige of the project leader could be enhanced by devices such as that in Toyota, where the new model project leader position was seen as the traditional stepping-stone to the leadership of the entire company. It attracted the company's better and more ambitious people (Womack et al. 1990: 112).

Many of the quality and productivity achievements of heavyweight teams are related to their superior ability to manage the general characteristics of large-scale projects. In order to understand the significance of the Japanese achievement, these need to be briefly introduced.

The problem of managing large-scale projects of work arises in many industries; the construction industry and software writing are obvious examples. Such projects consist of many defined tasks, some of which can be pursued simultaneously, in principle shortening project lifetime. Actual project lifetime is, however, determined by the organisation of the class of tasks that require the completion of other tasks before they are begun. A sequence of dependent tasks forms a path of work, and many such paths may run in parallel in a project. The longest sequence of dependent tasks is called the 'critical path' and it is this that effectively defines the minimum time to project completion. Any delay on the critical path delays the whole project and so it is this that deserves the closest management monitoring. An early, classic and relevant account of the problems of project management is Brooks's essay *The Mythical Man-Month* based on his experience as manager during the writing of the IBM Operating System 360 software (Brooks 1995).

Relevant to car design is Brooks's explanation of why project schedules always slip. The blame rests partly on the optimism of designers, who assume that each of the many sequenced tasks that comprise a major design project will only take the time that they 'ought' to take. Brooks also places blame on the unit of work used to schedule projects, the 'man-month'. This unit provides a temptation to assume that men and months are interchangeable, as if a doubling of workers on a task should halve the time taken. As Brooks expresses it, although project cost *does* vary as the product of the number of workers employed and the number of months worked, *progress does not*, because of communication costs (Brooks 1995: 17). Communication costs take a number of forms. New workers must be trained in the project design and typically by the already competent workers, but worse, if work is repartitioned to enable these new workers to be put to work, and if that partitioning raises the necessity for worker intercommunication to coordinate design, then the time required to communicate may outweigh the expected time savings from the rescheduling of the work.[6] In these circumstances a paradox arises: if management give in to the temptation to add more workers in an attempt to restore a slipping schedule, the result is instead to *lengthen* the schedule, to add to its expense, and because of the increased complexity of communication within the project, 'without a doubt' to yield a poorer product (Brooks 1995: 25).

The same description of dysfunctional project management is given by Womack et al. when they describe the management of car design within the functional departments of the car industry (Womack et al. 1990: 115). Car design shares with systems programming this characteristic of a complex inter-dependency of tasks that require significant worker communication as part of the necessary coordination. Brooks' description of the effects of attempting to restore a slipping project schedule makes sense of the outstanding feature of project management in car design; the heavyweight teams had *small numbers* of *multifunction* engineers which facilitated effective worker communication and so the efficiency and quality of design. In contrast, some of the US firms had overspecialised their engineers – one firm had specialist door lock design-ers (Clark et al. 1987) – and this drove up the size of design teams. Larger teams demanded disproportionately more coordination time and created more oppor-tunities for slippage – for diseconomies of scale – as coordination of interrelated tasks became difficult.

Now we are in a position to understand some of the more exotic means the Japanese firms developed to reduce project lead-time. Because the teams were smaller they carried a lower intrinsic time burden of task coordination. This allowed them to coordinate innovative efforts more closely to reduce, rather than simply manage, the critical path itself. The prime example of this given by Clark's researchers was the achievement of a great degree of 'overlap' of the most important critical path tasks, that is the achievement of a degree of simultaneous work on two tasks, where normally one would only begin upon completion of the other (Clark et al. 1987: 756ff.). The greater the degree of over-lapping of tasks that lie on the critical path, the shorter it is possible to make the project lifetime.

This idea really needs a concrete example and Clark's team give us the case of the design and development of the hardened steel dies used for pressing steel body panels (Clark et al. 1987: 760). Overlap took the form of starting the rough cutting of the dies *before* the final panel shapes had been decided, a practice that depended for its success on close communication between the panel design and die design engineers. The best practitioners had shortened the develop-ment time for dies from 2 years to 1 year[7] (Womack et al. 1990: 117).

Since Clark et al.'s article, this practice of overlapping project activities has become popularly known as 'simultaneous development' or 'concurrent engi-neering' and, divorced from its car industry context, is widely advocated as the means of reducing project lead-times – as a casual Internet search will reveal. It is obvious from the context that unless other technology design processes have disproportionately long tasks on the critical path, such as die formation, there will be no good targets for critical path reduction. If the effort is nevertheless made to overlap tasks on the critical path, there will be a much lower prospec-tive return in terms of time saved, yet an increased overhead of communication costs – and so a greater likelihood of slippage and diseconomies of scale. Given the normal problems of project management, design projects without an outstandingly long task on the critical path would be better advised to con-centrate on good project management: the effective definition of tasks and the identification of the critical path itself. In Brooks' account, this is difficult enough, and clever attempts to complicate project management result in more slippage and worse project quality. This argument for caution depends only on the task length characteristics of design project work. There were other organisational and expertise reasons why the Japanese achievement of simultaneous engineering was possible.

The key question is how the Japanese kept their design teams small and effective. In the first instance, small size was only possible because of the breadth of functional knowledge possessed by the individual engineers in the team; fewer individuals were needed to cover the necessary areas of expertise. This only begs the question of how this was achieved. So we turn to an examination of how the elite Japanese industrial company distinctively forms its engineering talent, something that nicely bears on its shop floor achievements.

The Distinctive Formation of Japanese Engineers and the Japanese Employment System

In contrast to western firms, Japanese graduate engineers typically begin their careers working on the shop floor. The historian Morikawa argues that this practice was originally supported by Japanese firms' urgent need to transfer technology from the west (Morikawa 1991: 138). The shop floor became something of a laboratory employing a high concentration of engineers as technology implementation problem-solvers. Long after Japan passed the acute phase of technology catch-up with the west, this practice of initiating engineers to the shop floor continued. Morikawa argues it had always formed part of the elite Japanese engineering schools' understanding of what an engineer should be (Morikawa 1991: 138). Whatever the explanation, the practice became so widespread that in Japan 'to work exclusively in the office or on research and development is not to be, by definition, an engineer at all' (Morikawa 1991: 136). Again, this stands in contrast to the west where immediate specialisation in such employment niches is an accepted career path for graduate engineers.

Shop floor experience is only the beginning of systematic job rotation through other functional departments that gives the Japanese engineer a broad experience of the industrial firm. This broad experience allows Japanese engineers to become the typical feedstock for promotion into the management hierarchy.

This distinctive use of the engineer helps makes sense both of the achievement of innovation on the shop floor and model design. In particular, such achievements as set-up time reduction, that involved the re-engineering of equipment, required engineers to cooperate with blue-collar workers over long periods of time. The achievement of the necessarily small, heavyweight teams in design was possible because the members already shared much experience of each other's nominal functional specialities. This was an obvious aid to communication, but it also allowed fewer individuals to represent the necessary fields of expertise.

The story of Japanese engineers' distinctiveness is not complete until the features of the Japanese employment system are described, for the work, motivation and careers of engineers are moulded by that system in the elite industrial companies. The dominant view within Japan is that there are 'three pillars' of the employment system that are intimately connected: lifetime employment, a seniority-plus-merit wage and promotion system and enterprise unions (Johnson 1982: 11; Whittaker 1990: 5).

Lifetime employment allows the company to invest heavily in costly forms of training – such as job rotation – and yet remain confident that it will reap the returns on this investment, rather than some employee-poaching competitor. Lifetime employment is itself facilitated by 'nenko' – the rewarding of employees over long, rather than short, time periods. Nenko describes the practice where employees cannot be promoted rapidly and above their seniors, but that as they age, if they are thought to have merit, they will gradually move up the 'career escalator', a term that captures the somewhat automatic nature of the process (Woronoff 1993: 85). Finally, enterprise unions represent the workforce of a particular Japanese company, rather than the workers in an industry (industrial unions) or an occupation (craft unions) (Whittaker 1990: 85). Some of the advantages of the Japanese employment system for the management of technology can be expressed as the avoidance of negative outcomes: of insufficient and inadequately broad training; of overspecialisation in some niche of technical expertise; of individual commitment to organisations other than the company, such as unions or occupational or professional groups.

The Japanese employment system represents an extreme form of reliance on internal management instead of the labour market for the skills, experience and expertise needed by a company. In DuPont or IBM we already have examples of western companies that have used market control to modify employment practices in a similar manner to the Japanese elite companies – by extending security of employment to their engineers and scientists. A difference is that when the source of that market control declined – when they lost their leading and innovative position in product markets – so did the ability of the company to extend security to its workforce.

There is a fundamental difference in industrial organisation between such leading western companies and the Japanese industrial firms – and so in the institutional basis for the control of the labour market. As introduced in the finance chapter, the elite Japanese industrial companies are invariably grouped around a big bank and a general trading company, a characteristic Japanese form of industrial organisation termed a keiretsu.[8] From the 1950s the economy became dominated by 20-odd large keiretsu that were effectively financially underwritten by the Bank of Japan itself (Johnson 1982: 206).

This distinctive industrial structure underpins the employment system. The number of keiretsu is small enough that they are able to enforce the norm of not poaching core employees from each other. Engineering graduates or core workers join a single keiretsu group for the life of their career and cannot 'job-hop' between companies affiliated to different keiretsu in search of promotion; in return the keiretsu offers the security of lifetime employment, extensive training and a broader range of employment and promotion opportunities *within the group* than any single western company could offer.

The obvious disadvantage of the extension of lifetime employment to a core workforce would appear to be its *inflexibility* with respect to changing economic conditions and company retrenchments. Dore, in his pioneering comparison of two British with two Japanese factories, certainly showed that in the face of such market shocks, extraordinary *efforts* are made by management to retain the core workforce (Dore 1973). Among the practices Dore identified in his comparison of Hitachi and English Electric plants: Hitachi's use of an expendable group of temporary workers; the minimising of in-house machining and fabrication in good times so that during demand troughs outside suppliers could be dropped; a greater ability to redeploy the very broadly trained core workforce between different types of work; a willingness to retrain any workers made redundant by changing technology; and because of its greater product range, a corresponding ability to transfer workers out of declining product areas while remaining within the Hitachi group of companies[9] (Dore 1973). In sum, Hitachi made great efforts to extend a high degree of security to its core workforce relative to ordinary workers in the economy, and to its own temporary staff. In exchange, Hitachi obtained a skilled and flexible workforce committed to the company. Even the scourge of 'Japanapologists', Jon Woronoff, would surely not disagree with this brief summary of the positive features of the system.[10] To be sure, this system transfers insecurity down the supply chain, but it retains security where it is most needed, in the companies that control final assembly and so the quality of the product delivered to the consumer. These elite companies are the ones that have the potential and the incentive to control product quality throughout the supply chain.[11]

This account linked the generation of distinctive Japanese shop-floor and design practices to the possession of an equally distinctive employment system that was itself embedded in a distinctive model of industrial organisation. Only now are we in a position to discuss whether the Japanese achievements in the management of technology are transferable as disembodied techniques to their western rivals.

The Question of Transferability of Innovative Japanese Technique in the Absence of Supportive Institutions

The discussion of such apparently alien institutions as keiretsu naturally throws doubt on any easy assumptions that full transfer of 'Japanese' practice is possible, or will be easy, or quick. No western country is likely to introduce keiretsu industrial organisation. On the other hand, as already argued, the organisational advantages that keiretsu structure spreads so widely within the elite Japanese industrial companies are not alien to western economies and companies, as the examples of DuPont, IBM and other innovating western companies demonstrate. Whereas a common source of the leading western firms' product market control has been through the maintenance of an innovative lead over their rivals, the Japanese industrial firms depend on an industrial structure that, like the financial institutions of over-loaning, was originally created to aid technological catch-up with the west. In other words, the institutional means by which leading companies in both Japan and the west gained a degree of control over product markets may have been very different: they are nevertheless similar in that a degree of product market control was obtained and again in the way that this control was exploited to extend such qualities as security to elite employees in exchange for their commitment to the company.

So it is possible to move the argument away from the significance of stark institutional differences. What matters is that wherever a degree of market control exists, the potential also exists for companies to learn how to emulate foreign best practice. That does not mean technological catch-up or overhaul of rivals can be taken for granted, or expected on any deterministic timescale.

The reasons are familiar from the car industry example: it took time to perceive the threat accurately, it took time for companies to experiment and find out what aided technique transfer. The quantitative evidence shows that by the end of the 1980s US–Japanese joint venture car plants in the USA had productivity and quality levels similar to Japanese plant located in Japan.[12] The car industry might be thought to be exceptional, because there are relatively recent published quality and productivity statistics – the result of an unusual degree of individual company openness and also partly the US government being willing to threaten Japanese producers with import tariffs.

If a general appraisal of the result of the US quality movement is wanted, it would be difficult to improve on the assessment of Robert Cole, an academic specialising in its study. Cole is upbeat about the progress that large US firms have made in closing the quality gap with the leading Japanese exporting firms (Cole 1998). After acknowledging the early 1980s' problems in identifying the scale and nature of the threat from Japanese quality practices, he describes the way that the US quality movement and principles developed:

> what we saw manifested in the large manufacturing firms were typically very partial versions of these principles, characterized by a bewildering mix of creative hybrids and degraded mutations. Such efforts tended to frustrate the quality zealots who railed against incomplete practices that failed to realize the true vision. The partial version also led scholars to dismiss the whole effort as a failed fad. Yet, such conclusions cannot account for the demonstrable improvement in quality across a broad range of manufacturing industries over the last decade and a half. A more plausible view is that many of these large firms adopted just enough of the various principles, often without the fanfare of faddish acronyms, and in a sufficiently pragmatic and creative way … that markedly improved their quality performance. (Cole 1998: 62)

In this view, as important as the initiatives of companies was the development of a rich infrastructure of non-market bodies, such as quality awarding bodies. Their origin can be traced to when the Japanese government had in the early post-war years invited US quality experts such as W. E. Deming, J. Juran and A. V. Feigenbaum to tour the country to speak to management audiences. Such was the Japanese enthusiasm that they created the world's first 'Deming Prize' for quality in industrial firms (Johnson 1982: 216). The idea that US experts had contributed to Japanese success was particularly galling in the USA: it also weighed in favour of the idea that US firms, by adopting the US experts' ideas, would be able to replicate the Japanese achievement. There grew an obsessive US interest in those experts most associated with quality such as Deming, Juran and Feigenbaum. There are now both Deming and Juran Institutes in the USA, multiple US quality prizes, and publications galore on 'quality', especially, by the 1990s, syntheses and derivations of the experts' practices and recommendations under the term 'total quality management' (TQM).[13]

Important also was the activity of consultants; by 1992 *Business Week* could estimate that US companies were paying out $750 million dollars a year to third-party suppliers of 'advice and materials on quality' (Cole 1998: 68). In this instance, then, consultants proved to be valuable diffusers of the knowledge of what worked between firms.

It is worth pointing out that what we might call institutional innovation external to firms was an important part of both the US and Japanese responses to their perceived technology gaps – the Japanese response was of course the more far reaching and the gap they sought to close was much the greater.

Modularisation and Outsourcing – the Japanese Lead Challenged?

At first sight some of the perceived trends in the organisation of production of the late 1990s, such as outsourcing and modularisation, might appear to contradict the identification of Japanese success with their superior modes of managing in-house design and production. Yet there is not necessarily a contradiction, because as argued by writers such as Pavitt, it is the advent of more effective information and communication technologies (ICT) that have made outsourcing a more available option for firms. ICTs allow geographical separation because they improve the ability of the firm to do such things as exchange detailed product specifications with suppliers and to maintain supervisory control over distant production sites (Pavitt 2003: 84). In Pavitt's schema the modularisation of production also aids the practice of outsourcing because if the specifications of subassemblies of components – modules – can be accurately specified and not changed over time, then they can be reliably outsourced. Pavitt is careful to point out that where personal transfers are necessary and tacit knowledge of production is important, there will be limits on the degree of outsourcing possible. Nevertheless, he is willing to speculate on where this trend might lead: towards an increasing split between firms specialising in 'systems integration' and manufacturing firms, with systems integration firms competing in 'innovations in the design, development, integration and marketing of increasingly complex products and systems' (Pavitt 2003: 88). The tendency for manufacturing to be outsourced to low-wage countries might accentuate and

> the high-skilled 'services' in which the high-wage countries specialise would not be 'immaterial' in the conventional sense. They would comprise high-tech machines (processing information rather than materials), mastery of the knowledge underlying manufacturing, and a capacity for designing, integrating and supporting complex physical systems, including simulations and modelling products and processes, production and logistic operations, monitoring and control and customer support: in other words, the skilled activities that manufacturing firms undertake except manufacturing itself. (Pavitt 2003: 88)

It is a grand future scenario, but one may doubt whether it is realisable or desirable in practice in all industries, and fear the consequences for firms if they persuade themselves that such a scenario is their immediate future. The story of modularisation in the car industry serves as a cautionary tale warning us not to jump to the conclusion that modularisation and outsourcing are inevitable trends of the future.

By 2003 the outsourcing of modules had become the favoured panacea for the problems of the car industry (Sako 2003: 229). She points out that 'in theory modularity captures the notion of a clear division of labour between the architect with architectural design knowledge and designers with knowledge of each module' (Sako 2003: 230). In other words, for modularisation to work, architectural designers with knowledge of how to design an entire system – for example, a range of car models – must have a dependable division of labour between their work and the work of those who design the modules which make up the total system. If the boundary between modules and the content of modules is unstable, then there is less scope for economies through the division of design labour: in the language of an earlier section, communication costs and coordination costs will be higher. So it is relevant that when Sako studied the composition of two proposed modules, the car cockpit and the car door, she found that the 'product architecture of a car differs substantially from model to model and that the notion of mixing and matching, or sharing and reusing modules across models, never mind across OEMs, is not generally possible due to large variations in modular boundaries' (Sako 2003: 234). The variability of module definitions meant the posited economic gains of modularisation would be largely unavailable.

Sako also noted that the large western car manufacturers were tending to retain their systems integration capability even as they sought to modularise and outsource to suppliers (Sako 2003: 239). She points out that this behaviour contrasted with the large manufacturers' rhetoric regarding modularisation and outsourcing, where they talked of a future reduction of their own ability to integrate systems in favour of their suppliers. She argues that it is not possible to tell whether the current state is stable or if it will prove a transition on the way to a more devolved future state. However, current behaviour in the car industry appears to be consistent with a number of other studies of outsourcing, for example of digital aircraft engine controls. So in this activity, as the digital control technology matured, aircraft engine makers increasingly outsourced the design and manufacture of components but kept in-house and further deepened their technological expertise relating to the outsourced components, for the purposes of systems design and integration (Prencipe 2000; Brusoni et al. 2001: 608). This suggests that the more extreme scenarios of modularisation and outsourcing that imply a loss of detailed technological capability on the part of the systems integrator may be both difficult and dangerous to realise in practice in the car industry – at least while technological change continues to be a feature of the industry.

The gap between rhetoric and practice as regards modularisation and outsourcing appears to prompt Sako to suggest something quite radical: that 'the popularity of the notion of modularisation in the United States and Europe may in part be due to the hope that it might enable the retention of, or reversion to, arms'-length trading with suppliers without being locked into any committed relationships' (Sako 2003: 241). In other words, faced with the Japanese organisational advantages in integrated design and committed supplier relationships, rather than face the difficult struggle to imitate these, US and European firms have preferred to believe that they can solve their problems with a package of practices that consolidate traditional practice.

This interpretation gains strength from Takeishi and Fujimoto's work on recent developments in Japanese car manufacturers (Takeishi and Fujimoto 2003). Their questionnaire to Japanese component suppliers revealed the greatest change in practice to have been a 'shift to integral architecture' (Takeishi and Fujimoto 2003: 263). The functions assigned to individual parts had become more complex and 'the need for... functional coordination with other components had increased' (Takeishi and Fujimoto 2003: 263). As the authors point out, these changes 'were in an opposite direction to modularization' (Takeishi and Fujimoto 2003: 263). On the other hand they did find some increase in the number of components assigned to subassemblies in the major manufacturers, but at the cost of some flexibility in manufacture. Overall, their picture is that the Japanese industry was interested in an increased degree of modularisation in production, but not as a means towards outsourcing, while the European industry has been motivated primarily by the predicted economies of outsourcing.

Sako goes on to list and query the motivation for modularisation in the car industry. Some of her 'strategic drivers' are not concerned with economics or the experience of customer demand in the car industry; they include dubious comparisons with customer demand in the PC industry and pure financial engineering, as in the desire to raise share values through shifting asset ownership to suppliers.[14] This is an obviously ephemeral motivation and suggests that the interest in modularisation in the car industry may prove something of a fashion.

The situation appears to be that the Japanese style of highly integrated design and manufacture continues to develop and compete with a more disintegrated western style of design and manufacture. Given that the Japanese are the leaders in effective design and manufacture it is suspicious that it is not they, but the follower western firms that are experimenting with modularisation and especially outsourcing. These are early days to make any definitive judgements on outcomes, but if Sako's suspicions are borne out, the experiments with modularisation and outsourcing may yet prove to be a distraction and a dead-end at this point in the industry's evolution. There is, then, no reason here to reject the idea from an earlier section that the Japanese keiretsu industrial organisation underpins the distinct employment practices that allow the development of a superior ability to design and manufacture products – superior systems integration ability. And this in turn raises some doubts about the realisation of Pavitt's grand future scenario of a division of labour between 'high-skilled service' firms in the industrialised countries and manufacturing in the low-cost countries.

Where the innovative gains through a cycle of design depend on knowledge of the entire manufacturing process, of comprehensive knowledge of the potential for redesign of all components, then modularisation may introduce inflexibility in design and production and manufacturing may need to be kept in-house as the better means of developing and retaining this comprehensive knowledge. While modularisation, *if successful*, has the obvious benefit of standardising and reducing the burden of knowledge required for complex product design, it is not necessarily and simply available by choice to management, but depends also on the nature of the innovative opportunities available in a technology at a particular time. In a nutshell, while modularisation is an established fact in certain industries, especially within electronics, there may be some industries, such as car manufacture, where the nature of technological change renders the imposition of a high degree of modularisation inappropriate at this time.

The Dematerialisation of Technique and the Generation of Management Fad – TQM 'diffusion'

There was an element of fashion in the adoption of modularisation in the car industry and this is a good point at which to address in greater depth the phenomenon of fashion in the adoption of practice. As mentioned above, one of the most visible reactions to evidence of superior Japanese practice was the institution of a 'quality movement' in western countries, with the aim of improving existing practice. TQM became one of the most widespread management 'innovations' of the 1990s and remains one of the most intriguing because of its origins in a genuine achievement within a known technology. The scale of claims of adoption and the volume of derivative advocacy texts remain impressive today. Some idea of the phenomenon can be quickly garnered: a Google search on 'TQM' yields 275 000 web entries and an Amazon.com search yields 135 entries for books. If we accept Cole's conclusion of gradual but haphazard improvement in US quality, we nevertheless want to know why the TQM method was so often subverted in practice and why claims of TQM adoption became unreliable indicators of the adoption of quality practices.

The Mutation of Technique Meaning during Implementation

The first and outstanding feature of TQM, shared with other management fashions, is that it is not an 'it', but a mutable entity, a package of concepts and techniques prone to change from one publication to another and from one organisation to another. One review of TQM research and practice begins:

> There is now such diversity of things done under the name "total quality" that it has become unclear whether TQM still has an identifiable conceptual core, if it ever did... Most organisational scholars who have responded to the call for TQM research have focused their theoretical efforts on refining definitions of TQM. (Hackman and Wageman 1995: 310)

> Had we attempted to organize our thoughts exclusively around contemporary TQM practice rather than use the philosophy and prescriptions of the TQM founders... it would have been impossible to write. (Hackman and Wageman 1995: 338)

This characteristic of mutable meaning in practice appears to be the outstanding feature of management fads. It is, for example, a feature of the business process re-engineering (BPR) fad that followed close on the heels of TQM. BPR consisted of various exhortations to restructure business processes, loosely linked to IT implementation. Jones and Grint are two examples of authors that could find little coherence of content between different BPR texts and no originality whatsoever in the constituent elements of the BPR 'concept' (Grint 1995; Jones 1995).

This is enormously significant, for if an entity has no stable content, serious analysis of its general value-claims is not possible. What is possible and interesting in its own right is to study the process by which the meaning of a fad like TQM mutated in implementation so that it became a pseudo-innovation. This is perhaps most clear for a single-technique fad like quality circles than for the mutable bundle of practices represented by TQM.

From Euphoric Adoption to Failed Implementation in Quality Circles

Quality circles (QCs) provide an early example of a technique that swept through western companies in the early to mid-1980s, as part of the search for production improvement but also as a means of improving employee relations through participation. Light is thrown on the process of their implementation by Hill's research into 13 British companies that had pioneered QCs in the early 1980s (Hill 1991: 544). By the end of 1989 only two of these companies still maintained a QC programme (Hill 1991: 545).

Hill's managers could not report that QCs had improved employee relations, evidenced by low and declining levels of voluntary worker participation and continuing hostility to QCs amongst a minority of workers (Hill 1991: 547). Every firm claimed some business improvements had resulted from QCs, but that the 'measurable return was seen as disappointing and as declining over time until circles contributed little' (Hill 1991: 548). The two firms that maintained programmes in 1989 did so for social rather than economic reasons.

An important reason for failure was 'organisational dualism', whereby QCs were established in parallel to the normal hierarchical lines of control. The middle managers responsible for making QCs work now had two lines of responsibility (Hill 1991: 549). Such managers were responsible for 'overseeing' circles, but were not empowered to choose circle members or to set the circle agenda and were not able to be rewarded for their circle work (Hill 1991: 550). Yet while circles analysed and proposed solutions to problems, they relied on these middle managers to decide to implement their solutions. Given the poor integration of circle responsibilities into middle manager roles, it is perhaps not surprising that obstructive and reluctant management was the most frequently mentioned problem by circle members (Hill 1991: 550).

This picture of QC implementation contrasted with a detailed study of Japanese QC practice that undermined the widely accepted idea that QCs were voluntary (Hill 1991: 550). In Japanese practice considerable pressure was brought to bear on both middle managers and workers to participate in QC activities. This picture of 'enforced voluntarism' certainly fits Kamata's description of his life as an autoworker on the Toyota shop floor in the mid-1970s (Kamata 1983). The British firms' inability to integrate QCs into the organisation 'was one of several signals that top management gave about their own apparent lack of commitment to the programme of quality improvement, after an initial period of enthusiasm' (Hill 1991: 550).

> The rhetoric of the early days of the boom, that circles would become a normal way of doing business, was hollow. Circles never really took hold in the great majority of firms, remaining both experimental and marginal throughout their lives. (Hill 1991: 551)

In sum, the British companies had implemented QCs for somewhat different objectives and in a manner distinct from the Japanese. A strong interpretation of this would be that what the British firms called QC programs were not true QC programs at all – and so they failed. Nevertheless, Hill remained optimistic about the prospects for QCs in the future, but as part of the package of practices represented by TQM.

Mutation of Meaning and the Logic behind Proliferating Claims of TQM Adoption

Both Zbaracki and Hackman found a tendency for the 'hard' or 'technical' element of TQM, such as the technique of statistical process control, to be lost during its implementation (Zbaracki 1998; Hackman and Wageman 1995). Yet as presented by authors such as Deming and Juran, such statistical techniques were the very element of TQM that provided the prospect of genuine quality and productivity change for adopters. Something similar appears to have happened with just-in-time (JIT) management, for survey evidence revealed that western firms which claimed to be adopting JIT had a distinct preference for certain techniques over others; that most radical and 'Japanese' of practices, kanban, attracted very little interest and in general the easier techniques were adopted, and not necessarily the most useful ones (Voss and Robinson's work cited in Oliver and Davies 1990: 556).

Zbaracki's study of five TQM-adopting organisations in the USA found that managers that had been trained in TQM, upon returning to their organisations, tended to drop the technical parts of TQM through ignorance and intimidation – they did not properly understand these parts of the programme and preferred not to stress them. What they did tend to stress were the inspirational success stories in similar organisations, for these seemed relevant and real to them (Zbaracki 1998: 615–16). His organisations included a government agency, a hotel, a defence contractor, a hospital and a manufacturing firm, but only the manufacturing firm properly implemented statistical techniques. Zbaracki concluded that at the hotel and hospital 'the TQM program is mostly rhetoric' – yet these two organisations were the ones with a TQM quarterly newsletter that reported *only* TQM success stories, whether the success stories were generated by TQM or not (Zbaracki 1998: 624). The TQM 'technique' that was most often claimed to be used by individuals in these organisations (95%) was brainstorming (Zbaracki 1998: 626), but Zbaracki's investigative interviews showed that what respondents called 'brainstorming' did not follow strict brainstorming rules.

Zbaracki found that what drove this mutation in all the programs was the organisation's internal need to demonstrate the success of the program. Although there were some successes, there were many more project failures and difficulties in finding problems for TQM 'solutions'. Such negative experiences were not useful in demonstrating the value of the TQM programme, and so when all Zbaracki's organisations began to send members out to other organisations, they would do so to describe only their success stories – this evangelical practice took hold before TQM had had a chance of affecting the behaviour of the organisation. It is ironic that it was exactly such positive stories that had motivated Zbaracki's organisations to 'adopt' TQM in the first place.

In this way a cycle was established where positive, inspirational stories generated new implementation attempts, then internal legitimation needs led to the active diffusion of a heavily edited, positively-slanted and misleading version of the implementation experience. This in its turn would stimulate other organisations into adoption. The hospital, whose programme Zbaracki described as mostly rhetoric, 'also spread news of its success. Administrators from the hospital took their best TQM story to a Juran conference. The hospital also became something of a model for surrounding hospitals' (Zbaracki 1998: 627). Zbaracki comments on the process by which TQM mutates during implementation:

> in using TQM, managers generate seemingly innocuous claims that TQM has provided their organisation with efficiency benefits. The evidence from the models shows that these claims, coupled with the various social psychological forces that drive out the technical content of TQM, generate increasingly inflated claims for the power of TQM and an imprecise technical meaning for TQM. (Zbaracki 1998: 632)

We are now in the land of symbolic rhetoric rather than measurable or demonstrable change. Zbaracki nevertheless argues that despite this mutation of meaning during TQM implementation, positive rhetoric may be useful in driving positive results where these do occur – and there were *some* positive results in *some* of the organisations. On the other hand, we might expect some positive change stories as a result of normal management processes in any organisation – TQM may have added nothing but its label to the otherwise unremarkable drip of occasionally positive results from management projects that are part of the background of any organisation. In addition, this cascade of positive stories creates a mounting pressure on the management of all organisations to conform to their peers: to start a TQM programme and then to *claim* successful adoption of TQM.

What else could be the explanation for otherwise astonishing facts, such as that within a population of 2700 US hospitals, 73% claimed to have adopted TQM by 1993, compared with a negligible number five years previously (Westphal et al. 1997: 380)? The survey evidence of Westphal et al. found that early adopting hospitals may have experimented with TQM and adapted it to their individual circumstances, but later adopters simply copied the mode of implementation of the early adopters and traded 'organisational efficiency benefits for legitimacy benefits' (Westphal et al. 1997: 388). By published efficiency measures, the later adopting hospitals were the least efficient, and as these authors argue, this finding is consistent with the idea that late adoption of TQM was primarily to *symbolise* that a hospital was a 'good' hospital.

The symbolic explanation of adoption also makes sense of the eventual abandonment of a fad, and the way that this occurs quietly and without public recantation. Once all organisations have *claimed* to have adopted a fad, its ability to indicate who is leading and progressive and who lagging and backward disappears, so interest suddenly dies.

Although management academics have begun to theorise about these organisational processes, they are familiar from other areas of our lives. Consider the following courtesy of the US Court of Appeals judge, Richard Posner, reflecting on the career of the public intellectual James Burnham:

> He was shameless in prophecy and undaunted by repeated falsification. In a recent study of public intellectuals, I listed the failed prophecies of a number of prominent contemporary public intellectuals, but I was wrong to think that such an exposé might discourage irresponsible prophesying. I now think the significance of such prophesying lies not in its truth-value but in its rhetorical effect. It does not really matter in the end whether the apocalyptic prophecies of 'The Managerial Revolution' turned out to be true or false; they form a breathtaking, if horrific, vision of the future, and they give the book an amplitude, a resonance, and a sense of urgency that it would otherwise have lacked. (Posner 2003: 41)

Burnham was an early exponent of a form of management cant, but if Posner's book on public intellectuals had been a book debunking management fads, he would surely have reached a similar conclusion. Management fads' claims of efficacy should not be judged in terms of truth-value, but as part of a package of activities that serves to symbolise, but not necessarily to contribute to, organisational progressiveness.

To sum up, it is not necessarily the case that TQM, as a synthesis of the quality experts' ideas, is without merit, but counts of claims of TQM adoption, as in the hospital case, are meaningless as measures of genuine change. It is right to be sceptical of the mass of claims based solely on 'success stories' but reasonable to believe claims on a case-by-case basis, if supporting and credible evidence is made available, as it was in the case of the car industry. Yet this cannot be expected to be usual when the object of study is current practice in organisations that have every reason to be secretive about the know-how on which their current and future profitability depends.

The kind of case evidence reviewed here, especially when carefully constructed as in Hill's work, raises the question of *why* senior management allowed the mutation of technique in implementation. It does not allow us a clear general answer: so it is both possible that senior management were *unable* to properly support implementation of the techniques in some cases and possible that they were not convinced of the value of the techniques – perhaps with some justice in the hospital case – or insufficiently motivated to manage effective implementation in others. While all these explanations may suit particular cases, a review of Prais' matched plant studies enables us to winnow some of them as *general* explanatory possibilities for management behaviour.

The Organisation of the Shop Floor – Anglo-German Matched Plant Studies of Operations Management

The major issue in this chapter has been the management problem of how to identify and then close a gap in technological practice with rivals. In the example of the quality gap between western and Japanese producers it was apparent that the very institutions that the Japanese had introduced as part of their effort to close their technology gap with the west were implicated in the 1980s in the reverse quality gap between western and Japanese firms. In the effort to close the US quality gap there was also a degree of what we might call 'institutional innovation' with the creation of quality bodies and an active consultancy industry diffusing novel practices between firms. The major object of the remainder of this chapter is to gain a deeper understanding of the relationship between institutional arrangements and the management of skills and work and this will be pursued through a discussion of Prais' empirical work on Anglo-German matched plants.

A General Quality and Productivity Gap Between British and German Manufacturers

The entire discussion of the nature, value and transferability of Japanese manufacturing techniques was motivated by Japanese firms' successful penetration of western markets. Whereas the Japanese case concerned a relatively recent achievement by large, export-oriented industrial firms, Prais was concerned to explain a long-standing, superior performance of middle-sized German firms over their British equivalents. More than this, Prais' object from the first was to demonstrate the detailed practical relationship between technical training and productivity on the shop floor (Prais 1995: xiii). His research was conducted with the knowledge that the formal German system of technical training was superior to the British informal one.

His method was to make detailed comparisons of productivity in British and German plants. These plants were matched for size – they typically employed between 50 and 300 employees (neither the smallest nor the largest plant were the object of study) and simple artefacts were chosen, ranging from biscuits to springs, valves and fitted kitchen cabinets, but also including hotel service (Prais 1995: 43ff.). Such simple artefacts could be expected to be manufactured in both countries and so facilitated comparison. The extent of the program of study is evident in the upwards of 160 plants investigated over the period 1983 to 1991 (Prais 1995: 44). As in the case of Japanese quality practices, we want to know the nature of the gap in performance, the differences in practical management of technology and then the arguments for the source of those differences in organisational and institutional forms.

These comparisons corroborated the existence of a large productivity gap between Germany and Britain evident from the national censuses of production. In the engineering examples of spring, screw and valve manufacture, German productivity was 63% higher than in Britain (Prais 1995: 48) and in furniture the German productivity advantage averaged 60% (Prais 1995: 51).[15] When comparable types and quality of clothing were compared, the German advantage was roughly double the production per employee hour (Prais 1995: 54), but such strict productivity comparisons fail to account for major differences in quality and approach to manufacture (Prais 1995). This was particularly evident for biscuit manufacture, where according to the productivity measure of 'biscuit-tons per employee hour' the German worker produced 20% *less* than the British (Prais 1995: 55), but the 'average German biscuit' commanded 2–3 times the retail price of the average British biscuit. The German biscuit manufacturers had moved their products so far up the quality scale that productivity was no longer a significant indicator of true difference.

Unlike manufacturing, hotel services are not subject to international competition, nor do they have the same scope for technological change. So it appears especially significant that when Prais used 'guest-nights' as a measure of output and full-time equivalent labour as a measure of input for hotels of the same Michelin star rating, he found that German and Dutch hotels required half the labour per guest-night of the British hotels (Prais 1995: 59). This example points strongly to some national influence on the nature and organisation of employees being at the heart of the international differences in productivity.

Practices that Explain the Anglo-German Difference in Productivity and Quality of Production

Prais' researchers were able to observe that these differences in productivity were all related to a distinct and superior ability for the German firms to manage operations. This ability was in its turn directly attributed to the superior quality and quantity of technical skills available in the German and other continental countries' plant. Examples of the connection between poor operations management and the availability of skills show precisely what it is about skills that proves valuable on the shop floor.

So a common source of low productivity in Britain was a greater frequency of breakdown of production machinery, even when this equipment was no older than in any other country. On their physical visits to English plant, Prais' researchers were able to observe and to measure the higher frequency of breakdowns; for example, emergency downtime of equipment was 10% of planned British biscuit production time, compared with an average of 3–4% in comparable continental plant – this, despite the more complex production processes in the continental plant (Prais 1995: 62). One explanation was that:

> Almost all continental plants carried out routine programmes of preventive maintenance – hardly any British did so; the consequences were apparent in the significantly lower rates of emergency breakdown [in the continental plant]. (Prais 1995: 71)

The inability to maintain machinery was symptomatic of a general British inability to 'manage' artefacts. This was so even for such an apparently mundane matter as equipment layout in hotels; furniture and washing facilities were more often built flush with the floor to enable easy cleaning on the continent (Prais 1995: 65).

Interviews with machinery suppliers confirmed

> that teething and subsequent heavy maintenance problems were significantly more prevalent among British than Continental users; British fitters' capabilities were sometimes so low that serious 'teething' problems had arisen simply because installation instructions had not been correctly followed. (Prais 1995: 64)

It is important that the British problem is hardly likely to be one of a lack of formal knowledge of best practice in these matters; rather, the firms were simply unable to deploy such practices. This view is supported by Prais' evidence that British management sought to *accommodate* their acknowledged weaknesses in operations. For example, British firms recognised their weak in-house ability to manage sophisticated machine models such as CNC machines. They accommodated this weakness by tending to buy simpler machine models and delaying purchase until they were sure suppliers had dealt with all the 'bugs' associated with advanced models.

Of distinct significance, superior German operational capabilities enabled a strategic response to strong international competition that was denied many British firms. For example, in clothing, German firms had shifted to short production runs of 150–300 high-value items that required, for instance, more complex stitching operations. This enabled them to quickly follow changes in high-value-added fashion items in their home market – changes that lower cost, but which distant overseas competitors found difficult to follow. In contrast, British firms relied on longer production runs (of the order of 15 000 items) for simpler products, with less stitching, tucking and fashion variations. This left them exposed to the developing country, low-cost operators that used the same operational practices.

The frequent changes in German textile design relied on operators being able to read directly from sketches; the British operators more often relied on physical demonstration by supervisors. The limited abilities of British operators were thereby implicated in the 'strategy' of long production runs. Despite the lower demands that long production runs should make on supervisory overhead, overall British clothing firms employed 2.5 times more supervisors and checkers than the German plants, to correct a higher rate of error (Prais 1995: 69). As in the US–Japanese shop-floor comparisons, here we find the contrast is between more managers, higher costs and lower quality on the British side and fewer managers, lower costs and higher quality production on the German side.

Hotels provided a demonstration of the importance of another basic operations function:

> Efficient work scheduling, we judged, was probably the single most important element in raising workforce productivity in German hotels. (Prais 1995: 72)

German hotel 'housekeepers', the equivalent to the shop-floor supervisor or foreman, spent more time on work-scheduling, stock control, purchasing, organising external services (laundry) and selecting labour saving equipment. (Prais 1995: 72)

In all, 75% of German housekeepers were qualified, but none of the British sample had attended any external exam (Prais 1995: 71). The result was to force a higher level of British manager to undertake these supervisory duties, but less effectively, and in turn this weakened the higher level managers' ability to plan long term for such matters as marketing campaigns and the purchase of computerised booking and invoice management tools (Prais 1995: 72).

In all these comparative studies, weak British management of operations was clearly associated with low concentrations of skilled workers, but especially, as in the example of hotels, of workplace supervisors. The poor qualification of the supervisor was perhaps the greatest weakness of British operations management; in British metalworking 85% of production foremen had been promoted without any qualifications, but in German metalworking 80% of foremen had attained the relevant *Meister* management qualification (Prais 1995: 71). In Britain, as in Germany, workers were promoted from the shop floor to supervisor positions and there is no doubt that in both cases they were somehow 'experienced' in their work. The difference is in the structure and content of this 'experience'. Prais' articulation of the practical difference between the British and German supervisors is worth emphasising:

> Someone who has followed a purely on-the-job route to learning a trade may be as competent in carrying out specified routine maintenance tasks as someone who has also attended college courses on theoretical aspects and passed written exams. But, in the modern more technically complex world, it seems he or she is less likely to be competent in knowing – sufficiently precisely and sufficiently quickly – what may have gone wrong, or is likely to go wrong, which the best way of putting it right and be able to do so in a way that ensures it will not soon go wrong again. (Prais 1995: 70)

In Prais' analysis, it is the weak British system of vocational training that explains the differences in operational performance between German and British firms.

It is not only Prais' research that draws this conclusion. An independent series of German–French matched workplace studies drew broadly the kind of conclusions as Prais' team – that the German vocational training system supports superior workplace organisation and performance. For example, comparisons of efforts to implement computer-integrated technologies found the French to be the less able (Lutz and Veltz 1992: 274). In general, Lutz found that French companies had more hierarchical levels, fewer shop-floor workers as a percentage of office workers and wider wage differentials between hierarchical levels. Management and supervisory functions in these French companies took 20% of wages and salaries compared with only 12% in comparable German companies (Lutz 1992: 261). Searching for an explanation of these differences, Lutz argued that the French employment system had adapted to the graded, hierarchical and highly academic output of the French educational system by providing an equally graded occupational structure matching the hierarchy of status that the exam system generates. Although France reformed vocational training from the 1980s, this result is consistent with Prais in its resort to the structure of the dominant educational institutions to explain otherwise inexplicable features of workplace organisation.

The matched plant research offers compelling evidence of persistent international differences in the quality of operations management associated with systematic differences in the availability of technically competent people at worker, skilled worker and supervisory levels. On the basis of this association we are asked to believe that deficiencies in vocational training are the cause of the poor management of operations.

I find the kind of examples cited above sufficient to accept that the nature and extent of vocational training has a strong influence on the options for managing operations on the shop floor.[16] However, it is obvious that simply because of the focus and detail of these studies, other features of the 'enterprise environment' are not considered. So, for example, different times of entry into the European Common Market might have left German and British plants with different experiences of opportunities for expansion, and therefore with different motivations towards the rationalisation of shop floor operations; conditions of access to finance might differ in the two countries in ways that alter the management of the enterprise long term. So the Prais argument is one of plausible causation, conditioned by the knowledge that other institutional features of the enterprise environment may also matter.

Changes in operations practice are plausibly outside management control in those countries with weak national vocational training institutions and if we wanted to close the gap in operations performance the obvious policy response would be to develop institutions of vocational training. Before this option is considered it makes sense first to understand how and why vocational training developed so differently in these European countries. A step in this direction is to consider the theoretical construct of the 'free-rider effect' that is usually offered by economists to explain how general skills shortages can develop in the market economy.

The Free-rider Effect in Skills Training

The empirical evidence suggests that the widespread British inability to manage operations well results from a general shortage of technical competence at both worker and supervisory levels. If an employer has unique skill requirements there should be no problem with the incentive to train because retention of the trained worker can be taken for granted. The free-rider problem applies to the labour market for *general* skills: skills that all employers seek. In this situation, an employer that bears the cost of general skills training cannot be sure that the trained worker will stay with the firm after being trained, especially if there exist rival firms that refuse to train, but seek instead to use higher wages to poach skilled workers from those that do train. The free-rider effect is usually described as if the important case concerns those general skills that are already developed, but the effect can be expected to be *worse* in the important case of innovation. New technologies often require new skill sets that will inevitably be in short supply compared with demand in the early stages of diffusion. When there is a mismatch between supply and demand of this sort for skills – even for established skill sets – the incentive not to train but to poach should be expected to be greater.

Direct empirical evidence of the free-rider effect should be expected to be hard to come by, since only a few cases of poaching need become known for firms to discipline their training behaviour. Nevertheless, examples of poaching behaviour are sometimes captured. James Fleck uses an example from his research into robotics implementation in the English West Midlands.[17]

In Fleck's study of four small engineering firms in the West Midlands, one firm spent two years training two workers to be able to maintain and reprogram a robot installation. Near the end of this period, a second firm approached the workers, offered higher wages and successfully poached them. The first firm had taken on the risk of increased debt in its attempt to pioneer the new technology. Without the workers it had trained, and in the absence of a market for robot-operator skills, it found itself unable to operate its robot installation effectively. It soon went bankrupt. The parasitical firm then bought the first firm's robot equipment cheap from the liquidators.

With the bankruptcy and sale of the pioneering firm's robots to the poaching firm, this is perhaps an extreme example. The lesson it would provide within the local engineering community would nevertheless be clear: do not pioneer early-stage technology and do not invest in the creation of rare technical expertise if the experts remain 'free' to move between employers in search of better employment conditions.

Here is the peculiar feature of the free-rider effect for innovation – it can be expected to do its damage not so much through such spectacular cases, but through the general dampening of entrepreneurial approaches to new technologies. Of course, this is what Prais' empirical research found: that British firms deferred implementation of new equipment and sought to purchase simpler models to reduce the complexity of implementation. Such practices can be understood as minimising the demand on shop-floor skills and can make sense if skills are in short supply and there is a general reluctance to train.

Apprenticeship as a Solution to the Free-rider Problem

The institutions of apprenticeship that had evolved in Europe by the medieval period can be judged as devices that overcome the free-rider problem in skills provision.

Industrial states have come to differ greatly in the structure and provision of technical and vocational training especially at the intermediate or sub-degree level – the level responsible for the creation of the skilled worker. This difference is greatest between the USA and Britain on the one hand and continental European states such as Germany and Switzerland on the other. The rather stark differences in provision today can be explained by the different fortunes of the institution of apprenticeship in these countries.

Apprenticeship offers the employer the advantage of sub-market prices for labour in exchange for training that labour in general skills over a fixed time period – the indenture period. So apprenticeship assumes an obligation on the employer to train beyond the immediate requirements of the job and willing-ness on the part of the apprentice to forgo current financial rewards for the future higher wages of the skilled worker. If the system works, a versatile, generally skilled workforce is created and the problems of parasitical poaching of scarce general skills will be avoided. The institution can be expected to be particularly important in industries like engineering that are skilled-labour intensive and provide a broad product range (Elbaum 1991). The trouble with this sketch is that every feature of apprenticeship may be subject to forms of labour and employer abuse.

As for new skills in innovative technologies, apprenticeship institutions can only be expected to be slow to incorporate new skill requirements (the question is raised of *who*, if anyone, will coordinate the change in apprenticeship con-tent) and tardy in providing sufficient stock of skilled workers in the new skills: for new technologies in high demand the institution may work as a barely adequate 'patch' over this particularly acute form of the free-rider effect.

Nevertheless, apprenticeship is the institutional mainstay of the German-speaking countries' supply of craft skills. The extent and character of the institution in Germany is in remarkable contrast to the situation in Britain – and the USA.

The Extent and Characteristics of the German Apprenticeship Institution

By the end of the 1980s, just over 60% of German 16–19 year olds took apprenticeship-level qualifications compared with 27% in Britain[18] (Prais 1995: 22). Prais makes a further division between a 'higher', technician level, involving more full-time training, and a 'lower', craftsman level, trained in large part at the workplace, because the difference between Britain and Germany lies entirely at the craftsman or workplace-based qualification level. Both countries have 7% of the workforce trained to technician level, but 57% of the German workforce is trained to craftsman level compared with 20% of the British[19] (Prais 1995: 18). As far as technicians are produced by and work within the large-company sector, the difference between the countries emphasises that technical training is a particular problem for small and medium-sized firms, where the free-rider effect is most likely to operate. These differences in raw numbers beg the question of how the institution is organised in Germany.

First, the devolution of coercive power to associations of firms is a feature of the German institution. In the nineteenth century the German state introduced *compulsory* membership of 'industry associations' for firms, then transferred a variety of state functions to these self-governing associations, comprising

> the entire system of vocational training, trade inspection, further training, company consulting as well as the activities of advisory committees and expert groups. (Weimer 1992: 317)

Second, the German state has been willing to periodically intervene to reform the number of authorised apprenticeships and their duration. This has helped the institution endure in white-collar occupations such as banking, the hotel industry and retail as well as in the blue-collar ones (Sengenberger 1992; Gospel and Okayama 1991). By the 1990s the number of authorised apprenticeship tracks leading to skilled worker or *Facharbeiter* status had been reduced to around 380 (Prais 1995: 31) and the period of apprenticeship had been short-ened to the now typical three and a half years.

Third, the value of a German certificate of practical, workplace-based attain-ment is secured by standardised, *externally* set and *externally* marked exams (Prais 1995: 32). This is in particular contrast to the practice of old *and new* British vocational qualifications: these have preferred workplace-based assess-ment, and worse, supervisor assessment that has the obvious defect of confusing the roles of teacher with examiner.

Fourth, in a German apprenticeship it is *compulsory* to continue one's education in public vocational schools, typically for two days a week. The classroom element of German apprenticeship deserves more attention. Two English dictionary definitions of 'academic' are 'excessively concerned with intellectual matters' and 'of purely theoretical or speculative interest' (Collins 1991) and together these capture the British sense that there is nothing practi-cal about academic study. In contrast, much of the German public vocational schooling consists of further study of the German language and maths. That there is in fact nothing more 'practical' than language and mathematical ability is supported by Wolf's review of the literature on the connection between the possession of skills and earnings. Some of the most reliable research derives from the British National Child Development Study (NCDS) where literacy and numerical ability were tested independent of any educational certificates an individual might possess, and then their earnings progress monitored over their lifetime (Wolf 2002: 33). This research confirmed that even after controlling for the level of formal qualifications achieved, high literacy and numeracy scores correlated with longer periods of employment and higher quality of work (Wolf 2002: 34). Furthermore, of all the possible subjects for study at the British Advanced-level exams for ages 17–18, only the mathematics qualification cor-related with significantly higher earnings, at about 10% more than those who did not take the exam (Wolf 2002: 35).

The significance of this is that although the pure academic educational route has the greatest prestige in Germany, as elsewhere, after the elite performers have been channelled away into pure academic study, the bulk of an age cohort continues to study the basic academic subjects of most practical value in schools that are organised around their individual, current attainments. This produces not only a workforce of high skill and education but also a 'very effective system for socializing young people and organizing the transition from school to working life' (Wolf 2002: 167).

A fifth feature of German apprenticeship is that its skilled worker or *Facharbeiter* status forms the basis for entry into higher level vocational quali-fications that lead into the management hierarchy. Most notable is that the *Facharbeiter* certificate is the starting point for a special workplace supervisor course that includes management content and that leads to the award of *Meisterbrief* (Prais 1995: 28). *Facharbeiter* is also a common entry route into the German Fachhochschule, or senior technical college, that grants qualifications in vocational subjects such as business economics or engineering at sub-degree level. This is an important source of qualifications for those who become German middle managers (Lawrence 1992: 79; Sengenberger 1992: 251). The mass entry of teenagers into the apprenticeship system in association with this structured hierarchy of courses and standards built around the achievement of *Facharbeiter* status gives many more people access to the higher level technical and management qualifications. The popularity of these courses also makes explicable the German tendency to conceive of 'management' as fundamentally based on technical competence – in their system, that is more often the reality than in the USA and Britain. This view Germany shares with Japan and it is notable that in both countries there are structured routes for promotion from the shop floor into the management hierarchy.

All five of these features of modern German apprenticeship were, and mostly still are, absent from the equivalent British institution of workplace training. A return will be made in a later section to the institutional origin of the German shop-floor route into the management hierarchy, but first the value of this and other characteristics of the German institution are reinforced through comparison of the German with the British institutional path of development. The contrast will add to the benefits of the German employer control of the institutions of training which also avoids the negative technology control outcomes that can arise when the workforce is left in control of skills.

Innovation, Technology Transfer and the Craft Control of Skills

The Craft Control of Skills and New Technology

Like Germany, Britain also inherited an institution of apprenticeship from the medieval period as the major means of providing shop-floor skills. It developed differently, without state regulation, without coherent employer control, but as a system of informal, worker–worker, on-the-job learning – the craft control of skills. This institution is interesting for the contrast it provides with the institutions of employer-controlled skills – through its defects, it demonstrates the advantages of the employer control of skills. Chief among these was that craft control became associated with a particular and potentially hostile position with respect to new technology on the part of the workforce. There is also evidence that it persists in the British economy today.

Lazonick is one of the notable authors who have gone on to argue that the persistence of this institution generally inhibited the adoption of twentieth-century innovations in textile and other technologies in Britain (Lazonick 1990), even going so far (with Elbaum) as to give it an explanatory role in his 'institutional' explanation of the relative decline of the British economy in the twentieth century (Elbaum and Lazonick 1984). The argument that inherited nineteenth-century institutional arrangements retarded technological change in the *cotton* industry has proved particularly controversial in the literature of economics history,[20] yet there is no doubt that craft control of skills persisted into many modern British industries and that it did tend to obstruct technological change, of a certain capital-intensive sort, and for at least a time, if it did not present an absolute block. One must take the evidence on a case-by-case basis. Lazonick's account of the development of craft control in the British cotton industry remains a useful demonstration of the underlying logic of this institution.

The origin of the craft control of skills in cotton lay in the only partial transformation of the old 'putting out', or cottage mode of production, into a more recognisably modern management-controlled form of production.[21] In the boom conditions of the late eighteenth century, British entrepreneurs wanted to attract skilled workers out of their homes into factories, where their work could be more closely overseen. These eighteenth-century handicraft workers were not disciplined for factory work and strongly resisted the organisation of work inside factories, and so to attract them British entrepreneurs were forced to make fundamental concessions over the control of work inside the factories. The compromise was the establishment of *internal subcontracting* throughout British textile factories (Lazonick 1990: 81).

Internal subcontracting was an incomplete form of owner control over production that left the skilled worker with control of operations; that is, of the recruitment and training of workers and of the supervision of the flow of work. From the handicraft workers' side, the retention of these forms of control over production was in exchange for moving into the more disciplined factory environment and dropping their aspiration to become independent craftsmen.

The subcontracting relationship was reflected in the pay scheme; in spinning, skilled mule spinners were paid piece rates while their unskilled employees were paid time wages. This practice left the skilled workers with every incentive to extract maximum effort from the unskilled workers – their source of income (Lazonick 1990: 80). They therefore retained great incentives to maintain the privileged position with respect to the unskilled aspect that internal subcontracting represented. This skilled elite early organised a powerful craft union to represent their sectional interests to employers. In textiles this elite would successfully preserve the internal subcontracting arrangement into the last stages of decline of the industry in the mid to late twentieth century.

It was typical of this privileged worker elite that its union expected to negotiate over any technology that affected its position of *power* in the internal hierarchy of the firm. These workers were therefore willing to accept new technology that stripped their role of any remaining skill, provided their position and power were recognised and maintained by their employers. In this way, the craft union came to represent those who controlled internal operations and the flow of work, rather than those directly in possession of artefact-related skills.

Craft control of skills and operations was the widespread means by which the shop floor in nineteenth-century Britain had come to be organised. The demand to control the conditions of new technology implementation was therefore also widespread, and even openly stated, as in this quote from documentation of the British boilermakers' craft union:

> When labour saving appliances … are introduced to work which is ours by right, by inheritance and by the fact that our members have served an apprenticeship, that whenever such a machine is introduced to work formerly

done by our members, it should be done by our members. (*circa* 1915, quoted in McKinlay 1991: 105)

This demand made sense given that the workforce, rather than the employers, were responsible for the transmission of skills. An employer's acceptance of an apprentice meant the firm was willing to legitimate the transfer of craft skills from one worker to another, but direct control of the process was typically left in the hands of a skilled worker. The result was that skilled workers saw *other workers*, and their own craft union, as the source of the skills which gave them power in the labour market, *not their employer*. Should new technology render existing skills redundant, the employer had no ability to retrain, since they had no control of training in the first place. Where this institution was prevalent and with the usual search for security of employment on the part of the workforce, it made sense for the skilled elite of workers to demand control of the conditions of implementation of new technology.

Since British apprenticeship was left under the control of craft unions, their prevalent interests, not the employers', shaped this institution. So craft unions had a natural interest in influencing apprenticeship conditions in the interests of *existing* skilled workers. For example, they tended to restrict the *number* of trainee entrants to skilled worker status below employers' needs, for this improved their bargaining position *vis-à-vis* employers. On the positive side they did have an interest in maintaining the *breadth* of their training, for the sake of job mobility. It is probable that the worst characteristic was that because workers trained other workers in the workplace, they tended to transfer knowledge of *existing practice in a particular firm*. As argued by Pollard, this was adequate enough if existing practice was also industry cutting-edge practice, as it was in Britain for much of the nineteenth century (Pollard 1989). But when a British industry fell behind the cutting edge of some foreign industry, there was no means by which craft-controlled apprenticeship could quickly adopt foreign best practice, and certainly no incentive for the craft union to retrain existing workers – the benefits would appear to accrue entirely to the employers, the costs to the craft union.

A spectacular example of the contrasting efficacy of craft control and internal, company-controlled training is the introduction of welding to ship production in 1920s' Britain and Japan. Mitsubishi shipyards instituted a large-scale in-house training programme that spread the technique rapidly through the industry. In Britain the boilermakers' craft union was allowed to 'capture' the right to transfer welding for the traditional apprenticeship system. With long indenture periods that existed partly to control entry to the craft union and no retraining of existing 'skilled' workers, the diffusion of welding skills was a decades-long process compared with Japan. The Japanese were welding ships for the Second World War, but even in the 1950s many British shipyards were still riveting steel plate (Fukasaku 1991).

The example of welding is one of pure technique transfer, but a more serious and increasingly general problem, as argued by Lazonick and others, was when new, typically capital-intensive and near-continuous process technologies appeared, that for their most efficient exploitation required major disruption, even the outright abolition, of existing occupations. How the transition to such technologies was managed had the potential to set management–owner interests hard against those of the craft union.

Some of the most interesting examples of the management of this transition in Britain concern the most skill intensive industries – the shipbuilding, engineering, car and steel industries. They also allow us to broach the question of how relevant variant forms of craft control continue to be in the Anglo-Saxon economies of today.

Managing Craft Control during the Transition to Radical New Technological Forms

Craft Control in Engineering

By the interwar years some British employers were becoming aware of the deficiencies of informal apprenticeship and began to make *ad hoc* and individual attempts to improve it, for example by including compulsory classroom education, on the German model. Gospel and Okayama in their overview of the British employer response argue that large British employers never fully controlled the system of skill transmission and failed to ensure organisational forms that would support *standards* of apprenticeship training:

> Out of 1573 engineering firms surveyed in an official inquiry in 1925, only 26 had separate "apprentice masters" whose sole duty was to train apprentices and even fewer had special training departments or schools. Only 10% of textile employers granted day release to apprentices and this was the most generous industry. (Gospel and Okayama 1991: 24)

In the absence of *collective* regulation or reform, the informal training system became subject to abuse. Employers increasingly sought to use apprentices as a source of cheap labour during recession, and even while during business cycle peaks firms remained unable to overcome shortages of skilled labour. Where they became interested in mechanisation (especially within engineering and shipbuilding) they attempted to limit the breadth of training for apprentices and to substitute large numbers of such cheap apprentices for the existing skilled workers (McKinlay 1991). According to McKinlay, the Engineering Employers Federation did attempt to coordinate industry-wide reforms of apprenticeship institutions to meet the agreed labour force needs of the industry, but such attempts foundered on the lack of interest of its member firms. Too many chose to continue to rely on informal apprenticeship for their skill needs and remained unwilling to back the proposals for collective reform. In this way the inheritance of informal apprenticeship practices undermined what had to be a collective effort to reform.

Lazonick chooses to place the responsibility squarely on British management for allowing the import and persistence of craft control into the most skill intensive industries.

> The exercise of craft control, even in the absence of craft skills, was a result of the historical willingness of British employers to relinquish what Americans and Germans came to define as the management function. (Lazonick 1990: 184)

Lazonick further develops the case of engineering as evidence of this British employer tolerance of craft control. Here, the craft union representing shop floor fitters and turners was weaker than in textiles and the employers always defeated its organised strikes. Such victories allowed engineering employers at times to free themselves of their craft union. They were therefore in a position of relative strength when compared with employers in other industries – for example, the employers in the cotton industry were never in a position to challenge that industry's craft union. Yet engineering employers never reorganised control of the shop floor and preferred to rely on experienced workers, or 'shop stewards', to maintain the flow of work (Lazonick 1990: 198).

In answer to Lazonick's tendency to blame management, one might suggest that he underestimates what was required for an adequate collective response. If we keep the free-rider effect and the coercive power of German industrial associations in mind, one can doubt that any *voluntary* effort at *collective* reform would have been much of an improvement on the sum of uncoordinated, voluntary and individual reform efforts. Of course, this shifts the explanatory burden onto the question of why the British employers did not seek state-derived coercive powers to support collective reform: not a question addressed in any detail by these sources.

Nevertheless, the form in which craft control persists into the twentieth century to plague the British workplace is in this reliance on shop stewards loyal to their workplace colleagues, rather than on the supervisor loyal to management.

In the case of engineering, a recent study of nine pairs of matched plant of British and Japanese machine tool manufacturers by Whittaker found, against his expectations, a continuing preference for craft control of work on the part of British management (Whittaker 1990). The Japanese, on the other hand, had what Whittaker termed a 'technical' approach to managing operations, one that involved a relentless search for operating efficiencies and no respect for custom and practice. One of Whittaker's illustrative examples of the contrasting approaches was the Japanese willingness to allow CNC machines to run unmanned and the British preference for the allocation of a skilled machinist to their CNC machines. This appears as something of an indictment: it can be said in defence of the British (and somewhat ironically) that their firms were in steep decline while the Japanese were rapidly expanding. The Japanese would have had every incentive to save on skilled labour, the British would have had a skilled labour surplus, with insufficient work. This example is better seen as suggestive, rather than definitive, evidence of a continuing problem with craft control in Britain in the small-firm skill-intensive sector.

Craft Control in the Car Industry

As Lazonick explains it, such was the preference on the part of British owner–employers for relying on craft control of the shop floor that they imported it into the new industries, such as car manufacture. Lazonick points out that this occurred in Britain in the 1920s, a time when US companies were conducting successful experiments with the organisational innovation of the personnel department as a means of winning worker loyalty and cooperation. It was also at a time when an example of how to use close supervision by foremen loyal to the firm was available in the example of Ford's 1911 assembly plant in Britain (Lazonick 1990: 201).

Once craft control had been imported into the British car industry, it endured to modern times and into the declining years of that industry. The role of the front-line supervisor was – and is – a critical one. This individual could represent management and its interests on and to the shop floor, but if power was allowed to shift to the shop floor, then this function could be considered discharged by the shop steward as worker representative. A vivid confirmation of Lazonick's indictment of British management is presented by Peter Wickens, the personnel manager for Nissan when the company built its first European car plant in Sunderland, Britain, in the late 1980s (Wickens 1987).

Wickens describes the process by which the supervisory function in the British car industry had been degraded over time. The familiar pattern was that supervisors tended to have been promoted from the shop floor on the basis of on-the-job performance, but had few formal qualifications. Their generally poor performance as supervisors had generated a short-run adaptive response by senior management. This involved the progressive stripping of management responsibilities from the supervisor and their allocation to professional specialists: so personnel specialists dealt with industrial relations; specialist layout and maintenance engineers had control over those functions; shop stewards controlled work organisation and the division of labour. Over time, this stripping of responsibility had helped to lose the supervisors their pay differential over shop floor workers. The result was a decline in the respect given to supervisors, a lessening in the authority of production managers within the firm and the inability of these managerial positions to attract capable people.

The description of degradation of the supervisory role makes Nissan's response intelligible. For its new plant, the company first doubled prevailing line manager pay to the level of a professional engineer. It then widened the supervisor's responsibilities to include: quality control; hiring of workers; maintenance; work area layout; design of the cost control system and work scheduling. In other words, all the basic operations and responsibilities were returned to the control of this pivotal management position. In the recruitment process itself it is significant that rather than using the tired British practice of appointing those with 'experience', Nissan used sophisticated psychological and personality tests to select people that could creatively discharge their responsibilities (Wickens 1987). When Nissan eventually hired 22 new line managers, only six would have the dubious privilege of having had management experience in the British-owned car industry. In this way, Nissan restored a functioning management hierarchy – a precondition for the restoration of systematic innovation on the shop floor.

Lazonick cites one piece of evidence which suggests that this picture of a degraded production supervisory function, with its implication of loss of management control over the shop floor, had become representative of a large swathe of British industry by the late 1970s. At this time, around 40% of British supervisors had joined trade unions, an act that signified their insecurity in employment (Child and Partridge 1982: 195).[22] Child and Partridge draw the familiar scenario of senior British management indifference to supervisor pay and conditions that had been allowed to deteriorate to the level of shop floor workers. At the same time senior British management had continued to try to hold their unsupported supervisors responsible for the flow of work on the shop floor. In other words, arbitrary and indifferent senior management treatment had driven British supervisors out of a dysfunctional management hierarchy and into unionisation (Child and Partridge 1982: 195).

The Conditions for the Elimination of Craft Control
in Large Companies – the Steel Industry

Some of the conditions necessary for British industries to manage successfully the legacy of craft control can be found in Owen's brief summary of the development and management of the British steel industry (Owen 1999: 115–50). The classic British industrial inheritance from the nineteenth century pertained in steel: a large number of fiercely independent, competing owners and a tradition of devolving the management and allocation of work to the shop floor. The difficulties of reform under a fragmented ownership structure are illustrated in the second period of private ownership from 1957 to 1967, when the efforts of one privately owned firm to tackle the overmanning of its plant foundered because of the excessively complex union negotiations (Owen 1999: 135).

While the industry was privately owned, and despite voluntary mechanisms set up to encourage merger and rationalisation, 'by the time of the Second World War the industry was only slightly less fragmented than it had been twenty years earlier' (Owen 1999: 121). The industry would be nationalised, then denationalised, not once, but twice. With regards to craft control, rather than either form of ownership having clear intrinsic advantages, what mattered was, first, whether under that form of ownership management had acquired a theoretical power to reorganise the industry, and second whether and how they chose to exercise that power.

The theoretical power to reorganise was certainly acquired in the second period of nationalisation beginning in 1967. The problem appears clear enough: in the year 1975 and measured in tonnes of crude steel per man year, British Steel had only 23.5% the productivity of the Japanese leader, Nippon Steel (Owen 1999: 136). Owen writes that the large number of obsolescent plants explained most of this gap, but 'the five largest works were also seriously overmanned'[23] (Owen 1999: 136). Yet when British Steel tried to negotiate on overmanning with national unions, it found it had to renegotiate at plant level, only to make little progress. This might not be surprising, given the 18 unions with which it had to negotiate (the number in 1979), each jealous of the conditions of work won by its rivals, and given that it was at the individual plant level that the power lay under the craft control of work (Owen 1999: 137). So as regards labour relations in these early years of nationalisation, if management had acquired the theoretical power to act, they had yet to find a way to apply it practically.

The last conciliatory effort by management was the attempt to interest the steel unions in radical plans for industrial democracy, an approach to win union backing for the necessary plant changes (Owen 1999: 139). It failed for lack of union interest. The increased pressure to reduce financial losses from the 1979 Conservative government led to a three-month strike that ended with the unions defeated and the highest annual loss in British Steel's history, of £1.8 billion (Owen 1999: 144). With power demonstrably shifted towards management and with £4.5 billion of continuing state support between 1980 and 1985 (Owen 1999: 144) management were able to press ahead with change at the plant level: they made multi-union agreements that reduced interunion disputes; they began to dissolve the strict demarcation of work between rival 'trades'; they began to introduce a version of the multifunction team, for both maintenance and production (Owen 1999: 145). By 1987 an industry report found British Steel to have the lowest costs per tonne of steel produced of any of the world's major steel makers (Owen 1999: 146). It was privatised in 1988 as a profitable concern. It continued the 'normal' search for productivity and quality improvements in private ownership so that by the end of the 1990s, France, Germany and Britain each had 'a major national steel company of roughly comparable size and roughly comparable levels of productivity' (Owen 1999: 148).

Owen concludes on steel that, 'if nationalisation had not taken place and the companies had been obliged to compete freely, it is likely that market forces would have brought about the necessary reorganisation of the industry' (Owen 1999: 149). For Owen it is the government that bears the major responsibility for the delay to the modernisation of the industry. This is rather strange, since by the evidence he assembles, when it had the chance, the private sector failed to manage either industrial restructuring or the elimination of craft control. Why did not 'market forces' work then?

The billions of state support in the 1980s were in part necessary to cover costs not incurred in earlier years. One may nevertheless suspect that the private sector would have baulked at the prospect of industrial action and the associated billions of heavy losses, necessary though these were if management control were to be reasserted over the shop floor. In this scenario, it might have proved economically rational to prefer industrial peace and continued craft control of the shop floor, at a price of lower operational efficiency and lower prospective returns on any new plant – as it had so proved in earlier decades. Of course we cannot know with certainty – we are playing the game of 'What if?' It seems to me at least as reasonable a scenario to the one preferred by Owen.

The larger significance of this example is that if it took a stream of blank cheques from the state and over 20 years from the inception of the capital-intensive technologies of the 1960s to assert management control over the shop-floor, what chance have industries such as general engineering?

Concluding Remarks on British Craft Control and the German and Japanese Institutional Means of Shop Floor Control

This short review of the British problem of craft control of the shop floor is interesting in its own right, yet it also deepens our understanding of the advantages of the German and Japanese institutional means to the end of shop-floor control. There are also questions, such as whether the craft control of work will continue to persist in Britain, whose answers depend on our understanding of how the German and Japanese institutions support employer control of the shop floor.

To recap, the institutions of reformed German apprenticeship and the keiretsu-supported Japanese employment system are quite distinct in form. Yet they are certainly comparable in terms of their effects on the management of shop-floor operations.

Central to the control of the shop floor is employer control of work and especially skilled work. The Japanese means to this end is the institution of lifetime employment. The market power of keiretsu organisation enables the elite Japanese industrial firms to extend security to the blue-collar workforce and remove the deterrent of a free-rider effect on training investments. The German means to this end is through the coercive collective power possessed by employer associations over the apprenticeship institution in their respective economic areas. Employers collectively control the supply and content of broad and general skill qualifications and in both countries there is a structured promotion route from the shop floor into the management hierarchy. A plentiful supply of general skills enables many German firms to establish their own version of the Japanese core–periphery model of employment; this is the arrangement where there exists a core of particularly valued and relatively protected skilled workers at the same time as a periphery of less secure, less valued workers (Sengenberger 1992: 248).

As viewed from Germany and Japan, the British institution of craft control is quite clearly linked to the employers' failure to take control of skills; so, for example, the German industrial sociologist Sengenberger states:

> [In West Germany] there is almost a complete absence of the Anglo-Saxon type of craft organisation or (competitive) craft unionism using occupational jurisdictions or demarcation lines as means of job control. (Sengenberger 1992: 248)

This pinpoints arbitrary on-the-job 'experience' and reward as the sources of the craft control problem. Because German skilled workers have a broad range of expertise, they are unlikely to be threatened by a change in technology that makes a particular task obsolete. British workers whose skill, status and pay pertain only to a narrow task are more likely to feel pressurised by technological task obsolescence. Even if this does not threaten them with redundancy, it will threaten task-dependent pay and status and so complicate the management task of worker reallocation to other tasks in the organisation. Such 'demarcation disputes' were a hallmark of the British car industry's labour relations problems in the 1960s and 1970s.

Now we are in a position to interpret the steep decline in the number of registered British apprentices that contrasts so markedly with Germany. Apprenticeship in Britain has declined steadily, from over 200 000 apprentices in the late 1960s to less than 90 000 in the late 1980s (McCormick 1991). It should now be evident that the British institution is so unlike the German that far from this decline constituting a disaster, it might not have been altogether bad *if* a more modern institution had been introduced to replace it.

Indeed, the British state made increasingly vigorous and expensive interventions into institutions of vocational training from the 1980s onwards. How and why the British state managed its attempted reforms of vocational training will be introduced in the next chapter, because the evidence is good that reform was becoming urgent at this time.

It is relevant to think again about the significance of the theoretical construct of the free-rider effect, incomplete as it is usually posited. It has already been pointed out that it is usually imagined only to apply to some *established* set of skills in the market economy, and that where new, highly valuable general skills are evolving around a new technology, such as robotics, any deterrent effect on employers' incentives to train can be expected to be greater.

Prais has argued that as an economy grows and local labour markets merge to become a national market, it is then that the free-rider effect on general skill training can be expected to become a problem (Prais 1995: xiii). This is a useful reminder that where labour markets are confined to local industrial regions, by geography or poor transport links, there is what can be called a natural constraint on the operation of the free-rider effect: for example, firms in small regions can engage in tit-for-tat discipline on known training cheats.

To continue with Prais' summary of the early stages of the evolution of the free-rider effect, as local, natural constraints are overcome and national markets are created, poaching behaviour begins and a tendency to reduce general skill training also starts to manifest itself. There is at first a reservoir of historically generated skills that buffers the effects of lower rates of training on internal organisation of operations, but with a long time lag in the order of a generation, the pool of available general skills decreases and a temptation to poach skills with the additional deterrent effect on training escalate together – a form of positive feedback, this time of a progressively damaging kind.

This further extends our introduction of a dynamic aspect into the free-rider effect when innovation is considered. The question that is opened now is whether, and how, the effect might continue to evolve; more specifically, whether the persistence of craft control in Britain is not merely an idiosyncratic hangover from that country's idiosyncratic past, but something akin to the 'end state' of evolution of the free-rider effect that threatens to occur in any country where the preventative institutions are absent. The problems of the free-rider effect would then consist not merely of insufficiency of supply of skills, but of the development of a worker interest that is pitted against a management interest in technological change. In this book, the question must remain an interesting and plausible possibility, for there is no space to work through the detailed arguments for and against.

It was remarked earlier that Lazonick blamed British management for failing to act on craft control and he even gave craft control an explanatory role in British relative economic decline. A case-by-case analysis partly bore out Lazonick's analysis and established that British management found craft control a convenient way of managing the shop-floor. Of our three cases, Lazonick's indictment of management appears most valid for engineering, for in that industry management was able to temporarily re-establish control over the shop-floor, but chose not to take control of skills. In the car and steel industries, management did not seem capable of this first step of re-establishing control over the shop-floor. State ownership and financing of the steel industry seemed to be the precondition for a management challenge to craft control.

My view is that to blame only British management is to underestimate the difficulty of transforming these organisational structures. The problem of expunging craft control wholesale was one that often required *some form* of coercive collective action to end the 'freedom' of a segment of the private sector to poach other firms' skilled labour. In their different ways, both German and Japanese institutions have this coercive element and it is of course available to private sector firms that have a high degree of market control by reason of their size. The establishment of this coercive power would, in industries of fragmented ownership, require the involvement of the state, as it did in Germany and Japan – and as it did in Britain in the instance of British Steel. If we wish to blame some agent for the persistence of craft control in Britain, the lack of involvement of the British state in a general reform of vocational training would seem to be as much a candidate for blame as 'British management'. As will be argued in the next chapter, the British state has usually required unanimity amongst the private sector community before it has been willing to act coercively. There is good evidence that the predominant style of state intervention in the private sector in Britain has hindered the resolution of the British problem with training.

Management Practice and the Divorce Between Management and Technological Education

This chapter has been about the ways in which innovative management has been enhanced by structured shop-floor experience combined with different forms of technical education. A marked feature of the German and Japanese institutions was the existence of a shop-floor promotion route into the management hierarchy. Associated with this was an understanding that management implied knowledge of technology, especially of the operations function. This section explores the significance of the widespread belief in the Anglo-Saxon world and increasingly in other industrialised countries that management should be understood as divorced from technical and operational knowledge.

Clues that this matters abound, in the history and contemporary debate of both engineering education and management education. In Britain the idea that if engineers are to be a source of managers then engineering curricula should include management teaching is an obvious one and an old one. Divall found that a consequence of the post-war attempt to make engineering curricula more scientific was the end of hesitant steps made in the 1920s to introduce management education into engineering curricula (Divall 1990: 94). These experiments can be understood as a continuation of the logic of engineering curriculum development during those years, that an engineering graduate should be a useful *technology manager*, not simply a specialist artefact designer. However, with the adoption of an engineering science curriculum, management education would be left to develop in the distinct institution of the business school. The Finniston report of the early 1980s found this unsatisfactory and once again urged the inclusion of management teaching in engineering curricula. Given the numbers of scientists and technologists that take MBA degrees today, it is possible that had engineering education continued in the path begun in the 1920s, this split between engineering and business education institutions would not have occurred to the extent that it did.

One of the consequences of the creation of distinct educational institutions for the teaching of technical and management subjects has been the gradual growth of an academic interest group that propagates a context and technology-free concept of management. As an engineer and a management academic, Armstrong calls those that hold such views the 'management education movement' (MEM). The MEM typically supports the teaching of a standard mix of management subjects including: business strategy; marketing; financial control and behavioural science (Armstrong 1992). Since this mix is free of technological context it can be claimed to be perfectly general in its relevance to practice, a position that maximises the potential market for such courses. According to Armstrong, the effect on students, particularly those who take lower level management courses such as undergraduate degrees and who therefore possess no operational experience, is to propagate a peculiarly British view of management, a view that management is

> something quite distinct from technical expertise; which, indeed in its more virulent versions, actually regards technical expertise as a disqualification for managerial positions. (Armstrong 1992: 42)

It is indeed common to encounter the idea that technical and management expertise are distinct in Britain. Not only formal education institutions propagate the idea: the phenomenon of craft control implies that operations and technology are identified with blue-collar workers rather than with the practice of management. There is empirical evidence that the split matters, and is embodied in other areas of management practice. Lam's comparison of the work organisation of 60 Japanese and 55 British R&D engineers in large electronics firms found a severe split between managerial and technical careers in the British firms and concluded that the

> mechanistically structured and functionally segmented organisation systems observed in the British firms have contributed to a vertical polarisation between technical and managerial roles, inhibited knowledge sharing and led to the gross under-utilization of engineers in product development and innovation. (Lam 1996: 206)

So the institutional polarisation of engineering and management education is mirrored in the polarised organisation of the work of engineers and management in these examples. It is mirrored again in the polarised organisation of management and skilled technical workers on the British shop floor.

A paradoxical form of evidence that the split matters comes from the efforts that technologically expert individuals make to bridge the educational gap through the acquisition of an MBA. When research was last conducted into the background of British MBA students it was found that 31% of students had an 'engineering background' from civil, structural, mechanical, electrical, chemical or IT engineering, while 19% had a science or maths background.[24] These two categories of MBA student are the first and second largest and their prominence does indeed suggest that the MBA has a major role in broadening the formal knowledge base of those who have technical and scientific qualifications. Again paradoxically, this function may even be the MBA's major contribution to management practice and best defence against recurrent criticism of the stand-alone and formal model of business education that it represents.

Evidence that this criticism is becoming sharper and more systematic in the USA is the launch in 2002 of a new journal in 'Learning and Education' by the US Academy of Management. Many of the criticisms of the business school *status quo* curriculum made in this rehearse issues already raised in this chapter.

For example, Pfeffer and Fong review the evidence on promotional and monetary returns to the acquisition of an MBA and conclude that 'there is little evidence that mastery of the knowledge acquired in business schools enhances people's careers, or that even attaining the MBA credential itself has much effect on graduates' salaries or career attainment' (Pfeffer and Fong 2002: 80). There *are* studies that find a positive correlation between career progression and possession of an MBA, but Pfeffer and Fong argue that those studies apply to the high-prestige programmes that screen for the very highest ability applicants. So in this analysis, later career success derives from the personal qualities of the individuals, not the possession of formal MBA business knowledge (Pfeffer and Fong 2002: 82).

What Pfeffer and Fong do not consider is my argument above that the MBA promises to add most value when it broadens existing technological and scientific expertise.

Pfeffer and Fong go on to observe that while business schools tend to teach through formal lectures and case studies, in reality management is a craft within a field of practice (Pfeffer and Fong 2002: 85). They then argue that if the object of the business school MBA is to improve management craft practice, it would be better to borrow the tried and trusted methods of other established vocational training schools, such as on-the-job training and clinical experience.

With the German and Japanese models in mind one might respond, first, that this is quite right, but that the appropriate setting for improving management craft practice is where practice occurs – within the firm, not the classroom. With the German and British models in mind, one would add that classroom education remains an important element of German 'vocational' training. A narrow concentration on individual craft skills is unlikely to improve the ability of management to *organise* skills for firm purposes. It also follows from these models that the formal content of classroom education does matter – it would not be surprising if there were scope for reform of the formal knowledge base of the management education curriculum, so that its content became more relevant to practice. Indeed, this is the gist of Donaldson's argument in this same issue of the Academy of Management's 'Learning and Education' journal (Donaldson 2002).

It is easy to become trapped by the implicit assumption in this kind of discussion that formal structures of 'education' have the full responsibility for producing effective individuals. Once again, reference to the German and Japanese models reminds us that the object of institutional arrangements in technical and management education is to enable individuals to combine knowledge to make them practically effective. The striking feature of those countries' educational models was the way formal and experiential education were integrated in individual careers from technology into management.

Despite the lack of formal structures with this object in the Anglo-Saxon countries, there were substitute means for this end. The exploitation of the MBA by scientists and technologists appears to be one means in these countries for achieving this integrative end – no doubt with weaker effect, for there is no standardisation of experience-in-breadth in these countries. Another example which reinforces that the goal is worthwhile and achievable, even if it proves difficult, is the idiosyncratic 'education' of entrepreneurs like James Dyson. Dyson studied interior design at London's Royal School of Art and had no formal engineering, science or management qualifications. If he became more of an engineer than most graduate engineers, it was because he successfully acquired the abilities that would allow him to achieve his ambitions through other means. These abilities included the full range of technological competences, from engineering and product design to finance, marketing, sales and management, but were acquired through an eclectic range of formal Royal Society of Art courses, mentors and project experience (Dyson and Coren 1997: 42). Entrepreneurs like Dyson show what is possible without the major engineering and management institutions if individuals are motivated to take responsibility for combining their educational and practical experiences. The abilities he lists are further confirmation of what makes an effective technology manager, engineer or entrepreneur. His individualistic means of acquiring them suggests, once again, that such people are more likely in the countries that have systematised the relationship between institutions of education and work experience.

Conclusions

Human beings evolved to be good at the acquisition, practice and development of skills: we are bipedal, with long arms free to manipulate, and with hands and opposable thumbs for precision in the manipulation of the things of this world. All individuals acquire and discard skills throughout their lives. The problems of the production, maintenance and alteration of a useful stock of skills in the market economy are those of organisation rather than individual ability. One problem is that because skills are acquired in a practical context and are an individual possession it is difficult to judge the quality of a stranger's skill without personal knowledge of that individual's performance in that practical context. And in industries that have fragmented ownership there is not only the problem of the lack of any agent able to coordinate supply and demand for skills but also the problem of the free-rider effect.

The institutions of German apprenticeship and Japanese keiretsu-lifetime employment appear to be different means to the same end of solving these problems of how to create and manage a useful stock of skills.

In the Japanese institution the binding of individuals to the keiretsu management hierarchy avoided any free-rider effect and gave management the ability to develop and judge individual ability in the long term. The reformed

German institution of apprenticeship enabled a labour market in skills to exist in industries that, unlike those dominated by the keiretsu, were characterised by fragmented ownership. Its system of independently-certified and related levels of craft skill were surrogates for the personal knowledge of individual experience upon which the informal British institution continued to largely rely, certainly for management-grades of skill and experience. The coercive power of industrial association levies largely solves the problem of coordinating skill supply and demand in a fragmented industrial ownership context.

In summary, the German and Japanese institutions discussed here are: means of eliminating any free-rider effect; of establishing confidence that a class of individuals possess a certain standard and breadth of skill; means of providing selection and promotion routes for the most able, skilled individuals into the management hierarchy. Last but not least, they provide the means to avoid the sectional interest problems that derive from the craft control of work.

Economists use the term 'market failure' to describe a market outcome as less good than the best attainable outcome. If we accept that some kind of institution of skill formation is a necessary part of a market economy, they would probably describe the British institutions of skill provision as responsible for a form of 'market failure'. This jargon might be useful if, from an Anglo-Saxon standpoint, one objected to the drastic and coercive institutional changes wrought in Germany and Japan suggesting that they are a curtailment of that decentralised decision-making which is the most valuable feature of the market economy. The answer would be that only certain 'freedoms' were coercively constrained and that this constraint created the opportunity for more creative operational and technological firm activity: collective coercion was associated with increased technological freedom of action for the decentralised decision-makers within the market economy.

It is tempting to write that the organisation of skills is a case where the more 'free' market solution produces a worse outcome than an institutionally constrained market. However, this chapter is a selective review and the German and Japanese coercive solutions were developed when those countries were committed to solving the problem of technology transfer as a means of economic catch-up with their leading, political rival states. So the lessons are most relevant to industries and technologies that lag best practice in other countries – in other words, where the problem is one of technology transfer rather than innovation. Britain represents the interesting case where technological and economic leadership has been gradually lost over many decades, in many industries, even as the country has become, like the other industrial countries, ever materially richer. It was argued at the end of Chapter 2, with reference to Edgerton's work, that in the process of falling behind other countries in economic wealth per capita, Britain has gradually exchanged the general problem of innovation for an increasingly acute technology transfer problem. The evidence in that chapter was the declining intensity of British industrial R&D expenditure and in this chapter it was contrasting Anglo-German operational practice. So industrial technology transfer policies are likely to be more important for Britain than those designed to promote groundbreaking innovation – although the latter remain one of the obsessions of the British government. As you might expect, the British state has turned to the problem of institutional reform of vocational training. Given that there appears to be a real problem of inadequate experience-based training in Britain, the conduct of the institutional reform effort deserves special attention in the next chapter.

Notes

[1] The book sold 400 000 copies in 11 languages, see www.leanuk.org/daniel_jon#25B6D.

[2] Kanban is an ingenious distributed information feedback practice that ensures parts are only manufactured in response to demand from final assembly. Kanban was used as a method of inventory control that involves attaching cards to each item of inventory and only allowing the manufacture of a part at a workstation when a returned and detached card indicates that its part has been used in a later production process. If the number of cards is reduced the number of items of inventory is reduced until problems matching supply and demand are encountered. The progressive solution of these revealed problems generates the more efficient production process (Monden 1983: 13ff.).

[3] See for example Schonberger (1982: 22–4).

[4] See Schonberger (1982: 27ff.).

[5] The evidence comprises a survey of car producers' *opinions* of which of their competitors were better at design for manufacture, and a General Motors study comparing one of its own plants with a Ford plant through 'tear down analysis' – the practice of dismantling a competitor's product to deduce the design of its assembly plant. A survey and one two-plant comparison are suggestive, but provide only a thin basis for confidence in exact percentage figures for the role of design on shop floor productivity and quality (Womack et al. 1990: 96).

[6] Brooks models this by pointing out that if each task must be independently coordinated with each other task, then the communication effort increases in proportion to the number of communicating worker pairs, or $n(n\ 1)/2$, where n is the number of communicating workers. So, for example, doubling the number of workers quadruples the effort spent communicating (Brooks 1995: 18).

[7] The process takes so long because the process of creep grinding is used to shape hardened steel blocks into dies. Creep grinding, as its name suggests, is a very slow process, slow in the order of years rather than months.

[8] Keiretsu organisation itself has antecedents in the zaibatsu, the giant shogun family-controlled industrial firms of pre-war Japan that were broken up by the post-war US occupying forces (Johnson 1982: 174). Once the occupying forces left, the Japanese economic bureaucracy MITI encouraged more than 2800 trading companies to group into 20 large companies each affiliated to a different keiretsu (Johnson 1982: 206).

[9] Nor does one have to believe that company-committed workers are necessarily 'happy' workers. On this question, Berggren's careful study of worker conditions in a lean production car plant in the USA drew mixed conclusions – workers were paid more, had more responsibility and control over their work, but worked more relentlessly and longer than in the conventional mass production car plants (Berggren 1993). They cannot be said to be clearly worse off in such a system.

[10] Woronoff allows that the Japanese system has created a first-class export-oriented production system; his attacks are directed against those that imagine that all aspects of the Japanese economy or society are equally marvellous.

[11] For reasons of space I have not explored Japanese management of the supply chain. Whether from Dore, Womack et al. or other authors, the essential feature of the techniques used in supplier relationships is that they are supported by a long-term and high-trust relationship between buyer and supplier. In essence, in exchange for continuity of orders, the industrial keiretsu company wields considerable power over the practices of its suppliers. If expanded to the level of practices, the argument would parallel the one assembled here with respect to design and shop floor practices.

[12] For example, by collaborating with its part-owned partner Mazda, Ford had achieved comparable results with the Japanese in new plant by 1989 (Womack et al. 1990: 86).

[13] A popular example is the book by Oakland. It is worth pointing out that despite the superficial appearance of comprehensiveness, like Schonberger and Monden, the details of innovative achievement in the Japanese organisation of new model design are simply absent, despite Clark's research being by then available (Oakland 1989).

[14] This motivation arose at the time of the dot.com boom, when the share valuations of the major car makers plummeted. There arose the belief that these companies could improve their *measured* return on assets (ROA) by shifting more of those underrated assets to suppliers – it was argued that this was necessary to raise their share values (Sako 2003: 245).

[15] It is important to compare productivity per worker *per unit of time*. If output per worker figures are used the gross German advantage dwindles to the order of 10% – because British workers work longer hours.

[16] An example of a determined critic of the NIESR research is Cutler, who works hard to deny this 'plausible argument' (Cutler 1992). One of his better arguments is that the capital equipment endowments of the compared plant were not always sufficiently comparable, so that productivity differences were derived to an extent from equipment differences. It is not a sufficient argument; it ignores the cumulative persuasive effect of the British examples of practice and in any case cannot be valid for the hotel comparisons. Cutler correctly identifies the NIESR research to have a degree of vulnerability at this point where vocational training is assumed to be the cause of poor operations practice, but all he can offer as alternative is the argument that British enterprises are to blame because they are uninterested in exploiting opportunities for profit through the organic growth of the enterprise. There are many problems with such an argument; the most relevant here is that even if this were to a degree valid, a good candidate to explain such behaviour would once again be the poor development of vocational training.

[17] The example has not been written up and published, except here. Readers, like myself, might believe in the veracity of the example because they trust James Fleck's reputation as an academic. Or if they have a predilection to believe only what they can read, they might prefer to see the example as a 'working through' of the theoretical device of the free-rider effect applied to innovation.

[18] This figure of 27% represented a fall in the proportion of the British workforce with apprenticeship-level qualifications, from 30% at the beginning of the 1980s (Prais 1995: 22).

[19] In Japan also, around twice as many 18 year olds complete the equivalent technical courses as in Britain, although the mode for doing so is not that of the German 'reformed apprenticeship' (Prais 1995: 34).

[20] There has been a long debate over what has been the precise institutional form that has blocked technological change and to what degree it has done so. The series of papers on cotton is instructive in its own right, for showing the shifting ground of what exactly the institutional retarding factors were. A good early formulation from Lazonick tends to pinpoint the vertical disintegration of the industrial structure of the British cotton industry as the institutional block to change (Lazonick 1983). Saxonhouse and Wright then cast doubt on vertical integration alone being the important factor (Saxonhouse and Wright 1984). Lazonick criticises their doubts of the inhibiting effect of vertical specialisation (Lazonick 1987) and they reply by expanding their criticism of his declinist argument (Saxonhouse and Wright 1987). A very useful clarifying review by Mass and Lazonick reworks their position to incorporate most of the points made in the debate (Mass and Lazonick 1990). In Lazonick's later book, the main source here, he has moved to the institution of craft control of skills as the central institutional explanation for relatively poor, twentieth-century British technological performance.

An excellent primer before entering this debate is the references to these authors in McCloskey's work on rhetoric and stories in economics and history (McCloskey 1990). For what it is worth I have adopted a compromise position: on cotton, with Saxonhouse and Wright, I cannot see why we should blame British management for making profits (and they did make profits) by adapting their inherited institutional form in the long period until the First World War, when both the labour- and capital-intensive technologies of cotton textile manufacture *coexisted*. There is a distinct echo here of the apparent paradox of coexistence of substituting technologies discussed in the chapter on competition and innovation. Reference to that chapter would prompt us to examine the nature of the markets served by the old and new cotton textile technologies. Indeed, in the Mass and Lazonick paper it appears that, in Japan, the old technology served the home market, the new the export market. Without further evidence from these sources, and with the knowledge that fashion is very important in textile markets, one may surmise that the old labour-intensive technology was better adapted to short production runs and rapid changes in textile design for the home market. The new technology concentrated on export markets because it needed long production runs in staple cloths to obtain its economic advantages – perhaps the scale of the home market for such production was too small, in Japan, at this time. If you object to this as conjecture, I would only say that while such conjectures can be made that are consistent with the evidence in this series of papers, as I believe this one is, they cast doubt on theories of irrational, profit-forgoing behaviour on the part of 'British management' and weaken theories of the strong blocking power of craft control on the adoption of new technology. Rose's recent book focusing on the history of the British and US cotton industries acknowledges that British institutional arrangements together with factors such as heavy-industry indebtedness inhibited technological change in the British cotton industry, but adds 'the role of shopfloor relations in inhibiting technological change in the interwar period should not be exaggerated ... diverging rates of re-equipment and productivity growth between the United States and Britain reflect far more the differences in the origin and magnitude of market difficulties confronting the two industries' (Rose 2000: 237); that is, the US industry produced largely for the US market, the British industry for export.

After the Second World War, but not in the textile industry, Lazonick's argument appears more convincing, as will be discussed above.

[21] It is a dangerous business associating forms of production with a series of ascending, progressive steps to modernity. The globally successful Swiss watch industry was organised as a putting-out mode of production until the advent of digital watches in the 1980s (Steinberg 1996).

[22] Lazonick cites this figure from Child and Partridge, who take a 1979 figure from the Institute of Supervisory Management (Child and Partridge 1982: 11).

[23] The situation for British Steel management was complicated over time by ad hoc government interventions. Two of the worst included: a Conservative government's decision to split a projected, economic plant into two plants sited to meet regional employment targets; and a Labour government's decision to cancel some of British Steel's desired obsolescent plant closures.

[24] These figures were obtained through personal communication with Peter Calladine of the Association of MBAs (AMBA), London. The report from which these figures derive was prepared for AMBA in the early 1990s and titled 'MBA – Reality and the Myth'.

8   The State and the Management of Technology

In its broadest sense technology policy concerns any activity by the state intended to promote or control technology. One could attempt a complete list of the many tools of intervention that could be adapted to promote a particular technology. This would be quite tedious and given the entwinement of technology with economic activity many of these tools would already be quite familiar to us: regulatory control, public ownership, tax and financial subsidies of various kinds.

In previous chapters state intervention was discussed where it happened to have an important bearing on events of technological change. Nor was it raised incidentally: the state had an unavoidable and important role in the reform and maintenance of the institutions of the market economy. Rather than analyse an abstracted set of policy tools, this chapter maintains the style of previous chapters and discusses the role of the state in the context of some process of technological change. What is different in this chapter is that the examples of technological change have been chosen to place the workings of the state policy process centre stage.

This completes the logic of the book, for the management of technological innovation will have been extended from the management of practice within the firm to the design of institutions and finally to the management of technologically relevant policy processes by the state.

The examples in this chapter have also been selected by the criterion that the policy outcomes appear to contrast with 'informed common sense' – this sharpens our question: why did the policy process proceed as it did?

The first example, of the reform of British professional engineering associations, establishes the idea of a preferred style of state intervention in a situation where there are established interests in the *status quo*.

The second offers the effort to reform vocational training in Britain in the 1990s as a particularly difficult reform problem: one where the intellectual case for reform was good, but where its implementation through the preferred consensual policy style ran into intractable difficulties. Comparison with the standardisation of the German shop floor promotion route into management offers insights into the difficulty of emulation of the German example, yet perhaps also offers a more hopeful, if modest, policy pathway.

Professional engineering and vocational training reform are like most examples in this book, because they concern socially valuable technologies developed primarily within the market economy. They include the difficulty of attempts to reform institutions with the purpose of encouraging better private sector management of innovation and technology.

For the sake of contrast, two examples are offered where state intervention was clearly needed to overturn the technology promotion activities of strong, active interest groups. The example of AC electricity supply adoption shows that even if there is a clear economic advantage in favour of change, the state may take decades to marshal the political resources needed to overcome conservative social forces favouring the technological *status quo*. The example of the economic millstone of plutonium reprocessing technology is that, despite the technology's many disadvantages and the loss of its original rationale, the British state proved unwilling and unable to frame a decision process that had the potential to block this technology's further development.

These examples show the limits of the state's ability to control technology on behalf of society. The examples contrast with politicians' positive rhetoric regarding innovation and technology and remind us that the political practicability of technological change is an issue. This has probably always been true and so the chapter ends with a brief, historical review of the forces that in recent history have encouraged the state to intervene actively in technological development.

Coercive Reform as a Temporary Departure from Normal Process – Engineering Association Reform in Britain[1]

The attempt to reform British professional engineering associations is an illustration of the great problem of reform when faced, not with a blank slate, but with institutions active on their own behalf. Difficult questions are what role such institutions should be given in the reform process and whether they should be coerced into reform, and if so, on what basis? As a political scientist, Grant Jordan calls the preferred and normal British policy style 'bureaucratic accommodation within policy communities' (Jordan 1992: 5) and for him the attraction of the study of professional engineering reform was exactly because the British government, in the form of the Department of Trade and Industry (DTI), appeared to have uncharacteristically abandoned this normal style of accommodation for a coercive intervention.

The historical roots of British engineering were in 'pupillage' (see page 237), a high-level form of the craft transfer of skills that characterised shop-floor apprenticeships. Like their shop-floor counterparts, engineers were also trained in relatively narrow skill bases associated with particular occupations. Recent times have received the significant inheritance from this nineteenth-century institution of 44 independent, professional engineering associations, each one originally founded to represent a high-level occupational or technological speciality, each one jealous of its independence and right to award 'engineering' status.

The sheer number of engineer accrediting organisations hindered both the standardisation of engineering technical qualifications and the ability of university engineering courses to redesign and experiment with their curricula, constrained as they were by their need to gain professional recognition from one or more of the institutions. Nor were 44 institutions likely to be as effective as one when it was necessary to represent general engineering interests in society at large.

An attempt to reform the professional engineering associations was mounted in the 1970s, after one of those periodic public debates about the role of manufacturing and engineers in British society. It was this debate and the perception that the engineering institutions had fumbled in their efforts to reform themselves in the mid-1970s that led the DTI to abandon its usual 'clientele' relationship with the engineering institutions, namely serving and representing their interests in government, in favour of an active reform position.

The result was the creation in the late 1970s of the Finniston Committee to investigate the organisation of engineering. The unusual reformist stance of the DTI meant that from its inception this committee represented a break with consensus politics by being staffed with members who were known to have reformist inclinations. The committee further broke with convention by deciding not to consult with the British professional institutions, but by investigating engineering practice worldwide. As is obvious when reading the report, they became especially impressed with the Japanese use of engineering expertise (Finniston 1980: 22).

The committee made the radical recommendation for reform that the government should create an 'Engineering Council' as a single body with statutory powers to oversee the registration and licensing of all engineers, but also with the remit to direct future engineering reforms. If it had been enacted, this recommendation would have constituted a direct challenge to the 44 existing professional institutions through the compulsory removal of some of their powers and activities.

But by the time the Finniston Committee reported in 1980, the government had changed. According to Jordan, this did not automatically seal the fate of the report, but it broke continuity and brought a new set of assumptions into government and especially the DTI, with the appointment of Keith Joseph as the new minister. Among the new assumptions were a suspicion of the professions, but also a dislike for government intervention and an absence of commitment to manufacturing. This created the opportunity for the existing institutions to provoke a lengthy and conservative renegotiation of the meaning of Finniston and the relevance of its recommendations that would advantage those parties with the 'stamina' to see the process through. Those with stamina would prove to be the old established engineering institutions, not the few committed reformers.

A key event in this process was the retirement in 1980 of Dr George Gainsborough, secretary of the Institute of Electrical Engineering (IEE) for 17 years. Gainsborough had been a prime mover in the creation of the Finniston Committee in the first place and was a strong supporter of the full implementation of its recommendations. With Gainsborough gone, the institutions began to appear united against statutory change. Within the DTI, the ministerial advisers who had portrayed the IEE as particularly enlightened and not at all self-interested, now had no case in favour of coercive reform. This mattered because the government did not have its own preference for change. In these circumstances the government responded to the active lobbying efforts of the existing engineering institutions by seeking a consensus between them and other interests over what to do about the report's conclusions. Without any discernible decision point, it soon became a working assumption that it was important to obtain the agreement of existing institutions to any reforms.

In this way Jordan shows that the 'Finniston assumptions' and the associated debate were successfully replaced by a discussion that centred on the importance of professional self-regulation, rather than government-imposed solutions. In Jordan's words,

> the explanation as to why the Big Four [the largest four engineering institutions] were so successful has much to do with the abdication of a theoretical role of decision by Ministers and civil servants. (Jordan 1992: 281)

In other words, as it reverted to its customary 'clientele' service function, the DTI would not judge the various potential outcomes of the post-Finniston debate in terms of revitalising British manufacturing or practice overseas, but in terms of finding the widest possible policy compromise among those constituting the selected 'policy community'.

Jordan has no view about what *should* have happened because he is not interested in the arguments themselves, but in the political process of government. A judgement of the outcome depends on what you choose to compare it with. One does not have to believe the entire Finniston argument, or to accept all Finniston's recommendations as valid,[2] to understand that part of that reform effort had some of the character of a simple standardisation, of the 'weights and measures' kind, for which modern nation states have always taken responsibility. So, for example, the existing institutions were acknowledged to be unable to agree common standards for the title of engineer (Jordan 1992: 126). Yet even on this most modest level of reform, Jordan concludes that 'it is difficult to envisage a fuller and more effective pattern of delegation on professional regulation to the institutions than they have secured under the formal position of Engineering Council control' (Jordan 1992: 256). When the government has no theoretical view on desirable outcomes, it has no basis for intervention and prioritisation of even the most basic of reforms.

Finniston himself famously dismissed the Engineering Council that the government and existing interests agreed to create:

> What I wanted was an engine for change. … Instead we have got a shunter moving along disjointed lines. (Sir Monty Finniston quoted in Sampson 1992: 80)

The Finniston story remains a good illustration of the limits of a political process that depends on 'bureaucratic accommodation within policy communities' when those communities are themselves the object of reform.

The State and the Reform of Engineering and Vocational Training in Germany and Britain

The evidence of the previous chapter pointed to a need to reform the British institutions of shop-floor vocational training. In the 1990s the British government did attempt reform, but by the way that the effort was first conceived and the interests that were allowed to become part of the accepted policy community, the reform would be comprehensively bungled. It is difficult to judge the inherent difficulty of even a necessary reform effort such as this one and so, in part to provide contrast, but also as interesting in its own right, this section begins with the story of how Germany legitimated the shop-floor promotion route into management and to engineer status in the early decades of the twentieth century. The contrasting conditions for the reform efforts suggest that the monolithic scale of the British attempt predisposed it to interest group subversion and failure.

Professional Engineers' Status Objectives and the Origin of the German Shop-Floor Route into Management[3]

The status objectives of the German professional engineer loom large in the story of the German shop-floor route into management. Pollard comments on the nineteenth-century Humboldt 'reforms' of German universities 'that it would be hard to imagine anything less likely to put Germany into the van of European progress than her universities as they emerged after 1815' (Pollard 1989: 145). They despised utility and resolutely excluded all practical subjects such as engineering. When higher engineering education was instituted in the 1860s, it had to be in separate institutions, the Technische Hochschulen. Because Prussia at this time was technologically backward, the industrial employers' interest was weak and the social status of engineering was low; the curricula of the Technische Hochschulen were largely designed to raise the status of engineers to the levels of the established professions and civil servants. This involved purging the curricula of practical content; the dismantling of college workshops and laboratories, and the introduction of 'cultivating disciplines' such as history and literature (Gispen 1989: 79). At their inception, then, Technische Hochschulen had a social and professional rather than an economic purpose and were intended to produce a social elite of engineers. The goal of creating an engineering elite would persist and would provide the shop-floor promotion route with both the opportunity for development and the chance to challenge the status of the Technische Hochschulen.

By 1880 the result was a rising level of industry criticism of the impractical Technische Hochschule graduate, who proved unable to adapt to industrial life where cost and production practicality considerations were a necessity of design (Gispen 1989). The increasingly influential industrial-employer interest blamed academic engineering professors and the abandonment of practical training in the engineering curriculum. The Prussian professional engineer association, the VDI,[4] became the mouthpiece for these views and the focus for a campaign to make engineering education more practical. Under this pressure, Technische Hochschulen began to restore practical content: the introduction of a year of compulsory workshop training; the adoption of laboratories for empirical research and training; the diffusion of compulsory drafting and design courses and the interesting 'anti-mathematician' movement, a deliberate attempt to expel complex techniques such as calculus wherever it was possible and to use simpler, easier – more practical – graphical methods (Gispen 1989: 153).

The transformation of the curriculum was dramatic as demonstrated by the change in laboratory and drafting hours in the Berlin Technische Hochschule: these rose from 35% of instruction time in 1881–2 to over 70% in 1898–9 (Gispen 1989: 156).

This was a true, practical engineering curriculum, but as a result of this change, the elite status of the Technische Hochschule graduate depended only upon the high *academic* standards for entry, the lengthy four-year period of instruction and the artificial restrictions on the number of students admitted.

It was these restrictions that enabled the rise of a fascinating private educational challenge to the intended engineering elite, in the form of graduates from a new kind of non-academic engineering school. These private forerunners of today's Fachhochschulen were initially founded to meet strong demand from employers and workers and were widely understood to be necessary to supplement the meagre numbers graduating from Technische Hochschule. Even the VDI supported the rise of the non-academic engineers, because in their ideal world the graduates of such schools would be subordinate to the Technische Hochschulen graduates and would release the elite from the burden of the more routine and tedious forms of technical work. Unfortunately for the would-be elite, by the First World War,

> the opposite happened. The non-academic engineering schools, established in part to remedy the ravages of the Industrial Revolution among the lower classes and to stabilise the social order at the point where it seemed to need shoring up most, eventually gave rise to indiscriminate intermingling of the two categories of engineers. The consequence was fierce competition, internecine warfare, resentment of all those who could remotely be blamed for this state of affairs, and a major career crisis in the engineering profession. (Gispen 1989: 160)

Employers refused to privilege the would-be elite Technische Hochschule graduate engineer over the non-academic engineers in the workplace. They argued that the individual abilities of non-academic engineers were under-estimated by the academics. The social crisis was made worse by one of the very features associated with economic success: the free market in education generated 'a continuously mounting glut of engineers' (Gispen 1989: 217). One estimate was that of approximately 250 000 people with technical education or expertise above foreman level 'no more than a quarter had true engineering functions'. The technically educated remained employed, but many were occupied in low positions little better than blue-collar work that did not fit their career expectations – a particular problem for the Technische Hochschule graduates (Gispen 1989: 198).

The government response to the social crisis generated by unstructured competition between different technical schools and colleges was the creation of a Committee on Technical Education.[5] This committee rationalised and standardised the competing and overlapping institutions on a national scale (Gispen 1989: 211). The academic engineering elite sought to have this committee buttress its status and privileges through devices such as legal restrictions on the forms of employment allowed to those promoted from the shop floor. Fortunately, the large employers dominated the committee and they were determined to defend the non-academic engineering schools, or Fachhochschule. The committee dutifully ignored the academic engineers and rationalised and preserved the workplace route into engineering and management to parallel the academic route represented by the Technische Hochschulen. The shop-floor promotion route has persisted as an outstanding characteristic of German technical education to the present day.

Comments on the German Reform

The obvious regulated order of German technical qualifications of today provides a temptation to see the late, *imposed order* on qualifications as the most significant event in this account. It was significant, but it was a response by a unified, employer-dominated professional association to the (now forgotten) preceding period of excessive supply and unchecked competition between different technical qualifications. This chaotic period of technical expertise oversupply gave the employers – and gives us – confidence that the Fachhochschule extension of the workplace promotion route into management produced technology graduates at least equal in quality to the elite stream of engineering education.

The greater lesson is that engineering is a practical art and that the ultimate criterion by which curricula should be judged is by their practical value in the workplace. The reason why Technische Hochschulen and Fachhochschule graduates were of similar practical value is suggested by a comparison of their essential curricula. The reformed and practical Technische Hochschulen combined academic study with a large dose of structured, practical experience in laboratories and workshops. Fachhochschule graduates that had come through the apprenticeship route combined structured workplace experience with academic classroom learning. In essence, both routes featured structured practical experience and a relevant academic program. Such effective curricula contrast with 'engineering curricula' that are, on the one hand, purely academic and scientific and, on the other hand, purely *ad hoc* experience.

Together with the story of the Finniston reforms, this account suggests that it is the range of final occupations for engineers that explains the fissiparous tendencies of professional engineering associations. To the occupational splits of the last section one can now add that between the interests of the academic and civil service engineers[6] and the industrial engineering employers.

The 'non-economic' programme of the academics was never permanently suppressed in Germany; they broke away from the VDI to form their own professional body and to campaign vigorously for Germany to follow Austria's decision in 1917 to restrict the prestigious title 'Ingenieur' to graduates of Technische Hochschule (Gispen 1989). Once again, German employers would successfully block the status-seeking stratagem.

In other countries too, it has been argued that professional engineers have distorted institutional paths of development for reasons of sectional interest. An important example for the USA is the result of the industrial relations trauma of the 1930s, when US engineers formed professional associations with the object of boosting their status and their identification with management through the exclusion of workers with lower level technical qualifications – and above all from manual workers (Meiksins and Smith 1992: 142). The long-lived legacy of this status-raising objective was the redefinition of engineering work to exclude routine work on the shop floor with blue-collar and unionised labour – a great contrast with the Japanese exploitation of engineers.

The example of the promotion of an impractical curriculum in the Technische Hochschulen may have relevance to today. Divall has argued that from the 1940s, in Britain, a consensus formed between university engineering academics and a small group of 'progressive' employers to move university engineering curricula in the direction of 'engineering science', the theoretical analysis of the physical aspects of engineering (Divall 1990: 94). Divall shows that engineering science curricula became dominant in the universities post-war and through the 1970s and that many of the civic universities dropped such practical activities as workshop training (Divall 1991). A British 'style' of graduate engineering education became established, where universities provided theoretical education, but left practical training to future employers.

Divall argues that an engineering science curriculum certainly met the engineering academics' need to be able to conduct research and so win esteem within the university hierarchy. It also appeared for a while to meet the aim of employers for whom 'particularly in electrical goods and aircraft, the graduate engineer came to be viewed as a talisman for economic growth through innovation'[7] (Divall 1990: 93). What it was not adapted to was the requirement that engineers should be able to be quickly and effectively deployed in the production environment in the private sector.

This is an example of a divergence between the practical needs of the firm and the status objectives of a 'profession'. Wickens' account of the degraded nature of production management in the British car industry pinned blame on a widespread tendency for functional specialists to aspire to 'professional' status and with it a 'hands-off' approach to production. And it has been argued that in comparison with Germany, there is a general tendency in Britain for those engaged in specialised work functions to aspire towards such status-seeking 'professional' objectives at the expense of practical attainment within the firm (Child et al. 1983). As in the extensive discussion of craft control in the previous chapter, the underlying suggested cause is the absence in educational institutions of an orientation towards and relationship with a field of practice.

The Mutation and Failure of the Radical Reform of British Vocational Training – the Fiasco of NVQs

The last section used the German reform of the shop-floor route into engineering and management to show that both the university curriculum and the practical role of the graduate engineer in the firm have typically been subject to political lobbying efforts over time and that this is one of the reasons for a number of significant international differences in the formation and use of such engineers. At the end of this section the German reform will again prove useful as a source of instructive comparisons with the British attempt to reform technical training in the 1980s and 1990s, not least because that reform was justified in part by reference to the German model. This section begins with an analysis of how, what were sometimes, *misrepresentations* of the German 'model' came to play a role in initiating the British 'reforms'.

The Role of Key Texts in Creating a Pro-reform Climate

Inadequate British technical training and technical education has been blamed so often for poor technological performance and by so many government-organised reports and commissions of inquiry that their individual discussion becomes beside the point. Wolf adopts the sensible expedient of simply listing the 23 inquiries published between 1867 and 1969 in a table (Wolf 2002: 65).

It is the number of such investigations and their common conclusion of the inadequacy of British technical training that raises a quite different question: if the inadequacy of training were so obvious and so persistent, why was there no successful policy response? In the 1980s a particular and forceful answer to such questions was delivered in Barnett's book *The Audit of War* (Barnett 1986: 282). Together with Wiener's tellingly titled book, *English Culture and the Decline of the Industrial Spirit* (Wiener 1987), this book came to have an unusual influence on policy:

> Correlli Barnett's Audit of War is undoubtedly one of the most influential books of the last decade [the 1980s]: it has greatly affected the intellectual climate in which questions of education, innovation, industry and trade unions are discussed in Britain today. It is that rare thing: a work of history which speaks even to those inclined to believe that history is bunk. Cabinet ministers cite it, and may even have read it. Practitioner symposia on the British economy are incomplete without a reference to, or an illustration from, the Audit of War. (Edgerton 1991: 360)

Wolf, in her explanation of why the British government began a series of uncharacteristically heavy-handed interventions in vocational training in the 1980s, cites the problem of 20% unemployment among young males in the early 1980s in Britain, but also the intellectual influence of Barnett's book and the series of research reports by Prais already cited in Chapter 7 (Wolf 2002: 4). It becomes interesting to ask why Barnett's book became influential when so many official reports into the issue of technical education apparently had no effect.

Edgerton is surely right that the attraction of Barnett's work is that it asks big questions and delivers big answers: Barnett's concern is British economic and technological decline and he lavishes blame on both the British government and civil service elite for their anti-technological bias and starry-eyed belief in the moral value of a welfare state and the *ad hoc*, on-the-job experience that represented the practical 'training' of British industrial leaders. Nor can the attractions of Barnett's characteristic style of writing be underestimated – it is a style of barely suppressed rage at the idiocy he must describe. So on the stultifying effect of the religious interest on the attempts to introduce technical schools and curricula as part of the 1944 reform of secondary educational schooling:

> The diaries of the President of the Board of Education and his Parliamentary Secretary … were stuffed not with meetings with industrialists and trade

unionists…or with the different engineering and scientific institutions, but with skull-emptying sessions with gentlemen of the cloth (of various persuasions) on the topic of how they could continue to run a large proportion of the nation's schooling thanks to the taxpayers' subsidy. (Barnett 1986: 280).

As a consequence the crucial question of providing the nation with an education *for capability* [author's italics] from primary school up to technical university equal to that of her competitors was squeezed away to the sidelines. (Barnett 1986: 282)

The German model is implicit in the last line of Barnett's judgement. In general, he breaks free of judgements limited by an acceptance of the *status quo* policy process – in contrast to Jordan, who accepted as normal process, bureaucratic accommodation within policy communities. As we will see later, his assurance that the lessons of the German model are clear and that this model is sufficient template for judgement of the British policy process can sometimes be faulted. Yet the contrast between Barnett and Jordan is striking – Barnett seeks to judge the policy process in terms of outcomes, Jordan is content to study the logic behind apparent deviations from typical process as an end in itself. The Barnett approach is naturally political and burdened with prior interpretation of the problem and the desirability of outcomes – it is complicated enough that it is naturally liable to error. The problem with the disinterested dissection of policy processes as an end in itself is – who cares?

In the example above, I find I cannot help but agree that what was important was what was *not* discussed and who were *not* included in the self-appointed policy community. Who else but the policy-makers could be blamed for these sins of omission? The example enriches our understanding – it is of a laissez-faire approach to the formation of policy communities: policy-makers have a theoretical power over who constitutes the relevant policy community and what should be prioritised as the important issues. In this example, there were other potential constituencies that could have been included – if they chose not to exercise this power, then it is hardly surprising when existing interests controlled the terms of debate.

In sum, Barnett's passion and his preparedness to judge the policy process from an openly stated viewpoint must account for some of the influence of his book.

The Faults of Corelli Barnett and the British Failure to Develop a Dyestuff Industry

Now Edgerton is able to demonstrate that elsewhere some of Barnett's facts are wrong – for example, on supposedly poor British aeroplane productivity during the war (Edgerton 1991: 373). And while Barnett is a ferocious critic of the 'practical man' running British industry he also dismisses the quality of engineering education of early twentieth-century British civic universities as not comparable with that from German Technische Hochschule (Edgerton 1991: 371). The discussion in the last section of the effective competition to Technische Hochschule graduates provided by the best German shop-floor workers should cause one to hesitate here; is not Barnett mistaking the claims of superiority made by the status-conscious academic German engineers as proof that they were indeed superior? On number of engineers he is nearer the mark with his accusations of a deficiency in numbers, but faulty in his estimates and wrong when he blames the British universities for inadequate supply. The role of British universities in engineering education deserves some explanation.

By the interwar years British civic universities had become enthusiastic providers of practical engineering education in response to demand from local industry and in return for the industrial gifts and endowments that were the essential features of a successful university in this period (Sanderson 1988). The number of British engineers did lag behind those in Germany. Pollard gives figures for 1914 of 250 teachers and 400 students in research bearing on industry in Britain, 673 and 3000 students in Germany (Pollard 1989: 195), but then notes that the difference mostly derived from the exceptional German research-driven dyestuffs industry. The number of graduating and active British higher educated technical professionals lagged Germany most plausibly because of weak, and to a large extent justifiably weak, British industrial demand, rather than a lack of interest in supply on the part of the British universities. The quality gap argument can probably be dismissed entirely. As Edgerton concludes, Barnett is not to be trusted as a historian of education (Edgerton 1991).

The problem with Barnett's judgement of the technical education institutions is that it takes no account of early starts and increasing returns to technological development. It is these that shift the responsibility for Britain's failure to develop dyestuffs to the institutions that supplied skills in the time *before* Germany established its scale economies in R&D.[8] The story is worth repeating here, both as an example of how a developing country can leap ahead of a richer nation (in one field of activity) and as a means of establishing the value of university engineering education to innovation.

In the critical period of the late nineteenth century it was the old British institution of 'pupillage'[9] that supplied demand for expert industrial engineers and managers. Pupillage differed from the lower level informal British apprenticeship in that middle-class parents paid a premium to bind their sons to an individual engineer for between three and seven years in return for a practical training. According to Divall, these premiums were sufficiently profitable for the leading engineering employers that they acted to *successfully discourage* alternative forms of engineering training into the early years of the twentieth century, whether that training was proposed to be through the professional engineering institutions or British graduate education (Divall 1990: 69). In the same way as for lower level informal 'training', while British industry represented best technological practice, this institution was adequate to maintain the established stock of skills, although not capable of large increases in supply. When new industries appeared that had new skill requirements, given that pupillage was rooted in existing practice, it simply could not respond at all. In Pollard's review of the role of science and technical education on British economic activity the one clear example where British deficiency and German ability to supply scientists made a clear difference in the establishment of an industry occurs in the rise of the German dye and organic chemical industry (Pollard 1989: 158).

The British chemist Perkins pioneered the industry in Britain with the discovery of a mauve coal-tar dye. Many German chemists began their careers in Britain, where they expected the major applications of their subject to develop in the textile industry, and because in Germany chemistry had initially fallen victim to the Humboldt reforms (see the previous section). However, in Germany the brilliant chemist Justus Liebig made so many practical contributions to agriculture and commerce that the prestige of the subject was raised sufficiently that the universities became willing to adopt it. The example throws light on the value of early starts: while Britain had had such a start in industrial *production*, Germany achieved the same in university chemistry *education and research* – it was the latter that counted. University departments drove the new discoveries and advances in understanding upon which the new field was based. Their graduates and PhD students were the basis of the key competitive advantage of the German dyestuffs industry: the industrial direction of a mass of chemists, for the management of which see page 51. Industrial concerns worked with universities to direct both young PhD students and 'the massed ranks of men of ordinary, plodding ability, thoroughly trained and methodically directed' in the R&D departments (Pollard 1989: 158). So here we have a stunning success in a new industry as a result of German universities' early willingness to stoop to the adoption of an evidently practical subject.[10]

Now we are in a position to 'do a Barnett' and apportion blame and responsibility for the 'failure' of the British to develop such a new industry. Comparisons between the two countries of absolute numbers of researchers, or science and technology graduates, will tend to mislead because Germany developed a dyestuffs industry and Britain did not and that industry by definition had a unique, high dependence on graduate chemists. Because the technology possessed increasing returns to scale of R&D, once Germany had established an early start in university-supported dyestuffs R&D, it rapidly came to dominate the global supply of these products – as it still does. If we want to apportion blame, then it should go to the leading British employers' efforts to preserve pupillage by blocking the foundation of more flexible institutions of education *during the period in which it was still possible to imitate the German development*. Once the German industry was established, it mattered not that there were British universities in existence willing to develop practical education – the window of opportunity for emulation in this industry had passed. Yet now we can see how tempting it was for Barnett and Wiener, when examining the situation in higher technical education in the interwar years, to resort to anti-technological cultural explanations in British universities to explain failure in this particular industry. We also have acquired a deeper understanding of the range of possible employer 'interests'. What established employers consider to be in their interest today may not be at all in the interest of future economic development.

Barnett and Wiener Misled the Policy-makers – or Did They?

So if Barnett and Wiener were influential, they were likely to mislead policy-makers – or were they used by policy-makers who were inclined to these beliefs in the first place? Sanderson commented on the obvious contradiction between the facts and the contemporary popularity of the view that British universities were anti-technological and anti-business, that:

> Politicians have found the argument attractive because it enabled them to create a climate of opinion hostile or indifferent to higher education as a milieu in which cuts could be imposed without resistance. (Sanderson 1988: 103)

Rather than the universities being the problem:

> It might be argued, both then and today, that the more insidious evil has not been anti-industrial attitudes in the universities so much as anti-intellectual, anti-academic attitudes in industry. (Sanderson 1988: 102)

The belief that universities should become, and can be made to be, ever more 'relevant' to business has persisted in government as countless speeches by higher education ministers and the recent British government science and technology White Papers demonstrate.[11] There is a striking contrast today between British governments' willingness to intervene in university governance however trivial the matter and their reluctance to intervene against the expressed wishes of private employer interests even when there is an economic as well as a social justification for so doing. As we have seen, in contrast to the universities, the case for poor quality and organisation at the skilled worker and foreman level of British technical and management training is a strong one.

The Development and Mutation of the British Vocational Training Reform Effort in the 1980s and 1990s

As far as Barnett's and Prais' work influenced government reform efforts, they offered the German system as a model, but when the British government sponsored the reform of vocational education in the 1990s, this soon bore little relation to the analysis of either Barnett or Prais, or the German model of vocational education. Wolf documents the mutation of policy into irrelevance in the critical period that followed the founding of the National Council for Vocational Qualifications (NCVQ) in 1986 (Wolf 2002: 72).

The grand object of the NCVQ was nothing less than a comprehensive national system of vocational training and a single set of related standards of such high quality that existing qualifications would wither away as students sought to acquire the appropriate National Vocational Qualification (NVQ). This system was expected

> to turn shop floors into centres of learning, creating a virtuous circle of training and productivity. (Wolf 2002: 72)

So in its scope and *ambition* the reform effort echoed Barnett's and Prais' laudatory appraisal of German vocational education. Ambition would prove to be the limit of valid comparison; NVQs in reality would be 'unlike any other vocational qualification the world had ever seen' (Wolf 2002: 80).

The fundamental difference between NVQs and other vocational awards was that NVQs were, in the accompanying jargon, 'competence-based', which meant that the achievement of an NVQ certificate required the demonstration of the ability to perform a specified task to an established standard. Assessment was typically by a trainer who would tick off the listed activities comprising the NVQ as they were achieved (Wolf 2002: 74). The 'competence-based' approach infected all the 1990s' 'reforms' to British vocational education; it was transferred to the 'modern apprenticeships' introduced in 1995 (Bierhoff and Prais 1997: 98) and even into classroom teaching for the GNVQ (General National Vocational Qualification): this was the classroom-taught 'vocational A-level', or the vocational equivalent of British 16–18 years' academic education. Unlike its academic, supposed equivalent, its outstanding feature was that the teachers had a large role in assessing what they themselves had taught (Wolf 2002: 92).

NVQs did achieve a theoretical comprehensive coverage of occupations. Although employers were supposed to define standards, in practice, and in order to achieve the aim of comprehensive coverage, specialist standard-writing consultants were used to define many NVQs so that they would win approval from the relevant government bureaucracies. By 1995, 95% of occupations had indeed been covered by standards and 794 distinct NVQs had been created. Slightly more than half of these standards had either one qualifying candidate or none at all (Wolf 2002: 75).

In the absence of an historical continuity of demand by employers and a willingness to train by young people, the mere creation of the option of a certificate made little difference to practice. Instead, it was striking that the most successful NVQs covered traditional craft occupations. Relative success at NVQ did not mean success as a standard of training: Bierhoff and Prais noted that in 1994, the number of British 19 years olds qualifying to the crucial NVQ level 3 in the five most important occupational categories (corresponding to the craft apprentice level) was only 9000, or 1.3% of British 19 year olds (Bierhoff and Prais 1997: 112). Even with allowances for some continuing award activity by the old certifying bodies like City and Guilds, Bierhoff and Prais estimate that no more than 2% of British 19 year olds were qualifying at NVQ level 3, compared with 9.2% under the pre-NCVQ system and 42% at the equivalent levels in Switzerland and Germany (Bierhoff and Prais 1997: 112). Although their results are for a recession year, by any terms this represents a catastrophic decline, mitigated only by Bierhoff and Prais' suggestion that an unknown number of employers had chosen to make internal arrangements for uncertified training rather than use the cumbersome NCVQ certification procedures. A good reason why employers might choose to do this would be the reasonable suspicion that NVQ standards were not reliable – they were, in fact, not standards at all, because they did not use independent assessment and validation of competence: by the mid-1990s accusations of low quality had attracted the attentions of the tabloid press (Wolf 2002: 116). In 1997, the failure of NVQs and of their sponsoring body the NCVQ was signalled when the NCVQ was effectively disbanded through merger with the schools curriculum and examinations body (Wolf 2002: 116).

What a contrast between the ambitious rhetoric that accompanied the 11 years of NVQ promotion and the reality of complete failure! Not only were 11 years of reform effort wasted, but so was the enormous commitment of time and effort of all those institutions and individuals foolish enough to believe the rhetoric and commit themselves to these qualifications.[12] At one point, even Wolf appears to conclude that attempts to transfer institutional arrangements are hopeless:

> The most unmitigated failures in public training policy occur when governments inspired by some other country's apparently efficient approach, attempt to transpose this approach wholesale. The favourite subject of such expensive and abortive transfers has been the German Dual System – an apprenticeship system which grew organically, which is run primarily by the companies themselves, and which, as we have seen, rests on a distinctive and complex network of labour-market institutions. (Wolf 2002: 159)

But as Wolf herself points out repeatedly in her text, the NVQ approach was *not* an attempt to make a 'wholesale transposition' of the German institution. On the contrary, the outstanding features of the German system had been dropped: there was no independent assessment, no compulsory classroom education element and no collective coercive control through industry associations. Summarised this way, it is clear that the outstanding feature of the British reform was the *absence of coercion*, either of employers or by employers. In this regard, the British 'reform' had continuity with the old-style, voluntary institutions of informal British apprenticeship.

In search of why the British reform, despite its accompanying rhetoric, bore little relation to the German shop-floor route to technical competence, we can compare it to the reform that generated the German institution (see page 230). Unlike the reform in Britain, this was organised by employers who had experienced a prolonged period of expansion, who effectively controlled the unified professional engineering association and who had experience of the worth of both graduate and shop-floor experienced engineers. The experience and motivation of employers at the time of reform is very different and it is surely this that explains why the German employers vigorously sought to shape the future training structures for engineers. In this analysis, the British failure to follow through with the Finniston reforms of professional engineering fatally weakened employer influence over the latter vocational training reform.

This analysis is reinforced by an examination of the management of the NVQ reform. First, the reform effort began in a period of exceptionally high youth unemployment. This was a social crisis and some kind of action became urgent. The solution of subsidised employment that was adopted in the Youth Training Scheme (YTS)[13] in 1983 can be seen as appropriate for the unemployment problem, but a truly rotten basis for the reform of vocational training. Yet this is what it became. It was in this scheme that the expedient was adopted of awarding a certificate for randomly acquired 'work experience'. The NCVQ organisation grew out of the YTS organisation, inheriting its managers and its characteristic approach towards training. The mistake here appears to be the attempt to 'economise' and hope that an existing institution could readily adapt to a new purpose. Little adaptation occurred and it is clear, in retrospect, that it would have been better to create a new organisation free from the influence of the subsidised-employment 'experienced' managers. The contrasting private sector equivalent to this is Nissan's careful creation of a new organisation to support its British car manufacturing activities.

Second, by Wolf's account it was critical that rather than real employers, it was the Confederation of British Industry (CBI) that was included within the relevant policy community as 'representative' of the employer interest. According to Wolf, the CBI faithfully represented general employer opposition to compulsory training in any form, successfully blocking such ideas when they were suggested by civil servants (Wolf 2002: 129). However, the CBI was also important as an enthusiastic supporter of workplace assessment, but here its position derived from internal political processes rather than some general employer view (Wolf 2002: 113). There is no instance in Wolf's account of the government adhering to some principle derived from research or history against the CBI and NCVQ preferences. The sustained government commitment to the peculiar voluntary and non-academic nature of the NVQ system of certificates can be understood as largely the result of its legitimation of the CBI's preferences and political influence.

So once again, how the state 'managed' the construction of the legitimate policy community was decisive to the policy outcome. The great mistake revealed by this account was the tendency of the state to seek administrative 'economies' by making use of the available organisations of the YTS administration and the CBI. These proved grotesquely false economies because these organisations had their own agendas.

Comparison with the German reform process suggests more productive paths of reform. The absence in Britain of an inherited, standardised apprenticeship institution and the lack of a strong professional engineering interest to form part of the reform policy community suggest modesty in ambition is appropriate. As Wolf points out, the state could have chosen to work with *real employers* in one of the skill-intensive, metal-forming or craft industries that stand to gain most from a form of communal compulsion in vocational training. Since these industries are characterised by many small employers, the administration of the relationship would have had to have been carefully thought out in advance. It would have been more costly and more difficult to administrate – but it might have worked.

A last point about the NVQ story is that it is the public policy equivalent of the management fad phenomenon. Both involve noisy public rhetoric about innovation and progress, both are rhetorically linked to good research, both symbolise that people are busy 'doing' and 'progressing', both result in little that is of practical value. People that accumulate experience of such abortive reform processes acquire legitimate reasons for a conservative resistance to future proposed 'change'.[14]

The State and the Management of Political Resistance to Technological Change

The Introduction of AC Electricity Supply in Berlin, Chicago and London

Hughes provides us with a wonderful comparative analysis of the implementation of alternating current (AC) electricity supply in three cities with very different political structures (Hughes 1983). Despite differences in the organisation and practice of city authority between Chicago and Berlin, these cities developed a modern system of electricity supply at the pace of technological advance. In contrast, by 1913 London had become technologically backward because the city's fragmented local government authorities would not cede their rights over their local electricity supply franchises – an essential precondition for the introduction of AC supply that exploited economies of scale.

In the last two decades of the nineteenth century there was widespread diffusion in many cities of Edison's pioneering direct current (DC) supply system, based on small-scale, local generating plants. These supplied direct current to users at some local voltage standard that need have no relation to adjacent supply station standards. However, from the mid-1890s it gradually became accepted that AC supply had superior characteristics to DC. The particular advantage of AC was that it allowed the use of transformers to change the voltage of supply between stages of generation, transmission and use. High-voltage AC could be transmitted over great distances with very small power losses.[15] It became possible to transmit power at standard frequencies from large generator plant possessing economies of scale and built at a distance from cities, but close to sources of water coolant and bulk fuel delivery. The realisation of these economies of scale in generation absolutely depended on securing scale in user demand, and this implied that the existing, small-scale DC (and experimental AC) suppliers would lose their markets and their investments. What complicated the realisation of the ideal economic scenario was the established role of city government as regulator of local electricity supply.

Berlin and Chicago – Alternative Political Routes to Implementation

Hughes writes that in Chicago 'venal, pliable politicians and ill-defined political institutions did not automatically frustrate the growth of public electric supply' (Hughes 1983: 202). They did not do so because they were outmanoeuvred by Samuel Insull, head of the Chicago Edison Company, and a close observer of Chicago politicians' behaviour towards the local gas industry.

When Chicago city politicians approached Insull to find out how much he would pay to block the grant of an electricity supply franchise to a rival company, Insull was prepared to refuse to deal. City politicians went ahead and awarded the franchise to a 'dummy company' for development (Hughes 1983: 206). The same manipulation in the gas industry had induced the city gas company to capitulate and buy the dummy company for $7 million (Hughes 1983: 206), but Insull ignored the city's dummy electricity company. Chicago city politicians then sought to develop the company, but found that Insull had obtained the exclusive rights to buy electrical equipment for the Chicago area from every US manufacturer (Hughes 1983: 206). Insull had effectively marginalised the corrupt city politicians and could proceed with the transformation of Chicago's electricity supply according to techno-economic criteria. This involved raising large blocks of finance to buy out 20 rival electric utilities (Hughes 1983: 207) and then converting their city generators to AC substations with one frequency standard. At the same time Insull built large-scale out-of-city AC generating plant to feed his unified, city electricity distribution network. The economies of scale of the new system paid down the debt needed to buy out the existing supply interests.

The city government of Berlin had a unitary political structure that enabled it effectively to represent city interests when contracts were negotiated with the city's electricity utilities. The largest of these was BEW, a subsidiary of the largest German electrical equipment manufacturer AEG, also based in Berlin. A pattern of contracting was established where the city would grant increasingly generous monopoly supply rights to BEW, for example for a given radius around the city centre, later for the city electric tram system, in exchange for a share in BEW profits and service guarantees to various classes of (poorer) consumers (Hughes 1983: 188). BEW prospered because AEG used BEW as the pioneering user of its new AC supply technology; in the 1890s this included AEG's new AC generator equipment, and in later years AEG's new steam turbines (Hughes 1983: 187, 195). BEW was therefore able to offer the city real prospective reductions in electricity prices, or the requested service guarantees, in exchange for extensions to its monopoly – the basis for further increases in BEW's economies of scale. In similar manner BEW was able to offer favourable prices to the important class of large industrial users, deterring them from installing their own generators and further diversifying the peak load demands on its own supply; the achievement of the ability to manage power loads in this way allowed higher utilisation of BEW generator capacity and so capital savings that provided further price reductions (Hughes 1983: 195). In contrast to Insull's buy-out of his rival Chicago utilities, BEW through its relationship with AEG maintained a leading position in the acquisition of economies of scale that enabled it to out-compete rival utilities. Although the city administrations of Chicago and Berlin were quite dissimilar and required different strategies for AC implementation, by 1913,

> Chicago and Berlin each had a centralised light and power system supplying the entire city from a handful of modern power stations; Greater London had 65 electrical utilities, 70 generating stations averaging 5.3 kw capacity [low], 49 different types of supply system, ten different frequencies, 32 voltage levels for transmission and twenty-four for distribution and about seventy different methods of charging and pricing. (Hughes 1983: 217).

This raises the question why it should have been more difficult to introduce AC supply in London than in Chicago or Berlin.

The Politics of AC Supply Implementation in London

London city government was largely in the hands of its many local authorities and, as today, the idea of a necessarily powerful, unitary city authority was highly controversial and strongly resisted (Hannah 1979: 44). Local London government authorities had acquired rights over supply in the 1882 Electric Lighting Act; for example, they could block the entry of a new supplier if a supplier already existed to users in that district (Hughes 1983: 230; Hannah 1979: 5). The granting of this power did not hinder the diffusion of the first phase of small-scale DC supply stations, for these stations had a regional span on the same scale as the local authorities. The 1882 Act is difficult to judge as bad, because at this early date, it was not clear how the supply technology would evolve. It also appeared to make sense to grant this power to local authorities because they, not private enterprise, had been the historic means of establishing networks of effective sewerage and water supply in Britain. There was a consequent, strong provincial civic pride in the record of such municipal enterprise (Hannah 1979: 23). Unfortunately, as local authorities became active initiators, owners and managers of DC supply schemes they acquired an interest in the retention of the old technological form.

The existence of local authority rights to regulate supply under the 1882 Act meant that all significant efforts to promote economies of scale in AC electricity supply in London had to be preceded by parliamentary bills to remove local authorities' blocking powers. Yet the necessity of following a parliamentary route in a democracy gave the local authority opposition plenty of opportunities to attack such bills.

There were other politically pertinent factors that disadvantaged the prospects for transformation in London. The city possessed no large electrical equipment manufacturer like AEG in Berlin that could have supplied some preferred utility with successive generations of leading-edge technology to rationalise supply incrementally. Nor did London possess, like Chicago, large and significant prospective users like the meat packing industry that could be tied into a political alliance to promote the transformation of electricity supply – although London was a major manufacturing city, activity consisted of a multitude of small establishments. The key political allies of potential 'system builders' were absent from London.

In contrast, conservative interests were highly organised, active and successful in delaying the implementation of this economically superior technology. The case becomes a rich study in the management of entrenched interests.

Hannah cites the electricity entrepreneur, Emil Garcke, complaining of the methods by which many private bills for areas of Britain had been emasculated in Parliament:

> The phrase 'organised opposition' is no mere euphemism. Local authorities possess the equivalent of a Trade Union in the Association of Municipal Corporations, by means of which pressure is brought to bear on every MP when desired...Many members find it difficult to resist this insidious local pressure and in this way divisions in favour of municipal ambitions are effectively produced. (Garcke in Hannah 1979: 26)

In London, 'successive years saw further Bills brought forward and dropped' (Hannah 1979: 44), until in 1905 Charles Merz, the Newcastle entrepreneur who had created a modern, integrated AC supply network in the north east of England, made the most serious private effort to remove local authority blocking power (Hughes 1983: 250). No one could have been better qualified to lead the effort; Merz had an unusual and carefully researched proposal demonstrating the techno-economic benefits of the new system. He had a petition in his support signed by some 250 London manufacturers. Against him were ranged some 100 local authorities represented by 35 legal counsel, London County Council and the Conservative Party, then in power (Hughes 1983: 254).

His opponents 'laid great stress on the rights conferred by their franchise' (Hannah 1979: 46) and raised the terrible idea that if it were passed, the bill

> would enable it to be said that business men could no longer rely on Parliament to protect the interest which it had itself brought into being, and on the faith of which public money had been subscribed and invested. (Cited in Hannah 1979: 46)

The bill became bogged down in select committee and Merz managed to antagonise Parliament with his committee tactics, to the extent that Lloyd George made an 'impassioned plea' to Parliament to teach Merz a lesson by stopping his bill (Hannah 1979: 47). After significant delay, the bill was eventually lost when the government fell.

A different approach developed from the short-lived enthusiasm for state intervention during the First World War. A report recommending radical action led the government to propose an Electricity (Supply) Bill in 1919 and the

> core of the government proposals lay in the District Boards, which would enable them to acquire generating stations and main transmission lines with or without the consent of their owners. (Hannah 1979: 71)

So coercion was finally proposed by the state – and scotched in Parliament, as a consensus emerged that 'government intervention had gone too far, and that with a changed atmosphere of cooperation the industry would be able to put its own house in order' (Hannah 1979: 73). Ironically, the onset of the Russian Revolution had increased business fears of these 'socialistic' proposals (Hannah 1979: 73) so that it could be said that fear of an anti-capitalist movement had helped block a proposal to restore the dynamic of capitalism.

The reform effort that would eventually prove successful finally began in 1924 when Prime Minister Stanley Baldwin appointed Lord Weir, a known enthusiast for electricity reform, to head the Ministry of Transport and to review policy options. In 1925 Lord Weir's committee duly recommended the creation of a national-owned Central Electricity Board (CEB) with coercive powers to rationalise the existing industry. Baldwin enthusiastically backed the bill and in 1926 it became law.

Why had reform worked this time? Hannah describes the 'real breakthrough' as the Weir report's avoidance of a recommendation for nationalisation: generators and distribution remained with the existing undertakings (Hannah 1979: 93) and so business fears of 'socialism' were largely neutralised. Political power would nevertheless lie with the new CEB; it would coordinate the planning of new generators, manage the operation of existing stock, be able compulsorily to close inefficient plant, and it would build and own a regional interconnection grid. All the 'blocking power' of fragmented and protected private ownership had been removed without the pain of forcible nationalisation.

The new organisation of the CEB finally allowed London – and Britain – to catch up with best practice AC supply in the rest of the industrialised world by the end of the 1930s (Hannah 1979: 148). Successful reform of the restrictive powers of local authorities had come some 30–35 years after the economic advantages of large-scale AC supply had been demonstrated in the early 1890s.

Cheap AC electricity supply was a transformational technology. One should expect to find the economic consequences of delay in the development of allied user and supplier industries. It was no accident that the development of electrolytic chemical technologies at this time was in the north east of England, the one region where Merz had established economies of scale in AC supply. Nor should it be surprising that Britain never established a lead in electrical equipment supply while home market development depended on ending this political gridlock. Electrical equipment manufacturers in Germany and the USA established economies of scale in production as part of the growth of economies of scale in AC power generation and supply. The British electrical equipment manufacturers faced not only halting supplies of finance (see Chapter 6) but also a backward, underdeveloped home market. Their retarded development in the crucial period where increasing returns to scale were obtainable meant the likelihood of their being permanently locked into a subordinate international position in the industry which symbolised modernity in the first half of the twentieth century.

Comments on the Political Process of AC Supply Implementation

What makes the British example special is the accidental creation of a powerful, conservative blocking interest through the 1882 Act of Parliament. Once the economies of AC supply were understood, the political story becomes one of repeated attempts to reverse this step down a dead-end path of development. The sources tell us that strong political actors in favour of reform were absent, but one cannot help but observe that the failure to develop the technology was itself a reason why a strong private interest in development failed to form. It is significant that it was Merz, the one actor to have relevant technological experience, who would mount the most serious private reform attempt. In a sense we can judge the political significance of the absent actors by the unrealised potential economies of the new technology – these latter were insufficient to induce coordinated political change.

So the reform of British electricity supply could not be a story of the disinterested management of a policy 'community'. Reform only occurred because Stanley Baldwin sought to engineer the outcome. The minister he appointed knew what result was required, but was prepared to choose as a means to his end whatever provoked least resistance in Parliament. There can be little doubt that it was the unrealised economies of scale that acted as the lure in the government's choice of this more coercive path. One might call it 'a path of state leadership'. We can conclude that when an outcome is sufficiently well established as socially beneficial, the policy process itself will be bent to achieve the desired end.

This case strongly supports the idea that the state *should* be prepared to adapt the policy process to achieve such desirable ends. It is a big intellectual and practical step to adopt this as a general principle. It makes the process by which a reformist end is decided to be desirable very important indeed. The plutonium reprocessing case demonstrates what can happen when the state convinces itself that intervention is necessary and beneficial, and must be brought about *somehow*. And the general issue of the state's willingness to intervene in the technological *status quo* is reviewed in the last major section of this chapter.

This chapter – and indeed this book – have been about changes that advantage technologies of potential or actual social benefit. This is largely because we are concerned with technologies developed by the private sector. Whilst even the private sector on occasion develops technologies with disastrous social effects – one thinks, for example, of the thalidomide disaster – it is amongst the class of technologies developed by the state that we can find the best examples of socio-economic waste and hazard. The story of the development of plutonium reprocessing is an excellent example of how a technology that the state initiated, for national strategic reasons, acquired a life of its own so that it became uncontrollable.

Plutonium Reprocessing – a Technological Mission Out of Control?

The state-initiated technological 'mission' is some large-scale exotic technology project that the private sector either will not or cannot fund or manage. The mission has been the mode of development of much military, space and nuclear technology, technologies developed for 'national–strategic' rather than economic reasons. Since the spectacular success of missions like the Manhattan Project to develop atomic weapons in the Second World War, the military victors of that war have been convinced of the decisive military advantage to be gained from planned scientific and technological advance – above all the USA, where defence and space R&D accounted for at least 60% of US federal R&D in the 1960s and 1970s and as much as 80% in the 1980s (Mowery and Rosenberg 1989). Much of this R&D is subcontracted to private defence companies. Such unprecedented and successful military technology development underpins the US role today as lone superpower.

Missions often involve the creation of a sustaining organisation dedicated to the fulfilment of the development goals. This organisation's continued existence is obviously dependent on the continuance of its development mission, and should the desirability of that mission change, political problems of control may occur. Walker's account of British government decisions to build and then operate the British plutonium reprocessing plant (THORP, standing for Thermal Oxide Reprocessing Plant) is an excellent analysis of the problems of control of an organisation, British Nuclear Fuels (BNFL), that remained dedicated to the fulfilment of its mission regardless of the changing rationale for that mission (Walker 1999).

The Technological Vision of the Plutonium Economy

The origin of THORP is in the oil crises of the 1970s and the sudden desire of many industrial countries to obtain *secure* energy supplies, independent of Middle East states and oil cartels. The nuclear industry seized the opportunity to offer its vision of secure future energy supplies based upon the realisation of an independent 'plutonium economy'. The technological future of the plutonium economy deserves a little explanation.

The neutron flux in the core of a normal, uranium-burning reactor converts a portion of the uranium in the fuel rods into plutonium. Like uranium, plutonium is an element capable of fission. So this plutonium can be recovered from the spent fuel rods through chemical 'reprocessing' and then in theory this plutonium could be burnt in a new design of fission reactor, the 'fast breeder' reactor (FBR), to produce electricity. 'Fast breeder' reactors are so called because they can be operated to generate more plutonium than they burn: the core of an FBR can be surrounded by uranium that will partly convert to plutonium through the capture of neutrons emitted by the core. The technological vision was that once a country owned a network of FBRs *and* a plutonium reprocessing industry it would be able to manufacture its own plutonium *while it produced electricity* – and so it would possess security of energy supply.

The technology was large scale, capital intensive, novel and high risk with major decision points divided by decades. It would only be worthwhile if the 1970s' assumptions of high future energy prices and political vulnerability to the OPEC oil cartel remained valid. Nevertheless Britain, but especially Japan, France and Germany, were prepared to develop this technological vision for the sake of security of energy supply.

The plutonium economy required development of both the novel FBRs and the plutonium reprocessing plant. In Britain the state-owned BNFL was already involved in plutonium reprocessing at its Sellafield site and was a keen advocate of a new large-scale plutonium reprocessing plant to support the future plutonium economy.

The Decision to Construct THORP and the Subsequent Loss of an Operational Rationale

This decision to construct the plant was first taken in 'private' in government and apparently backed by the British Cabinet (Walker 1999: 14). Untimely disclosure of leaks of radioactivity from the BNFL Sellafield site then persuaded the government to allow a public inquiry, the 'Windscale inquiry', that would run for 100 days in 1977. Walker comments on the chairman of the inquiry's final report:

> Reading the report today, twenty years after it was prepared, it is more apparent than ever that Justice Parker and his advisers set out, for whatever reasons, to construct a case for giving unequivocal consent to the proposal that BNFL had submitted. (Walker 1999: 17)

This was largely achieved by, first, the acceptance of BNFL's projections for the scale and imminence of future plutonium demand and, second, the inquiry being framed to exclude consideration of a 'mixed' strategy of both reprocessing and the major alternative technology to reprocessing, the dry storage of used uranium fuel rods.

At the time of approval there already existed a reprocessing plant for British Magnox reactor fuel and there was already a growing stockpile of plutonium, so by BNFL's own forecasts, THORP would only be necessary to guarantee plutonium supplies after the *eighth* FBR was constructed (Walker 1999: 18). If a lower level of plutonium demand had been accepted and/or a mixed strategy considered, there would have been no case for THORP because the 1970s' level of reprocessing could have supplied the plutonium for a small FBR programme. Walker writes that although the actual outcome of no FBR construction was not strictly foreseeable in the 1970s, even then the scale of the FBR building programme imagined by BNFL was understood to be exaggerated by scientists and engineers outside of the nuclear industry (Walker 1999: 18–19). Nevertheless, the inquiry approved BNFL's proposal for THORP, construction began in 1985 and in 1991 the plant was complete and awaiting approval to operate.

By 1991, the 1978 forecasts for plutonium demand were patently false. FBR development programmes were in trouble everywhere: Britain had abandoned its FBR program in 1990 and built none; Germany abandoned its program in the mid-1980s; France was struggling to operate its prototypes safely (Walker 1999: 32). Without FBRs, there could be no commercial demand for plutonium. In addition, oil prices had dropped far from their 1970s' peaks, new sources of oil and gas within industrial countries' territory had been discovered (the North Sea) and the fear of being held hostage by an oil cartel had receded.

So not only was there no demand for plutonium, but it was increasingly unlikely that there ever would be. Nevertheless, ever larger stocks of plutonium were accumulating outside the existing reprocessing plants to add to the stocks derived from decommissioned nuclear weapons. In these circumstances, to allow the operation of THORP would be to incur large decommissioning costs of the order of billions of pounds (the plant would immediately become contaminated by plutonium) and to add to the burden of radioactive waste discharged in to the environment. And all these costs would be incurred in order to produce a product already in oversupply and without any prospect of future demand. Indeed, as a stockpiled 'waste' plutonium was a security risk because of the danger of theft and terrorist nuclear bomb manufacture. Nevertheless, when BNFL, faithfully pushing its favoured technology project, wanted to begin operation of the completed THORP plant, the British government would grant permission to operate in 1994 (Walker 1999: 1).

Entrapment – Walker's Analysis of the Policy Process for Approving the Operation of THORP

The government created an organisational vehicle to consider the authorisation of operation decision – the Interdepartmental Government Committee (IGC). With the background outlined above, it at first appears incredible that the IGC would approve operation. However, this committee was not created to conduct a full review of whether the plant was needed. Its remit was limited to ensuring that operation would breach neither European nor British environmental regulations and that THORP would bring some benefit to Britain (Walker 1999: 79). It is therefore understandable that Walker speaks of the 'tyranny' of the original 1978 decision to build the plant: 'once taken, it seemed fixed and beyond further consideration or challenge' (Walker 1999: 141).

Given the real costs of the technology, the criterion that the operation of THORP should bring some benefit to Britain is interesting. The IGC obtained a cost–benefit analysis from the consultants Touche Ross that found THORP would have large positive financial benefits of £1.8 billion for BNFL and £1 billion for Britain at a discount rate of 8% (Walker 1999: 87). Walker points out that this calculation ignored the interests of foreign utilities that were locked in to THORP by 'binding', cost-plus contracts – the German and Japanese utilities were liable for most of the cost of constructing, operating and even decommissioning THORP. In other words, THORP was neither economic nor useful, but other countries, not Britain, would incur the measurable and massive losses.

The cost–benefit calculation also put a low value on the option of *not operating* THORP, by assuming that if these contracts were broken by *Britain*, it would be liable for the massive financial penalties stipulated in the contracts. Touche Ross did not consider the option of 'collective withdrawal' by all parties to the contracts, although collective withdrawal was certainly a prerequisite for 'non-operation', given the nature of these BNFL–utility contracts.

According to Walker the larger problem was that the government in 1992–4 was seeking to justify continued engagement with the development of the technology. This is demonstrated by what the government did *not* do, so that as Walker comments:

> As far as I can discover, no attempt was made to weigh the likelihood of achieving the outcomes that must follow THORP's operation [for example, the approval of waste disposal sites]; no studies were carried out by the British Government of the situations in Germany or Japan, nor of the political realism of transportation [of plutonium] and of MOX [mixed plutonium oxide] recycling proposals; there were no consultations on feasibility with the German or Japanese Governments; and there were no assessments of the political and administrative costs attached to future problem-solving. (Walker 1999: 93)

All the way through this story the DTI, with theoretical authority over the state-owned BNFL, chose instead to be the representative and champion of BNFL's mission within government. This excessively close relationship was one of the chief sources of what Walker calls the government's *entrapment* in this technology. Other contributions to entrapment were the failure to develop the alternative technology of dry storage – so that policy options in 1992–4 were reduced – and the adherence to the international political principle of non-interference in the internal affairs of other sovereign nations; the latter inhibited unilateral withdrawal by any of the international partners, despite the appalling economics and social costs of plant operation (Walker 1999: 132–6).

With the DTI's capture by BNFL, there was no other body with the expertise or *responsibility* to review critically or fully the risks that the technology of THORP represented. Other government departments would be wary of stepping onto the DTI policy 'turf', for the precedent it might establish. Naturally, many of Walker's suggestions for future technology policy concern the creation and use of independent project review ability, whether within government, or outside, perhaps in the form of a standing body (Walker 1999: 149–50). Had such a standing body existed, one of the more critical branches of government such as the Treasury would at least have had the option of commissioning a full and independent review of the technology.

Independent Review Bodies and the Management of Technology Missions

The technologies of plutonium processing and AC electricity supply are polar opposites in terms of public and private welfare, yet there are common elements to the policy problem in each case. In comparison with the cases involving the management of policy communities, in both these cases the state was essentially alone in facing the policy problem. In both cases there was an undesirable outcome that posed the problem of how the state could legitimate a corrective policy intervention in the name of the public interest. In neither case could appeal be made to some normative policy procedure to remedy the problem.

Although some British MPs feared entrapment in THORP and suggested the possibility of instituting review in the original 1978 parliamentary debate, nothing was done and Britain has yet to experiment with an independent science and technology review institution. Nor is the call for such a capability in Britain new – it is one of Henderson's chief recommendations in his 1977 review of the state policy decisions that generated the economic disasters of the Anglo-French development of the supersonic aircraft Concorde and the British Advanced Gas-cooled Reactor (Henderson 1977: 193). Henderson's analysis of the gestation of these technologies has much in common with Walker's analysis of the THORP decisions: secret government decision processes with strictly limited organisational participation sustained these projects, even more so than with THORP. In the Concorde case we even have binding contracts that played a similar role to the contracts in the THORP case; only this time, it was the British Labour government of 1964 that found that there was no provision for unilateral withdrawal in the contracts, signed two years previously by its predecessor government (Henderson 1977: 162). The Labour government felt it had no choice but to continue to pour in billions of pounds of development money with no prospect of economic return.[16]

Independent review bodies of the type that Walker advocates for Britain have been created in other countries, and in response to similar problems of technology and science mission control as outlined here. Two examples are the US Congress for Technology Assessment, founded in 1974 (but later abolished for political reasons), and the European Parliament's STOA (Scientific and Technological Options)[17] program begun in 1987 (Ford and Lake 1991: 42). Both were intended to enhance the respective elected chambers' ability to scrutinise science and technology policy issues.

The value of STOA was shown by one of its first acts, the critical appraisal of the economic viability of the European fusion research program (Lake 1992). This was one of a number of inherited, drifting technology missions from the idealistic, early days of the European Union, when it was thought that mutual cooperation on science and technology might cement the union. Instead, it has generated weakly controlled programs of work that persist through decade after decade with, in the fusion case, no change in the prospect for a workable, let alone economic, technology. The common problem with the collectively controlled European Union science and technology programmes appears to be that no one state has responsibility for either their full cost or their direction.

The problem of uncontrollable mission drift is epitomised by the European Joint Research Centre (JRC). It is telling that this was originally the European Joint *Nuclear* Research Centre, and that with the failure of its fission reactor designs to be adopted, rather than a strong decision to close the centre, it was given a new, civil research mission – and name (Ford and Lake 1991). A 1987 European parliamentary appraisal of the 'new' JRC found that at the main research site of Ispra in Italy, 700 of the 1600 personnel were said to be research staff; of these most were technicians, leaving 250 with academic qualifications.

> The actual scientific life of the centre is the business of 30 to 40 people. They are faced in the programme directorate with 40 people including 20 with academic qualifications. (European Parliamentary Document A2-174/87 cited in Ford and Lake 1991: 40)

Once again the European Commission passed over the chance to close the centre in favour of another 'reorganisation' of the JRC. Ford and Lake commented that despite the potential for the JRC to be revitalised,

> the main barrier to this is probably cultural rather than organisational. It will take remarkable management skills to 'turn around' the listlessness, apathy, lack of direction and lack of conviction which have characterised too much of the JRC's history. (Ford and Lake 1991: 40)

With STOA the European Parliament has a powerful investigative tool to pursue like abuses of the public purse.[18]

The Historical and Equivocal Interest of the State in Technological Change

A theme that is emerging from the examples developed so far is that the state has an equivocal interest in technological change. Where technological change involves the overriding of interests hostile to change, the temptation is variously: to capitulate (professional engineering reform); to buck the issue (vocational training reform); to do nothing (THORP, and for several decades also AC electricity supply). This brutal summary does not imply that the choice of a course of action was either always obvious or easy to take: the detailed analysis of the examples shows otherwise. It does serve to raise the radical question: why should the state take on the burden of the 'public interest' in these matters?

The answer to this question will show that the expectation that the state take on such responsibilities is a modern phenomenon, but one intimately involved in the evolution of the state itself. If we take the long historical view, we can consider ourselves lucky that the state is prepared to struggle with the political problems of technology management. That does not mean that its ability to do so can be taken for granted. On the contrary, with such suggestions as the institution of independent technology review, we are continuing to live and make the history of innovation in the state's capacity to manage new forms of technological change.

Two Examples of Successful, State Suppression of 'Useful' Technology

An example of a state hostile to a technology and intent on its reversal, rather than its advance, is given by the Tokugawa Shogunate of Japan. The Tokugawa regime was not only hostile in intent, but it succeeded in suppressing the European technology of guns in Japan. How this was achieved and what conditions made this possible are discussed below.

The Portuguese introduced European guns to Japan in the middle of the sixteenth century. They proved indispensable in the warfare that led to the unification of the country under the Tokugawa Shogunate. By the end of the sixteenth century, Japanese gunsmiths had produced so many guns that there were 'almost certainly' more guns in Japan than in any other country in the world (Perrin 1979: 25).

With so many guns in circulation there was no question of an immediate and outright ban on their production and use. But with political unification, the Tokugawa Shogunate took steps to concentrate gunsmiths and gun manufacture in controllable geographical locations. State orders for guns were gradually reduced and permanent annual salaries introduced for the surplus gunsmiths in order to remove the necessity of their having to work for money (Perrin 1979: 62). Although small numbers of guns continued to be produced, after the last significant military engagement in 1637, the major use was in ceremonial processions (Perrin 1979: 63). Technological development of guns altogether ceased, so although the Japanese became aware of the advance of the flintlock, they never introduced it into use. There would now follow two centuries of technological stasis in gun technology.

Guns were suppressed so that the Japanese samurai could indulge their cultural and aesthetic preference for the sword. There was also a general Japanese reaction against things foreign that included the gun at the end of the civil war period.

This cultural choice was evidently only *possible* because of the achievement of political unification with its concomitant internal peace and state monopoly on weapons production and use. The choice of technology suppression could have been challenged by foreign intervention, but the country's relative isolation and, ironically, its reputation for waging effective war, meant there were no foreign attempts to intervene in Japanese affairs until the mid-nineteenth century (Perrin 1979: 35). Just as internal factional war had enforced adoption of the gun and the perfection of tactics based on its use, so a monopoly of political control *allowed* the expression of values hostile to its use. There is no reason to think that the suppression of gun technology would have reversed through any internal Japanese political dynamic – when change came it would be in reaction to foreign intervention in Japanese political affairs from the mid-nineteenth century onwards.

Fourteenth-century Sung-dynasty China offers several examples of indigenous technological development that rapidly outpaced the level of European technology development, but subsequently stalled, allowing Europe to become the leader in technology development. One indicator of development is total iron output. This had reached 150 000 tons per annum in China as early as the end of the eleventh century – a similar level to that achieved in Europe 600 years later, by the year 1700, yet per capita Chinese output was higher than European output for these dates (Jones 1987: 202). China had also invented printing, gunpowder and the magnetic compass by the end of the eleventh century, and as Basalla comments, these were the three inventions that the English philosopher Sir Francis Bacon

> identified as the source of great changes in Renaissance Europe ... responsible for revolutionising literature, warfare, and navigation. If these discoveries were of monumental importance in the making of the modern Western world, why did they not exert a similar influence in China? There is no wholly satisfactory answer to this question; and the search for an explanation will take us into an exploration of the cultural values of the Chinese elite. (Basalla 1988: 170)

Whereas the fate of these three technologies in China was relative stagnation compared with Europe, the fate of Chinese navigation after 1430 offers another impressive example of the state reversing technological development.

Before 1430 Chinese maritime trade with faraway East Africa had been regular and fleets had even visited Kamchatka and Zanzibar (Jones 1987: 203). The benefits of such trade were not obviously great, while the Chinese state was very concerned with illegal attempts by the Japanese to trade through Chinese ports and to engage in piracy and smuggling with Chinese coastal settlements (Jones 1987: 205). The Chinese banned maritime trade in an attempt to control these illegal activities because the existence of shipping was easier to monitor than its usage. Chinese merchants soon stopped building large vessels to avoid government accusations that they engaged in long-distance trade. The strategy of suppression of maritime trade was obviously at risk as long as there remained written records of the achievements of the trading period: but in 1480 the records of previous voyages were deliberately destroyed by a government faction intent on the strangulation of an internal proposal to use the records to recommence voyages. Now it became possible that 'by 1553 it was admitted that the art of building large ships had been forgotten' (Jones 1987: 205).

These examples of technological halt and reversal establish a lesson valuable in itself – that technological 'progress' is not inevitable, even when possible and for some social factions desirable: the discovery or invention of 'advanced' artefact forms does not in itself ensure their subsequent diffusion and continued development. At least in principle, technologies that are thought to be socially undesirable, even if superior in some manner of use, may be controllable.[19] Yet what would bring these experiments in suppression to an end would be foreign intervention. It is no coincidence that it was in Europe, where foreign intervention was a permanent threat to the viability of the constituent states, that technological development would outpace that in China and Japan.

The European States System – Learning to Promote Technological Change

Jones gives the European 'states system' great weight in his comparative review of explanations for the economic and technological rise of Europe (Jones 1987: 107). Technologies that enhanced the power of the state could not be suppressed for long, but, more importantly, European states in proximity and rivalry to one another gradually acquired the habit of active promotion of technological change. In Jones' review, this general search for useful technological change within the states system fostered that associated phenomenon unique to Europe: the nation state, with its elaborate internal institutions and services for its population.

Over the time span of the last 500 years, then, the history of European national attempts to advance technological change is not fully separable from the evolution of the nation state itself – and modern history in general, as European states fought one another to control access to the resources of the New World.[20]

> The general explanation of change lies at the intersection of technological change, increasing market size and the ambitions of a system of nation-states. To review this from another angle, the European experience is not properly captured by a simple opposition of 'mercantilism' and laissez-faire, nor by the phrase 'the rise of capitalism'. This is not just because these terms are vague. It is because the rise of the nation state and its programme of services was equally vital.[21] (Jones 1987: 149)

This kind of long-range historical review shows that the search for technological change is not an obvious and enduring good in itself. The conditions for its pursuit depend on the establishment of the institutional fabric of the nation state, and this includes functions often taken for granted, such as law and order. In sum, it is an acquired value within the social context of a system of state rivalry.

This gives us an opportunity to develop a new perspective on this chapter's detailed examples of state policy interventions. First, it makes sense to develop further our idea of how the state learns to promote internal technological change. An easy way to explore the nature of the state's incentive to promote technological change is through the dramatic military–technological shocks that in recent modernity served as the catalyst for the destruction of pre-modern regimes.

Military Shocks to the State and the Equivocal Valuation of Internal Technological Development

It was when US gunboats sought to enforce the opening of trade with Japan through a demonstration of superior weapon technology in 1853 that the Japanese elites with a shock realised that they were powerless to resist. This began a process of experimentation in technological catch-up with the west that was highly contentious internally to the elites. A turning point came in the Satsuma Rebellion of 1877, when conservative elements of the samurai class chose to take up arms against those who had changed the course of their society's development (Perrin 1979: 73). The devastating defeat of the rebellion left the Japanese elite fully committed to technological catch-up with the west.

Even in the Japanese case it is clear that the shock of external military force did not deterministically result in a full commitment to technological catch-up. Other states had even more wrenching and incomplete experiences of internal change in response to external intervention. Pipes nicely describes the shock experienced by the old Russian Tsarist regime of the seventeenth century:

> The reason why the Russian monarchy found it necessary to tamper with the closed and self-perpetuating system which had cost it so much trouble to establish has mainly to do with Russia's relations to western Europe ... she was the first (country) to become aware of the inadequacies of her rigid, regulated system when confronted – especially on the field of battle – with the more flexible and 'scientifically' managed institutions of the west. Russia was the earliest of the non-western countries to undergo that crisis of self-confidence which other non-western peoples have experienced since; a crisis caused by the realisation that inferior and odious as it may appear, western civilisation had discovered the secrets of power and wealth which one had to acquire if one wished successfully to compete with it. (Pipes 1990: 112)

The modernisation of Russian society forced by Peter the Great was incomplete and with the break in the Tsarist dynasty after his death, it would never be driven with such single-minded ferocity again, at least under the Tsarist system.

The scale of internal change, the associated power struggles and breaks in the pace of development are naturally a subject of such countries' modern history. That does not mean that all modern history seeks to explain and compare national stories of progress and development; too often descriptive narrative predominates in which what is the case is assumed to have been inevitable. Anderson's work represents an exemplar of the comparative historical method deployed in search of answers to the question: why did these states develop (Anderson 1992)? One of the valuable insights he offers is that the supposed vanguard role of the middle class in industrialisation is significant by its absence – another myth – and instead it is to the internal coherence and belief of a society's elite that we must look for the determination to initiate industrialisation (see especially Anderson 1976). Yet in his work too, it is clear that it was rarely the case that an elite was able to make a collective and coherent choice in favour of industrialisation; rather the process involved complex internal political struggle over discrete reforms, it was characterised by stops and starts and compromise and it sometimes broke down into political revolution and violence.

Even within Europe where the technological and institutional gap between states was far less, the desire to possess the superior technology of another state did not lead automatically and comfortably to possession of that technology. Landes' account of Prussian and French reaction to the British ascendancy in textile and other technologies in the eighteenth and nineteenth centuries conveys the persistence and experimentation necessary to transfer another country's advanced technique within the states system.

Continental Emulation – Prussian and French Imitation of British Technological Achievement

The striking feature of Landes' account is how Prussia and France moved from artefact transfer to social and institutional experiments as a means of obtaining control over technology. As it became suspected that Britain possessed leading-edge technology, the first phase of continental reaction was to inspect and monitor the symbols of the technology gap: the leading-edge artefacts in production. From the middle of the eighteenth century European government representatives and business people began to come to Britain on tours of inspection to learn about superior British industrial technique – until the British government realised that the country had a technological advantage and moved to discourage such visits (Landes 1969: 125). Given that European states had already learnt that it was important to close such gaps, these observations established copycat industrialisation as a political imperative. Even for these most technologically advanced countries of the time, several human generations would pass before some of the continental countries could close the gap. Landes makes the point well:

> Why the delay? Surely the hardest task would seem to have been the original creative acts that produced coke smelting, the mule and the steam engine. In view of the enormous economic superiority of these innovations, one would expect the rest to have followed automatically. To understand why it did not – why even the quickest nations marked time until the third and

fourth decades of the l9thc – is to understand not only a good part of the history of those countries, but also something of the problem of economic development in general. (Landes 1969: 126)

Landes' analysis of the blocks to technology transfer that these countries experienced ranges over the elements of the technology complex.

It includes the more straightforward physical aspects such as relatively low population density and poor internal transport, but also political fragmentation, which implied the fragmentation of markets. Germany was then 'a patchwork of kingdoms, archduchies, duchies, bishoprics, principalities, free cities and other forms of sovereignty, each with its own laws, courts, coinage and above all, customs barriers' (Landes 1969: 127) and even France, unified politically, remained divided into distinct trade zones based on the constituent, once sovereign states. There was a more unequal distribution of wealth than in Britain and the mass of the continental population was almost outside the money economy (Landes 1969: 127). Associated with this and at the cultural end of the complex was a predisposition on the part of the continental elites to believe that luxury consumption promoted economic wealth.

The task of closing the technological gap was therefore as much a job of social and institutional reform as of importing the skills and artefacts of the British Industrial Revolution. Of course, the import of advanced artefacts and skilled personnel was an early and important part of the acquisition of British technological abilities: there were at least 2000 skilled British workers on the continent by 1825 (Landes 1969: 148). A familiar set of state policies were directed at this level of technology transfer: direct subsidy, patents and low interest rate loans all sought to ameliorate the lack of a market incentive for such investments.

It was as these states successfully transferred production artefacts and skills and yet observed that they were not transferring the technology or closing the technology gap with Britain, that they embarked on the more difficult reforms that challenged tangled internal interests.

So the pre-revolutionary French state tended to enforce the authority of the medieval guilds over production because it relied on the guild organisation to collect taxes. It would always remain ambivalent about freeing production from the guilds and only after the revolution was the guild system abolished with the Loi Le Chapelier of 1791 (Landes 1969: 145). A final example, of obstructive social structure, concerns the excessive social prestige of the aristocracy in France that tended to drain capital out of trade as successful businessmen sought to purchase land and the other trappings associated with the aristocracy. Even the pre-revolutionary French state had attempted, not very successfully, to stem this flow by making patents of nobility conditional on a family continuing in the line of business for which it won its honour (Landes 1969: 130).

Yet even this is insufficient description, for the continental states' determined focus on the closure of the technology gap with Britain was not only a matter of clearing away social obstructions and outdated practices – and this was no simple matter – but also the generation of institutional innovation. Nineteenth-century technical education institutes were originally introduced and understood as a 'compensation for a handicap', the handicap being the absence of best practice skills (Landes 1969: 151). Later they would be perceived to be an asset in their own right. The industrial investment bank was another innovation intended to solve the characteristic problem of an insufficiency of capital supply in the technologically backward countries, compared with Britain (Landes 1969: 206–9).

Continental states had been forced to greater and greater degrees of social engineering to close the gap with Britain, but by 1870:

> As a result of a generation of drastic institutional changes and selective investment, the nations of western Europe now had the knowledge and means to compete with Britain *in certain areas* on an even plane. (Landes 1969: 230)

Today the most obvious evidence of this period of struggle is the continued existence of distinctive institutions such as industrial investment banking and technical training.

The thrust of this material is that a state sufficiently determined to acquire the useful technologies developed in other countries will directly challenge internal social interests where necessary and be innovative with regard to new institutional means of promotion. It should not be a surprise to learn that such a powerful and historic shaping force should remain active today. To discuss it further, and for the first time in this book, it seems necessary to invent an 'ism' – 'techno-nationalism' – as a convenient shorthand representation of this value: the determination *by reference to practice in other countries* to mould society to enable technology adoption and promotion.

The British Technological Reform Process Reconsidered – a Need for the Benign Form of Liberal Techno-nationalism

Techno-nationalism offers fresh perspective on the halting examples of British reform described in this chapter. Although the example of foreign technological practice appeared as a source of motivation for the British reforms, such ideas were prone to subversion by the preferred process of administering the reform effort.

There were striking similarities between the analyses of British administrative style by Henderson, Walker, Jordan and Wolf. They tended to stress: its secretive nature; the limited number of included organisations and the sometimes dubious basis for inclusion; the scrupulous regard for consultative procedure at the expense of ideas and therefore outcomes. In sum, the administration of the reform efforts quickly became closed and internally orientated, with the result that the *status quo* tended to be preserved. The 'techno-nationalist' imperative described above was notable by its absence.

It was said that the state's determination to promote technological change was an acquired value within the social context of a system of state rivalry. In the expression of this idea there is clearly nothing about state determination that is eternal or automatic. Not only may determination decay or perhaps warp into a damaging form (THORP), but, on the contrary, we have had enough evidence of what internal conservative forces can achieve that we might *expect* the determination of the state to tire and to falter over time. With regard to the cases presented, this appears to be what has happened in Britain.

The way in which I defined techno-nationalism insists on judging the techno-logical, institutional *status quo* of our society by reference to practice elsewhere. From a British techno-nationalist perspective, the examples of reform in this chapter are a disgraceful series of bungles; and now the language is appropriate, for we judge the quality of the reform by its failure to achieve desirable goals.

It should go without saying that a techno-nationalist position is not a crude pro-technology position – one only has to think of THORP, Concorde and the British AGR programmes. Special technological interest groups with a history of state dependency, such as the aero-industry and the nuclear industry, are often quick to claim their latest project is vital to the national interest, to increase the likelihood of state funding. Only fools would take such self-interested claims at face value. An unfortunate side effect of British state support of these industries may be that techno-nationalism has become associated with the administrative device of the technological mission.

On the contrary, a techno-nationalist position can be a strong liberal position. Friedrich Hayek, that great defender of the liberal economic order, has argued passionately that a vigorous liberal state is one that challenges internal conser-vative forces in defence of progressive change (see the chapter 'Why I am not a Conservative' in Hayek 1960). The most obvious form of progressive economic change is technological, and so internal institutional reform in pursuit of technological capacity may be both a liberal and a techno-nationalist policy. All of the British reform efforts in this chapter fell into this category of liberal techno-nationalist reform; they were liberal economic reforms that were legiti-mated by reasoned comparison with overseas practice and that required the overriding of internal interest groups for their enactment. With the failure of most of these reform efforts a classification of the British state suggests itself – a laissez-faire liberal state rather than a strong liberal state.

Contemporary Rivalry-driven Techno-nationalism and Political Ambivalence towards Associated Social Change

The liberal form of techno-nationalism is essentially benign, unless one is a cultural conservative. It is a short expression of the idea that reference to foreign technological practice should be used to stiffen the liberal economic reform process.

When the development of desirable foreign technologies suggests that the very structure of national sovereignty itself should be 'reformed', it might seem inappropriate to use the term 'techno-*nationalism*'. Yet this is only because we are in the habit of taking the national unit for granted. In the historic examples the creation of the national unit and effective possession of technological ability were entwined. Techno-nationalism continues to play an important role in the renewal of the European states' efforts to strengthen collective political and economic institutions.

A perception that there was a growing technological gap between Europe on the one hand and Japan and the USA on the other helped generate the collective political will of the European states to strengthen the powers of the collective institutions of the European Union in the 1980s. The renewal of European Union institutions began with the passing of the Single European Act of 1987. With this Act, European Union member states collectively agreed to reduce individual states' ability to veto collectively binding policies in the Council of Ministers (Sharp 1991). Major collective policy initiatives that promised net collective gain, but with the gain spread unevenly between member states, had proven vulnerable to the veto. With the limitation of the veto, such policy initiatives became worthwhile once more.

Sharp gave the European telecommunications industry as an example of the kind of technological weakness that motivated the renewal of collective European institutions. In the 1980s there were six European 'national champion' companies that developed digital switching equipment to modernise their analogue telephone exchanges (Sharp 1991). However, the cost of R&D for this technology was so high that it necessarily raised the price per unit unless it could be spread over the entire European or world market. Since there were six rival companies and projects in Europe, each subsidised and cosseted by its respective government, not one of them could win the entire European market. Without a large-scale European market, each 'European' digital exchange could expect to be higher priced in the non-European markets than the equivalent technology developed by the single Japanese (NTT) or US (AT&T) developer (Sharp 1991). For 'Europe' to compete with the USA and Japan in similar technologies the lesson was clear: Europe needed a genuine European market and an end to national protection of 'technology champion' companies within that market.

The consolidation of a single European market 'space' on the scale and with the institutional qualities of the US market has proceeded apace since that time, with notable steps being: the reduction of European customs barriers in 1992; the Schengen agreement on the free movement of people within the European economic area; the single European currency, the euro, introduced in 2002.

The relinquishment of national control over national champions has been a much more hesitant and political business. The litmus test of the European states' determination to support a genuine European market in high-technology industries is whether they are willing to allow their national technology champions to fail in that market – and by some failing, as they must, others gain the chance to become the pan-European, coordinating technology developers that Europe needs to rival the USA and Japan.

By 2003 most of the European telecoms companies had acquired life-threatening debts as a result of rash expansion during the 2000 Internet bubble. Now, if ever, was the time for some rationalisation of the former national champions. The British and German states had privatised their respective national telecoms companies and these had had to resort to private capital markets for debt restructuring solutions. However, the French state remained majority shareholder in France Telecom and the French government confirmed in December 2002 that through a special state holding company it would 'lend' the massive sum of 9 billion euros to France Telecom to help it with its 70 billion euro debt (BBC 2002). This would ensure the company would not default on a scheduled 15 billion euro debt repayment. The European Commission has a mandate to regulate state aid and in a matter of months it launched a formal investigation into the status of the French state's 'loan'. Even if the loan is ruled illegal and fines imposed and extracted, it is clear that when the necessary crisis approaches, the French state is not willing to stomach the loss of French national control over what is – now it is clear enough – still very much a French national technology champion company. The signal is obviously very damaging to the idea of a common European market and promotes the thought: why should other states relinquish control of their national champions only to see the subsidised French champions take their markets and become the pan-European developer companies? (See TV Broadcasting Standards on page 81 for another example of French national champion favouritism.)

If the hopes are that pan-European developer companies will appear, the fears may have a better press and concern the loss of national state discretion over economy and society. The *tension* between these hopes and fears is the familiar tension of modernity, between the desire to acquire greater power, wealth and national security through the acquisition of new technologies and the wish to preserve existing institutions and avoid social dislocation. In Europe we are 'living the history' of this process and it is not clear how far it will be either encouraged or allowed to progress.

If in this European example techno-nationalism provided an incentive to build market institutions, in certain circumstances techno-nationalism can provide an incentive to suspend the free operation of the market. An outstanding example of the latter is the adaptation of trade policy by the US federal government to hinder foreign – especially Japanese – high-technology exports into the US market.

With the rise in Japanese high-technology exports to the USA in the 1980s the USA began to more actively manage trading outcomes. Some spectacular Japanese technology development projects, such as the design of VLSI semiconductor chip technology, generated trade friction in their own right (Anchordoguy 1989: 143). High-technology industries like semiconductors have never been 'free' of government support and if support can bring forward development in technologies with increasing returns then there is a prospect of an early and permanent lock-out of foreign competition – a winner-takes-all prospect. With the VLSI chip technology, the Japanese ran a more efficient design project and produced a better commercial product earlier than the USA,[22] achieving a high degree of domination of the important commercial DRAM[23] chip market.

The real significance of such examples is that, whereas after the Second World War only the USA was a significant high-technology developer, from at least the 1980s there were two major rivals. Now, the issue of fairness is acute, given the ubiquitous, increasingly rivalrous subsidies for high-technology development combined with winner-take-all outcomes. It is not clear whether a general set of rules ever could be agreed to govern the outcomes of rivalry between high-technology subsidising states. In practice, the Americans and the Europeans have resorted to trade policy interventions in response to significant national producer market share losses.

Tyson writes that had it not been for US trade policy intervention after 1985, the Japanese keiretsu-controlled semiconductor manufacturers would probably have moved to a permanent position of global dominance of semiconductor chip markets (Tyson 1992: 104). US semiconductor producers initiated anti-dumping and unfair-trading practice suits against Japanese competitors in 1985 and this led to negotiation between the US and Japanese governments and the announcement in 1986 of the ground breaking semiconductor trade agreement (SCTA) (Tyson 1992: 106). This essentially suspended the US unfair-trade and anti-dumping legal actions in exchange for a managed market outcome. The Japanese agreed first to open access to their market with a five-year target that US producers should gain 20% market share. Second, Japanese producers were required to reveal cost data so that the USA could monitor the significance of pricing practices in the US and Japanese home markets (Tyson 1992: 108). In other words, fair trade outcomes would be insisted upon, regardless of comparative product quality. Since US penetration of the Japanese market had only reached 14% by 1991, in the negotiation of a second five-year plan in that same year it was more strongly insisted upon by the USA that Japan must enable the 20% target for US producer share of the Japanese market (Tyson 1992: 128).

This kind of stipulation of outcomes has nothing to do with any kind of economic theory, orthodox or otherwise. Nor had economic theory anything to do with the creation of Sematech, a consortium of US producers formed to re-establish US technological parity in the various branches of semiconductor equipment manufacture. The largest share of the $200 million annual budget would be paid by a US defence agency: the US military had become prepared to fund civilian technology development to maintain their security of technology supply.

These US actions are best seen as symbolic of the US state's determination, for reasons of national power, to maintain US ownership of leading-edge semiconductor technology. Tyson expresses this determination when she warns that: 'If the United States fails to choose the semiconductor industry as a winner, American producers may well become long-run losers in the rigged game of international competition' (Tyson 1992: 154). In other words, what mattered here for the USA was not some particular set of rules for trade – these were mere means to an end – but the will of the state that it should retain control of this most important of technologies. The case serves as a reminder that relatively open international markets are only one way of organising relations between states: if a state finds that open markets are not producing the control of technology it perceives to be in its national interest, and if it is powerful enough and its markets valuable enough, as in the case of the USA, markets can be selectively rigged to produce the preferred technology-possession outcome.[24]

Conclusion – the Continued Relevance of Techno-nationalism within an Enlarged States System

Techno-nationalism within an enlarged states system of rivalry continues today to be a motivation for state intervention in the institutions of the market economy. As the European and US examples show, it can generate dissatisfaction with the current outcomes of the market economy and thereby becomes a motivation to change those institutions: in the European case to extend and enlarge the market economy, in the US case to reverse an undesired 'market' outcome. Schumpeter might have liked this conclusion, for he thought innovation to be the most important feature of capitalism. Here we have states intervening in the institutions of capitalism in search of better innovation outcomes.

None of this is to argue that techno-nationalism is a wholly benign influence in the world today, as it would be plain foolish to argue that it has been entirely benign in history: one only has to look at the twentieth-century histories of Germany and Japan to realise that the political purpose to which technological development is put matters.

Nevertheless, from the material above, a positive value of national techno-logical rivalry has been its historic role in motivating a political elite to confront internal national interests and procedures that obscure and obstruct internal technological development and that would never otherwise be confronted. I understand the British examples of reform to show just how important an external technological and institutional template remains when it comes to internal institutional reform. Reform efforts that make reference to status quo institutions may be sufficient when a country is a world technology leader, but it is unlikely to be sufficient when a country is either determined to catch up with the leaders or determined not to fall further behind them.

This does not imply that reform will be easy if only the state looks overseas. In the examples in this chapter, for the reform of its institutions the state's problem is how to manage plural institutions when certain among them have compromised interests, are unrepresentative, ineffective or, in some cases, simply absent. The abandonment of a laissez-faire acceptance of the institutional inheritance implies that the state must exercise judgement of the participants in the reform process itself. Here the overseas template of practice can prove useful. The active management of pluralism is difficult, requires determination, but is necessary if a society wishes to maintain the vigour of its capitalism.

Notes

1 The authoritative source for the detail of the reform process on which I have relied is Jordan (1992).

2 In particular one can doubt the idea that professional engineering reform alone could be the effective means of reform of British manufacturing. The Japanese comparisons that inspired Finniston were as much about the nature of organisation and the use made of engineers as the qualifications and standards of the engineers themselves. Then some of the report recommendations read as a familiar attempt to raise the status of 'engineer' both by restricting entry to the profession and through the creation of a new elite stream of engineers (Jordan 1992: 142). If one bears in mind the lessons of Prussian engineering reform, discussed later in this chapter, it can be doubted that this kind of professional stratification would have proven useful to British employers or would have contributed to the regeneration of manufacturing.

3 This account is derived from Gispen (1989).

4 Verein Deutscher Ingenieure.

5 The DATSCH (Deutscher Ausschuss für Technisches Schulwesen).

6 The objective of creating a legally protected engineering elite also in part represented the desire for status within a hierarchical and legally ordered society of the sort that existed in pre-industrial Prussian society.

[7] It remains difficult to draw an 'economic obstruction' conclusion from the move towards engineering science, because of the existence – as usual – of 'compensating institutions'; in Britain's case, a system of municipal technical colleges offering sub-engineering qualifications with a greater practical content and tending to recruit students with working-class backgrounds (Divall 1990: 98). For unclear reasons these colleges were in decline by the 1960s (Divall 1991: 193) and as they declined so engineering science curricula attracted increasing criticism from factions of academics and employers for their lack of practical design content (Divall 1991: 193). One might imagine that employers were performing their usual and benign role of correcting academic drift towards status objectives. Divall argues that this time the critics were too weak to shift majority opinion and engineering science retained its grip on the curriculum. In his opinion, then, a gap in the provision of practical British engineering education had opened, associated with the inability of industrial interests to maintain the practical relevance of university engineering curricula. One wonders, once again, if this situation would have occurred if there had been a unified, employer-controlled professional engineering body active on the national scene.

[8] These economies and related patenting strategies ensured would-be imitators faced massive entry costs – as DuPont found to its cost (Chapter 3).

[9] The institution survives in British legal training for the bar, as an Internet search will reveal.

[10] Pollard likes to note that since total British chemical output remained larger than the German one even as late as 1900, the dyestuffs industry and its relationship to science remained a 'wholly exceptional example even within the chemical industry' (Pollard 1989: 160). One might respond to this that the example of German dyestuffs was nevertheless a better indicator of profitable future development paths in the chemistry industry than established practice of the time.

[11] The great theme of the 1993 White Paper was a search to increase the commercial orientation of public-funded science. Knowledge of the then-prevailing belief that universities were anti-commercial organisations helps make sense of the document. Research councils were given mission statements reorienting them to support 'wealth creation', business schools would be encouraged to set up management of innovation teaching modules, and, more seriously, public science and technology projects should be funded only if they met what I termed the 'appropriability criterion': 'the likelihood that they can be appropriated by firms and organisations' (Department of Trade and Industry 1993: 16). There was never much chance of applying this criterion rigorously. The most recent science and technology White Paper continues to demonstrate the government's determination to make universities play a larger commercial role and to fund science as a means of achieving innovative, commercial and national success; the subtitle was 'a science and innovation policy' (Department of Trade and Industry 2000).

[12] As an undergraduate admissions tutor in a British university I began to deal with a small number of earnest students, parents and teachers who believed in the rhetoric promoting the new qualifications but did not seem to understand the consequences of an absence of independent assessment. One can only wonder about the degree of disillusionment of such people when the NVQ effort was finally abandoned.

[13] Wolf refers to the YTS scheme as if it was a genuine training scheme, but after a review of its origin and content others have concluded that it was 'hard to avoid' the hard truth that it was an unemployment reduction measure rather than a meaningful training scheme (Ashton et al. 1989: 143).

[14] Wolf notes that CBI resistance to any form of compulsion was in part based on employers' collective memory of another British government experiment in training, from the 1960s, that involved compulsory industrial sector levies.

[15] The power loss in a line carrying current I is given by $P = IV$, with V as the voltage drop across the resistance R of the length of line. Ohm's law ($V = IR$) can be used to substitute for V, giving power loss, $P = I^2 R$. At the point of generation a transformer raises the voltage and drops the current preparatory to transmission – and clearly the current far more than the transmission line resistance determines its rate of loss.

[16] One reviewer, Henry Ergas, has argued that for the large technological missions which the USA, France and Britain have favoured as a means of technology promotion, the scale of the US economy allows the US federal government to exploit competition between defence contractors to maintain relatively efficient management; France has been relatively successful at closing down drifting proects; which leaves Britain as the poor performer, disadvantaged by the limited number of potential developers (there is only one BNFL) and its secretive policy practices (Ergas 1987).

[17] See http://www.europarl.eu.int/stoa/.

[18] The private sector parallels to these public sector cases are complex technologies for which adopting firms lack sources of advice independent of the technology vendors. Robertson, Swan and Newell found that informal sources of knowledge about the inventory control technology MRP/MRPII were strongly influenced by the technology vendors. Since the strongly propagated ideas of best practice were not always in the interests of all adopters, one of the researchers' recommendations was that better adoption would result if more independent sources of technology assessment were available to the private sector (Robertson et al. 1996). There is a role for a private sector equivalent to the STOA.

[19] Today, it is only because there is a near international consensus amongst states that nuclear proliferation technologies are undesirable that it is even conceivable that these technologies might be suppressed.

[20] There are fashions in history, as in other subjects, so it should be said that this description of modern history is recognisable as the classic liberal history of modernity, as presented for example by Lord Acton in his *Lectures on Modern History* (Acton 1960).

[21] With this perspective on long-range technological and economic change, Jones reasonably comments that: 'Arguably, there is more relevance for the less-developed world in the history of this kind of provision than in those staples of Industrial Revolution history, canals and cotton mills' (Jones 1987: 238). Another instance, then, of the argument that the artefact is generated and sustained by a less visible, but vital and complex, social and institutional fabric.

[22] The Japanese VLSI project ran from 1976 to 1979 and cost $360 million, of which the Japanese government supplied 50%. The comparable US project ran from 1979 to 1984 and received a similar subsidy from the US government of $200 million but this money was spent by the Pentagon (Anchordoguy 1989: 141). The result was a design compromised by military design objectives: a chip with great heat and shock tolerance but at a higher cost than required for civil applications. For the first time Japanese chip producers gained a significant share of the US chip market, while Fujitsu became the first computer manufacturer in any country to have greater mainframe sales than IBM (Anchordoguy 1989). The reasons for the better Japanese management of design are similar to those discussed in this book for the car industry.

[23] DRAM – Dynamic Random Access Memory.

[24] The hostility between the USA and Japan at this time was dramatised by a group of Congressmen taking sledgehammers to smash a Toshiba radio-cassette player on the steps of the Capitol (Ishihara 1991: 42). On the Japanese side there was the book by Ishihara, the LDP politician and mayor of Tokyo, *The Japan that Can Say No*. This advocated that Japan use its superiority in technology to get tough with US demands to access Japanese technology, especially defence technologies (Ishihara 1991).

9   Concluding Comments

This book has been organised as an in-depth development of the intellectual device of the technology complex that was introduced in the first chapter. One of the advantages of this organisation is that while it confirms the utility of the broad conception of technology, it shows clearly how technology is an integrated part of society. It does so by capturing the diversity of organisations and institutions that are involved in accommodating and controlling new technologies. It becomes obvious that innovation is not something restricted to the private sector alone; rather private sector management exists within a web of institutions that support, enable and sometimes constrain its decisions, and that are themselves subject to the influence of private sector management. It is also evident how individual innovation decisions are rooted in the past, the past of individual experience and of inherited organisational and technological forms. This extension of subject scope to include the role of institutions of intellectual property, of finance, technological education and the state accepts that many disciplines offer insights into innovative processes. In other words, if one wants to understand how innovation may be managed, some form of multidisciplinary review is necessary.

It is nevertheless worth remembering that by the standard of the technology complex, this book is *deliberately* skewed towards the elements of technology that in principle are managerially and socially controllable. The role of, for example, geographical and physical endowments is ignored: they are important, but they are beyond our scope here. This is perhaps no surprise in a book that is directed at students of the social and managerial control of technology.

Another aspect has been the selection of sources for the detail they give of the framing of innovation decisions. If the technology complex gave structure to the breadth of topics included, here was a choice in favour of depth of analysis. This was a necessary choice if one wished to understand how rational people could commit themselves to courses of action that – in retrospect – could prove fruitless and sometimes destructive of their organisations and declared purpose. A complementary result of this choice in favour of depth was the discovery that many of what purported to be straightforward accounts of innovation could be astray in their interpretation of events.

In sum, if the object is a general understanding of the innovation process, it is necessary to cover both the breadth of topics implied by the technology complex and the detail of particular decision events. This approach provides a surrogate experience of the innovation process that has a good chance of informing actual innovation practice.

Bibliography

Abrahams, P. and Griffith, V. (2001) 'Slow-Acting Medicine', *Financial Times*, 6 April: 14.

Acton, L. (1960) *Lectures on Modern History*, London: Collins Fontana Library.

Adam, R. (1989) 'Tin Gods – Technology and Contemporary Architecture', *Architectural Design Magazine*, October.

Anchordoguy, M. (1989) *Computers Inc: Japan's Challenge to IBM*, Cambridge, MA: Harvard University Press.

Anderson, P. (1976) 'The Notion of a Bourgeois Revolution', in Anderson, P. (ed.), *English Questions*, London: Verso.

Anderson, P. (1992) *English Questions*, London: Verso.

Anderson, P. and Tushman, M. L. (1990) 'Technological Discontinuities and Dominant Designs: A Cyclical Model of Technological Change', *Administrative Science Quarterly*, 35: 604–33.

Andreski, S. (1974) *Social Sciences as Sorcery*, Harmondsworth: Penguin.

Armstrong, P. (1992) 'The Management Education Movement', in Lee, G. L. and Smith, C. (eds), *Engineers and Management – International Comparisons*, London: Routledge.

Armstrong, P. (2002) 'The Costs of Activity-Based Management', *Accounting, Organisations and Society*, 27: 99–120.

Armstrong, P. and Jones, C. (1992) 'The Decline of Operational Expertise in the Knowledge-base of Accounting', *Management Accounting Research*, 3: 53–75.

Arrow, K. (1994) 'Foreword', in Arthur, W. B. (ed.), *Increasing Returns and Path Dependence in the Economy*, Ann Arbor, MI: University of Michigan Press.

Arthur, W. B. (1994) *Increasing Returns and Path Dependence in the Economy*, Ann Arbor, MI: University of Michigan Press.

Ashton, D., Green, F. and Hoskins, M. (1989) 'The Training System of British Capitalism: Changes and Prospects', in Green, F. (ed.), *The Restructuring of the British Economy*, London: Harvester Wheatsheaf.

Axelrod, R. (1984) *The Evolution of Cooperation*, Harmondsworth: Penguin.

Barnett, C. (1986) *The Audit of War*, London: Macmillan, Papermac.

Barrett, A. (1998) 'The Formula at Pfizer: Don't Run with the Crowd', *Business Week*, 11 May, http://www.businessweek.com/1998/19/b3577005.htm.

Basalla, G. (1988) *The Evolution of Technology*, Cambridge: Cambridge University Press.

BASF FAZ Institut (2002) 'Du Pont Separates its Textile Unit', 12–18 February, http://www.chemicalnewsflash.de.

BBC (2002) 'France Telecom in 9bn-euro bailout', BBC News, 4 December, http://news.bbc.co.uk/-1/hi/business/2542297.stm.

Bell, D. (1973) *The Coming of Post Industrial Society – A Venture in Social Forecasting*, New York: Basic Books.

Berggren, C. (1993) 'Lean Production – The End of History?,' *Work, Employment and Society*, 7, (2): 163–88.

Bierhoff, H. and Prais, S. J. (1997) *From School to Productive Work – Britain and Switzerland Compared*, Cambridge: Cambridge University Press.

Bimber, B. (1994) 'Three Faces of Technological Determinism', in Smith, M. R. and Marx, L. (eds), *Does Technology Drive History?*, London: MIT Press.

Braun, E. and Macdonald, S. (1982) *Revolution in Miniature*, Cambridge: Cambridge University Press.

Bright, A. (1949) *The Electric Lamp Industry*, New York: Macmillan.

British Society for Human Genetics (1997) *Patenting of Human Gene Sequences and the EU Draft Directive*, http://www.bshg.org.uk/, September.

Bromberg, J. L. (1991) *The Laser in America 1950–1970*, Cambridge, MA: MIT Press.

Brooks, F. P. (1995) *The Mythical Man-Month*, London: Addison-Wesley.

Brooks, M. W. (2000) 'Introducing the Dvorak Keyboard', http://www.mwbrooks.com/dvorak/.

Brummer, A. and Cowe, R. (1994) *Hanson – The Rise and Rise of Britain's Most Buccaneering Businessman*, London: Fourth Estate.

Brusoni, S., Prencipe, A. and Pavitt, K. (2001) 'Knowledge Specialisation, Organisational Coupling, and the Boundaries of the Firm: Why Do Firms Know More than they Make?', *Administrative Science Quarterly*, *46*: 597–621.

Callon, M. (1991) 'Techno-economic Networks and Irreversibility', in Law, J. (ed.), *A Sociology of Monsters – Essays on Power, Technology and Domination*, London and New York: Routledge, 132–64.

Cassidy, J. (2002) *Dot Con – The Greatest Story Ever Sold*, New York: Harper Collins.

Cassingham, R. C. (1986) *The Dvorak Keyboard*, Arcata, CA: Freelance Communications.

Chandler, A. D. (1977) *The Visible Hand – The Managerial Revolution in American Business*, Cambridge, MA: Belknap Press.

Chandler, A. D. (1990) 'The Enduring Logic of Industrial Success', *Harvard Business Review*, *68*: 130–40.

Child, J. (1969) *The Business Enterprise in Modern Industrial Society*, London: Collier-Mac.

Child, J., Fores, M., Glover, I., et al. (1983) 'A Price to Pay? Professionalism and Work Organisation in Britain and West Germany', *Sociology*, *17*, (1): 63–78.

Child, J. and Partridge, B. (1982) *Lost Managers: Supervisors in Industry and Society*, Cambridge: Cambridge University Press.

Christensen, C. (1997) *The Innovator's Dilemma*, Boston, MA: Harvard Business School Press.

Clark, K. B., Chew, W. B. and Fujimoto, T. (1987) *Product Development in the World Auto Industry*, *3*, Washington, DC: Brookings Institute.

Clark, M. (1993) 'Suppressing Innovation: Bell Laboratories and Magnetic Recording', *Technology and Culture*, *34*, (3): 516–38.

Clark, R. W. (1985) *The Life of Ernst Chain – Penicillin and Beyond*, London: Weidenfeld and Nicolson.

Cohen, S. S. and Zysman, J. (1994) 'The Myth of the Post-Industrial Economy', in Rhodes, E. and Wield, D. (eds), *Implementing New Technologies*, Oxford: Blackwell.

Cole, R. (1998) 'Learning from the Quality Movement: What Did and Didn't Happen and Why?', *California Management Review*, *41*, (1): 43–73.

Collins (1991) *Collins Dictionary*, London: Harper Collins.

Cooper, A. C. and Schendel, D. (1988) 'Strategic Responses to Technological Threats', in Tushman, M. L. and Moore, W. L. (eds), *Readings in the Management of Innovation*, 2nd edn, London: Harper Collins.

Cooper, A. C. and Smith, C. G. (1992) 'How Established Firms Respond to Threatening Technologies', *Academy of Management Executive*, *6*, (2).

Coopey, R. and Clarke, D. (1995) *3i – Fifty Years of Investing in Industry*, Oxford: Oxford University Press.

Cordis News (2002) 'Commission to Investigate Patentability of Human Gene Sequences and Stem Cells', *EUBusiness*, 11 October.

Coren, G. (1997) *James Dyson – Against the Odds, An Autobiography*, London: Orion.

Crane, J. (1979) *The Politics of International Standards – France and the Color TV War*, Norwood, NJ: Ablex.

Cringeley, R. X. (1996) *Accidental Empires*, Harmondsworth: Penguin.

Cusumano, M. A., Mylonadis, Y. and Rosenbloom, R. S. (1992) 'Strategic Manoeuvring and Mass Market Dynamics: VHS over Beta', *Business History Review*, Spring: 51–94.

Cutler, T. (1992) 'Vocational Training and British Economic Performance: A Further Instalment of the British Labour Problem', *Work, Employment and Society*, *6*, (2): 161–83.

Daniel, C. (2003) 'Industry Bets on the Future of Air Travel', *Financial Times*, 2 May: 21.

Danielian, N. R. (1939) *AT&T – The Story of Industrial Conquest*, New York: The Vanguard Press.

David, P. (1985) 'Clio and The Economics of QWERTY', *American Economic Review*, *76*: 332–7.

David, P. A. (1986) 'Understanding the Economics of QWERTY – The Necessity of History', in Parker, W. N. (ed.), *Economic History and the Modern Economist*, Oxford: Blackwell.

Davies, A. (2003) 'Integrated Solutions: The Changing Business of Systems Integration', in Prencipe, A., Davies, A. and Hobday, M. (eds), *The Business of Systems Integration*, Oxford: Oxford University Press.

Davis, E. P. (1994) 'Whither Corporate Banking Relations?', in Hughes, K. (ed.), *The Future of UK Competitiveness and the Role of Industrial Policy*, London: Policy Studies Institute.

Dearden, J. (1969) 'The Case Against ROI Control', *Harvard Business Review*, May–June: 124–35.

Dearden, J. (1987) 'Measuring Profit Centre Managers', *Harvard Business Review*, September–October: 84–8.

DeLamarter, R. (1986) *Big Blue: IBM's Use and Abuse of Power*, New York: Dodd, Mead and Co.

Dennett, D. (1996) 'The Scope of Natural Selection', *Boston Review*, October/November.

Department of Trade and Industry (1993) *Realising Our Potential*, Cm 2250, London, HMSO.

Department of Trade and Industry (2000) *Excellence and Opportunity – A Science and Innovation Policy*, Cm 4814, London, HMSO.

Divall, C. (1990) 'A Measure of Agreement: Employers and Engineering Studies in the Universities of England and Wales, 1897–1939', *Social Studies of Science*, 20: 65–112.

Divall, C. (1991) 'Fundamental Science versus Design: Employers and Engineering Studies in British Universities', *Minerva*, 29, (2): 167–95.

Djeflat, A. (1987) 'The Management of Technology Transfer', *International Journal of Technology Management*, 3, (1/2): 149.

Donaldson, L. (2002) 'Damned by Our Own Theories: Contradictions Between Theories and Management Education', *Academy of Management Learning and Education*, 1, (1): 96–106.

Dore, R. (1973) *British Factory – Japanese Factory*, Berkeley, CA: University of California Press.

DTI 'Future and Innovation Unit' and Company Reporting Ltd (2001) 'R&D Scoreboard', http://www.innovation.gov.uk/projects/rd_scoreboard/database/databasefr.htm.

Dunford, R. (1987) 'The Suppression of Technology as a Strategy for Controlling Resource Dependence', *Administrative Science Quarterly*, 32: 512–25.

Dyson, J. and Coren, G. (1997) *Against the Odds – An Autobiography*, London: Orion.

Eaglesham, J. (2000) 'The Politics of Plagiarism', *Financial Times*, 17 November: 16.

Easterbrook, G. (2003) 'Axle of Evil', *The New Republic*, 20 January: 27–32.

Edgerton, D. (1991) 'The Prophet Militant and Industrial: The Peculiarities of Correlli Barnett', *Twentieth Century British History*, 2, (3): 360–79.

Edgerton, D. (1993) 'Research, Development and Competitiveness', in Hughes, K. (ed.), *The Future of UK Competitiveness and the Role of Industrial Policy*, London: Policy Studies Institute.

Edgerton, D. (1996) *Science, Technology and the British Industrial 'Decline' 1870–1970*, Cambridge: Cambridge University Press.

Edgerton, D. (2004) '"The linear model" did not exist: Reflections on the history and historiography of science and research in industry in the twentieth century', in Grandin, K. and Wormbs, N. (eds), *Science and Industry in the Twentieth Century*, New York: Watson.

Edgerton, D. and Horrocks, S. (1994) 'British Industrial Research and Development Before 1945', *Economic History Review*, 47, (2): 213–38.

Editors (2002) 'Deal Breathes New Life into Napster', 17 May, http://www.napster.com/latest_news.html.

Elbaum, B. (1991) 'The Persistence of Apprenticeship in Britain and its Decline in the United States', in Gospel, H. F. (ed.), *Industrial Training and Technological Innovation – A Comparative and Historical Study*, London: Routledge.

Elbaum, B. and Lazonick, W. (1984) 'The Decline of the British Economy: An Institutional Perspective', *Journal of Economic History*, 44, (2): 567–83.

Ergas, H. (1987) 'The Importance of Technology Policy', in Dasgupta, P. and Stoneman, P. (eds), *Economic Policy and Technological Performance*, Cambridge: Cambridge University Press.

Faberberg, J. (1987) 'A Technology Gap Approach to Why Growth Rates Differ', *Research Policy*, 16: 87–99.

Farrelly, P. (2000) 'Europe? Invest? It's a Bit Late, Lord Hanson', *Observer*, 9 July.

Feynman, R., Leighton, R. B. and Sands, M. (1965) *The Feynman Lectures on Physics – Quantum Mechanics*, Reading, MA: Addison-Wesley.

Fingleton, E. (1999) *In Praise of Hard Industries*, London: Orion Business.

Finniston, M. (1980) *Engineering Our Future – Report of the Inquiry into the Engineering Profession*, London, HMSO.

Fisher, F. M., McKie, J. W. and Mancke, R. B. (1983) *IBM and the US Data Processing Industry*, New York: Praeger.

Fleck, J. (1987) 'Innofusion or Diffusation? The Nature of Technological Development in Robotics', PICT Working Paper 4, Edinburgh University, Edinburgh.

Fleck, J. and Howells, J. (2001) 'Technology, the Technology Complex and the Paradox of Technological Determinism', *Technology Analysis and Strategic Management, 13*, (4): 523–31. www.tandf.co.uk/journals.

Fleck, J. and White, B. (1987) 'National Policies and Patterns of Robot Diffusion', *Robotics, 3*, (1): 7–22.

Ford, C. (1996) 'A Theory of Individual Creative Action in Multiple Social Domains', *Academy of Management Review, 21*, (4): 1112–42.

Ford, G. and Lake, G. J. (1991) 'Evolution of European Science and Technology Policy', *Science and Public Policy, 18*, (1): 38–50.

Forrest, J. E. (1991) 'Models of the Process of Technological Innovation', *Technology Analysis and Strategic Management, 3*, (4): 439–53.

Foster, R. (1986) *Innovation – The Attacker's Advantage*, New York: Summit Books.

Foster, R. N. (1988) 'Timing Technological Transitions', in Tushman, M. L. and Moore, W. L. (eds), *Readings in the Management of Innovation*, 2nd edn, London: Harper Collins

Franks, J. and Mayer, C. P. (1996) 'Hostile Takeovers and the Correction of Managerial Failure', *Journal of Financial Economics, 40*: 163–81.

Freeman, C. (1982) *The Economics of Industrial Innovation*, London: Pitman.

Fruin, M. (1997) *Knowledge Works: Managing Intellectual Capital at Toshiba*, Oxford: Oxford University Press.

Fukasaku, Y. (1991) 'In-Firm Training at Mitsubishi Nagasaki Shipyard, 1884–1934', in Gospel, H. F. (ed.), *Industrial Training and Technological Innovation – A Comparative and Historical Study*, London: Routledge.

Galbraith, J. K. (1975) *The Great Crash 1929*, Harmondsworth: Penguin.

Garud, R. and Ahlstrom, D. (1997) 'Technology Assessment: A Socio-cognitive Perspective', *Journal of Engineering and Technology Management, 14*, (1): 25–48.

Garud, R. and Karnøe, P. (2003) 'Bricolage versus Breakthrough: Distributed and Embedded Agency in Technology Entrepreneurship', *Research Policy, 32*, (2): 277.

Garud, R. and Rappa, M. A. (1994) 'A Socio-cognitive Model of Technology Evolution: The Case of Cochlear Implants', *Organization Science, 5*, (3).

Gilfillan, S. C. (1935) *The Sociology of Invention*, Cambridge, MA: MIT Press.

Gispen, K. (1989) *New Profession, Old Order: Engineers and German Society 1815–1914*, Cambridge: Cambridge University Press.

Goldstein, D. (1995) 'Uncertainty, Competition and Speculative Finance in the Eighties', *Journal of Economic Issues, 29*, (3): 719–46.

Gompers, P. A. and Lerner, J. (2000) *The Venture Capital Cycle*, Cambridge, MA: MIT Press.

Gospel, H. F. and Okayama, R. (1991) 'Industrial Training in Britain and Japan: An Overview', in Gospel, H. F. (ed.), *Industrial Training and Technological Innovation*, London: Routledge.

Gould, S. J. and Lewontin, R. C. (1979) 'The Spandrels of San Marco and the Panglossian Paradigm: A Critique of the Adaptationist Programme', *Proceedings of the Royal Society of London, Series B, 205*, (1161): 581–98.

Graham, M. B. W. (1986) *RCA and the Videodisc*, Cambridge: Cambridge University Press.

Granovetter, M. and Swedberg, R. (ed.) (2001) *The Sociology of Economic Life*, Cambridge, MA: Westview.

Grant, J. (2003) 'A Burst of Speed from the Japanese – Book Review of "The End of Detroit" by Micheline Maynard', *Financial Times*, 30 September: 16.

Green, K. and Morphet, C. (1977) *Research and Technology as Economic Activities*, Oxford: Butterworth–Heinemann.

Grint, K. (1995) 'Utopian Reengineering', in Burke, G. and Peppard, J. (eds), *Examining Business Process Reengineering*, London: Kogan Page.

Guenther, A. H., Kressel, H. R. and Krupke, W. F. (1991) 'Epilogue: The Laser Now and in the Future', in Bromberg, J. L. (ed.), *The Laser in America, 1950–1970*, Cambridge, MA: MIT Press.

Haber, L. F. (1958) *The Chemical Industry During the Nineteenth Century*, Oxford: Oxford University Press.

Hackman, J. R. and Wageman, R. (1995) 'Total Quality Management – Empirical, Conceptual and Practical Issues', *Administrative Science Quarterly, 40*: 309–42.

Hamilton, D. P. (1999) 'Microsoft Case Highlights Role of Browsers', *Wall Street Journal Europe*, 8 November: 2.

Hannah, L. (1979) *Electricity Before Nationalisation – A Study of the Development of the Electricity Industry Supply Industry in Britain to 1948*, London: Macmillan.

Hannan, M. T. and Freeman, J. (1989) *Organisational Ecology*, Cambridge, MA: Harvard University Press.

Hare, R. (1970) *The Birth of Penicillin*, London: Allen and Unwin.

Hartford-Empire Co. vs. *US* (1945) US Supreme Court, 323 US 386, http:/www.ripon.edu/faculty/bowenj/antitrust/hart-emp.htm.

Hayek, F. (1960) *The Constitution of Liberty*, Chicago: Chicago University Press.

Hayes, R. H. and Abernathy, W. J. (1980) 'Managing our Way to Economic Decline', *Harvard Business Review*, July/August.

Hayes, R. H. and Garvin, D. A. (1982) 'Managing as if Tomorrow Mattered', *Harvard Business Review*, May/June: 264–73.

Hayter, E. W. (1939) 'Barbed Wire Fencing, A Prairie Invention – Its Rise and Influence in the Western States', *Agricultural History*, *13*: 189–217.

Hecht, E. and Zajac, A. (1974) *Optics*, New York: Addison-Wesley.

Henderson, P. D. (1977) 'Two British Errors: Their Probable Size and Some Possible Lessons', *Oxford Economic Papers*, *29*, (New Series, 2): 159–205.

Heymann, M. (1998) 'Signs of Hubris – The Shaping of Wind Technology Styles in Germany, Denmark and the United States, 1940–1990', *Technology and Culture*, *39*: 641–70.

Hill, S. (1981) *Competition and Control at Work*, London: Gower.

Hill, S. (1991) 'Why Quality Circles Failed but TQM Might Succeed?', *British Journal of Industrial Relations*, *29*, (4): 541–68.

Hill, S., Martin, R. and Harris, M. (2000) 'Decentralisation, Integration and the Post-Bureaucratic Organisation: the Case of R&D', *Journal of Management Studies*, *37*, (4): 563–85.

Homburg, E. (1992) 'The Emergence of Research Laboratories in the Dyestuffs Industry, 1870–1900', *British Journal for the History of Science*, *25*, (March): 91–111.

Hopper, T. and Armstrong, P. (1991) 'Cost Accounting, Controlling Labour and the Rise of Conglomerates', *Accounting, Organisations and Society*, *16*, (5/6): 405–38.

Hounshell, D. A. and Smith, J. K. (1988) *Science and Corporate Strategy – Du Pont R&D, 1902–1980*, Cambridge: Cambridge University Press.

Howells, J. (1994) *Managers and Innovation – Strategies for a Biotechnology*, London: Routledge.

Howells, J. (1995) 'A Socio-Cognitive Approach Towards Innovation', *Research Policy*, *24*: 883–94.

Howells, J. (1997) 'Rethinking the Markets-Technology Relationship', *Research Policy*, *25*: 1209–19.

Howells, J. (2002) 'The Response of Old Technology Incumbents to Technological Competition – Does the Sailing Ship Effect Exist?', *Journal of Management Studies*, *39*, (7): 887–907.

Howells, J. and Hine, J., (eds) (1993) *Innovative Banking – Competition and the Management of a New Network Technology*, London: Routledge.

Hughes, T. P. (1983) *Networks of Power: Electrification in Western Society*, Baltimore, MD: Johns Hopkins University Press.

Hutton, W. (1995) *The State We're In*, London: Macmillan.

Ibison, D. (2002) 'Total of Bad Loans Understated by Y13 000 bn – FSA', *Financial Times*, 10 November: 3.

Ishihara, S. (1991) *The Japan that Can Say No: Why Japan will be First Amongst Equals*, New York: Simon and Schuster.

Johnson, C. (1982) *MITI and the Japanese Economic Miracle: The Growth of Industrial Policy*, Stanford, CA: Stanford University Press.

Johnson, H. and Kaplan, R. S. (1991) *Relevance Lost – The Rise and Fall of Management Accounting*, Boston, MA: Harvard Business School Press.

Jones, E. L. (1987) *The European Miracle*, 2nd edn, Cambridge: Cambridge University Press.

Jones, M. R. (1995) 'The Contradictions of Business Process Reengineering', in Burke, G. and Peppard, J. (eds), *Examining Business Process Reengineering*, London: Kogan Page.

Jordan, G. (1992) *Engineers and Professional Self-Regulation – from the Finniston Committee to the Engineering Council*, Oxford: Clarendon Press.

Jorde, T. M. and Teece, D. J. (1991) 'Antitrust Policy and Innovation: Taking Account of Performance Competition and Competitor Cooperation', *Journal of Institutional and Theoretical Economics*, *147*: 118–44.

Kamata, S. (1983) *Japan in the Passing Lane: An Insiders' Account of Life in a Japanese Auto Factory*, New York: Random House.

Kaplan, R. S. (1985) 'Financial Justification for the Factory of the Future' Working paper, Harvard Business School, Boston, MA.

Kehoe, L. and Denton, N. (1997) 'Microsoft Could Face Federal Fines of $1 m a Day', *Financial Times*, 21 October: 1.

Kennedy, W. P. (1987) *Industrial Structure, Capital Markets and the Origin of British Industrial Decline*, Cambridge: Cambridge University Press.

Kindleberger, C. P. (1996) *Manias, Panics and Crashes – A History of Financial Crises*, 3rd edn, New York: Wiley.

Kitch, E. W. (1977) 'The Nature and Function of the Patent System', *Journal of Law and Economics*, 20: 265–90.

Kitch, E. W. (2000) 'Elementary and Persistent Errors in the Economic Analysis of Intellectual Property', *Vanderbilt Law Review*, 53: 1727–41.

Klopfenstein, B. C. (1989) 'The Diffusion of the VCR in the United States', in Levy, M. R. (ed.), *The VCR Age*, Newbury Park, CA: Sage.

Kortum, S. and Lerner, J. (2000) 'Assessing the Contribution of Venture Capital to Innovation', *RAND Journal of Economics*, 31, (4): 674–92.

Kotler, P. (2000) *Marketing Management*, London: Prentice Hall.

Krohn, S. (1999) 'The Wind Turbine Market in Denmark', Danish Wind Industry Association, http://www.windpower.org/articles/wtmindk.htm.

Krohn, S. (2002) 'Danish Wind Turbines: An Industrial Success Story', Danish Wind Industry Association, http://www.windpower.org/articles/success.htm.

Krugman, P. (1994) *Peddling Prosperity – Economic Sense and Nonsense in the Age of Diminished Expectations*, London: Norton.

Lake, G. J. (1992) *Repatriating Refugees from Reality: Fusion and Fantasy in the European Community*, Joint 4S/EASST Conference, Gothenburg, Sweden.

Lam, A. (1996) 'Engineers, Management and Work Organisation: A Comparative Analysis of Engineers' Work Roles in British and Japanese Electronics Firms', *Journal of Management Studies*, 33, (2): 183–212.

Lancaster, G. and Massingham, L. (2001) *Marketing Management*, 3rd edn, London: McGraw-Hill.

Landes, D. (1969) *The Unbound Prometheus: Technological Change and Industrial Development in Western Europe from 1750 to the Present*, Cambridge: Cambridge University Press.

Langenfeld, J. and Scheffman, D. (1989) 'Innovation and US Competition Policy', *Antitrust Bulletin*, 34, (1): 1–63.

Langreth, R. (1998) 'Pfizer Pins Multibillion-Dollar Hopes on Impotence Pill – Aimed at Heart, Research Leads to a Surprise', *Wall Street Journal*, 19 March: B1.

Latour, B. (1991) 'Technology is Society Made Durable', in Law, J. (ed.), *A Sociology of Monsters – Essays on Power, Technology and Domination*, London and New York: Routledge.

Lawrence, P. A. (1992) 'West Germany – A Study in Consistency?', in Lee, G. L. and Smith, C. (eds), *Engineers and Management – International Comparisons*, London: Routledge.

Lawrence, P. R. (1987) 'Competition: A Renewed Focus for Industrial Policy', in Teece, D. J. (ed.), *The Competitive Challenge*, Cambridge, MA: Ballinger.

Lawrence, P. R. and Dyer, D. (1983) *Renewing American Industry*, New York: Free Press.

Lazonick, W. (1983) 'Industrial Organisation and Technological Change: The Decline of the British Cotton Industry', *Business History Review*, 57: 195–236.

Lazonick, W. (1987) 'Stubborn Mules: Some Comments', *Economic History Review*, 40, (2nd series): 80–6.

Lazonick, W. (1990) *Competitive Advantage on the Shop Floor*, London: Harvard University Press.

Lazonick, W. (1992) 'Controlling the Market for Corporate Control: The Historical Significance of Managerial Capitalism', in Scherer, F. M. and Perlman, M. (eds), *Entrepreneurship, Technological Innovation and Economic Growth*, Ann Arbor, MI: University of Michigan Press.

Lessig, L. (1999) *Code and other Laws of Cyber Space*, New York: Basic Books.

Lessig, L. (2000) 'Europe's "Me-too" Patent Law', *Financial Times*, 12 July: 17.

Lessig, L. (2002) 'Time to end the race for ever longer copyright', *Financial Times*, 17 October: 17.

Liebowitz, S. J. and Margolis, S. E. (1990) 'The Fable of the Keys', *Journal of Law and Economics*, 33: 1–25.

Liebowitz, S. J. and Margolis, S. E. (1994) 'Network Externality; An Uncommon Tragedy', *Journal of Economic Perspectives*, 8, (2).

Liebowitz, S. and Margolis, S. (1995) 'Path Dependence, Lock-In and History', *Journal of Law, Economics and Organisation*, *11*, (1): 205–26.

Liebowitz, S. and Margolis, S. E. (1996) 'Typing Errors', Reason Magazine Online, June, http://reason.com/9606/Fe.QWERTY.shtml.

Liebowitz, S. and Margolis, S. E. (1999) *Winners, Losers and Microsoft*, Oakland, CA: The Independent Institute.

Lindert, P. H. and Trace, K. (1971) 'Yardsticks for Victorian Entrepreneurs', in McCloskey, D. N. (ed.), *Essays on a Mature Economy: Britain after 1840*, London: Methuen.

Lipsey, R. G. (1989) *Positive Economics*, 7th edn, London: Weidenfeld and Nicolson.

Lutz, B. (1992) 'Education and Job Hierarchies – Contrasting Evidence from France and Germany', in Altmann, N., Kohler, C. and Meil, P. (eds), *Technology and Work in German Industry*, London: Routledge.

Lutz, B. and Veltz, P. (1992) 'Mechanical Engineer versus Computer Scientist – Different Roads to CIM in France and Germany', in Altmann, N., Kohler, C. and Meil, P. (eds), *Technology and Work in German Industry*, London: Routledge.

MacDonald, S., Lamberton, D. and Mandeville, T. (eds) (1985) *The Trouble with Technology – Explorations in the Process of Technological Change*, New York: St Martin's Press.

MacFarlane, G. (1984) *Alexander Fleming – The Man and the Myth*, London: Chatto and Windus.

MacKenzie, D. (1992) 'Economic and Sociological Explanations of Technical Change', in Coombs, R., Saviotti, P. and Walsh, V. (eds), *Technological Change and Company Strategies*, London: Academic Press, 25–49.

MacKenzie, D. and Wajcman, J. (eds). (1985) *The Social Shaping of Technology*, Milton Keynes: Open University Press.

MacKenzie, D. and Wajcman, J. (1999a) 'Introductory Essay: The Social Shaping of Technology', in MacKenzie, D. and Wajcman, J. (eds), *The Social Shaping of Technology*, 2nd edn, Buckingham: Open University Press.

MacKenzie, D. and Wajcman, J. (eds) (1999b) *The Social Shaping of Technology*, 2nd edn, Buckingham: Open University Press.

MacLeod, T. (1988) *The Management of Research Development and Design in Industry*, London: Gower.

Majumdar, B. A. (1977) 'Innovations, Product Developments and Technology Transfers: an Empirical Study of Dynamic Competitive Advantage: the Case of Electronic Calculators'. Unpublished PhD thesis, Cleveland, OH, Case Western Reserve University.

March, J. G. and Simon, H. A. (1973) *Organisations*, Oxford: Oxford University Press.

Mark, R. (1996) 'Architecture and Evolution', *American Scientist*, July/August: 383–9.

Mark, R. and Billington, D. P. (1989) 'Structural Imperative and the Origin of New Form', *Technology and Culture*, *30*, (2): 300–29.

Marsch, U. (1994) 'Strategies for Success: Research Organisation in German Chemical Companies and IG Farben until 1936', *History and Technology*, *12*: 23–77.

Mass, W. and Lazonick, W. (1990) 'The British Cotton Industry and International Competitive Advantage: The State of the Debates', *Business History*, *32*: 9–65.

Matthews, D., Anderson, M. and Edwards, J. (1997) 'The Rise of the Professional Accountant in British Management', *Economic History Review*, *50*, (3): 407–29.

Mayer, C. P. (1997) 'The City and Corporate Performance: Condemned or Exonerated?', *Cambridge Journal of Economics*, *21*: 291–302.

Mayer, C. P. and Alexander, I. (1990) 'Banks and Securities Markets: Corporate Financing in Germany and the United Kingdom', *Journal of the Japanese and International Economies*, *4*, (4) 450–75.

McCloskey, D. N. (1971a) 'Editor's Introduction', in McCloskey, D. N. (ed.), *Essays on a Mature Economy: Britain after 1840*, London: Methuen.

McCloskey, D. N. (ed.) (1971b) *Essays on a Mature Economy: Britain after 1840*, London: Methuen.

McCloskey, D. N. (1990) *If You're So Smart*, London: University of Chicago Press.

McCormick, K. (1991) 'The Development of Engineering Education and Training in Britain and Japan', in Gospel, H. (ed.), *Industrial Training and Technological Innovation – A Comparative and Historical Study*, London: Routledge.

McKinlay, A. (1991) '"A Certain Short-sightedness": Metalworking, Innovation and Apprenticeship, 1897–1939', in Gospel, H. F. (ed.), *Industrial Training and Technological Innovation*, London: Routledge.

Meiksins, P. and Smith, C. (1992) 'Engineers and Trade Unions: The American and British Cases Compared', in Lee, G. L. and Smith, C. (eds), *Engineers and Management – International Comparisons*, London: Routledge.

Meyer-Thurow, G. (1982) 'The Industrialisation of Invention: A Case Study from the German Chemical Industry', *ISIS, 73*, (268): 363–81.

Miller, P. and Napier, C. (1993) 'Genealogies of Calculation', *Accounting, Organisations and Society, 18*, (7/8): 631–47.

Miller, R. and Sawyers, D. (1968) *The Technical Development of Modern Aviation*, London: Routledge.

Misa, T. J. (1994) 'Retrieving Sociotechnical Change from Technological Determinism', in Smith, M. R. and Marx, L. (eds), *Does Technology Drive History?*, Cambridge, MA: MIT Press.

Monden, Y. (1983) *Toyota Production System*, Norcross, GA: Industrial Engineering and Management Press.

Morgan, P. (2003) 'Excess Corporate Debt, Not an Ageing Population, is Japan's Problem', *Financial Times*, 22 January.

Morikawa, H. (1991) 'The Education of Engineers in Modern Japan: An Historical Perspective', in Gospel, H. (ed.), *Industrial Training and Technological Innovation – A Comparative and Historical Study*, London: Routledge.

Mowery, D. C. and Rosenberg, N. (1979) 'The Influence of Market Demand upon Innovation: A Critical Review of Some Recent Studies', *Research Policy, 3*: 220–42.

Mowery, D. C. and Rosenberg, N. (1989) *Technological Change and the Pursuit of Economic Growth*, Cambridge: Cambridge University Press.

Myers, S. and Marquis, D. G. (1969) *Successful Industrial Innovation*, Arlington, Virginia: National Science Foundation.

Nelson, R. R. (ed.) (1993) *National Innovation Systems*, Oxford: Oxford University Press.

Nelson, R. R. and Winter, S. (1982) *An Evolutionary Theory of Economic Change*, Cambridge: Cambridge University Press.

Nelson, R. R. and Winter, S. G. (1977) 'In Search of a Useful Theory of Innovation', *Research Policy, 6*: 36–76.

Oakland, J. (1989) *Total Quality Management*, London: Butterworth–Heinemann.

Oliver, N. and Davies, A. (1990) 'Adopting Japanese-Style Manufacturing Methods: A Tale of Two (UK) Factories', *Journal of Management Studies, 27*: 555–70.

Ormerod, P. (1994) *The Death of Economics*, London: Faber and Faber.

Owen, G. (1999) *From Empire to Europe – The Decline and Revival of British Industry Since the Second World War*, London: Harper Collins.

Pascale, R. (1990) *Managing on the Edge*, Harmondsworth: Penguin.

Patel, K. (2000) 'Too Much Theory, Laments Inventor', *The Times Higher Educational Supplement*, 3 March: 6.

Pavitt, K. (2003) 'Specialisation and Systems Integration – Where Manufacture and Services Still Meet', in Prencipe, A., Davies, A. and Hobday, M. (eds), *The Business of Systems Integration*, Oxford: Oxford University Press.

Perrin, N. (1979) *Giving up the Gun – Japan's Reversion to the Sword, 1543–1879*, Boston, MA: David R. Godine.

Pfeffer, J. and Fong, C. T. (2002) 'The End of Business Schools? Less Success than Meets the Eye', *Academy of Management Learning and Education, 1*, (1): 78–95.

Pfizer vs. *Lilly ICOS* LLC (2000) High Court of Justice, Chancery Division, Patents Court, Case No. HC 1999 No. 01110, 8 November.

Pipes, R. (1990) *Russia Under the Old Regime*, Harmondsworth: Penguin.

Pollard, S. (1989) *Britain's Prime and Britain's Decline*, London: Edward Arnold.

Porter, M. E. (1980) 'Competitive Strategy: Techniques for Analysing Industries and Competitors', *Research on Technological Innovation, Management and Policy, 1*: 1–33.

Porter, M. E. (1985) *Competitive Advantage*, London: Macmillan.

Posner, R. (2003) 'The Oligarch – Book Review of "James Burnham and the Struggle for the World: A Life" by Daniel Kelly', *The New Republic*, 23 December.

Prais, S. J. (1995) *Productivity, Education and Training – An International Perspective*, Cambridge: Cambridge University Press.

Prencipe, A. (2000) 'Breadth and Depth of Technological Capabilities in CoPS: the Case of the Aircraft Engine Control System', *Research Policy, 29*: 895–911.

Prencipe, A., Davies, A. and Hobday, M. (eds) (2003) *The Business of Systems Integration*, Oxford: Oxford University Press.

Primrose, L. and Leonard, R. (1983) 'Financial Evaluation of Flexible Manufacturing Modules, Department of Total Technology, UMIST, Manchester.

Primrose, P. L. and Leonard, R. (1984) 'The Financial Evaluation of Flexible Manufacturing Modules', *Proceedings of the First International Machine Tool Conference, Birmingham*.

Primrose, P. L. and Leonard, R. (1987) 'Financial Aspects of Justifying FMS', *Proceedings of 2nd International Conference on Computer-Aided Production Engineering*, Edinburgh.

Primrose, P. L. and Leonard, R. (1988) 'Predicting Future Development in Flexible Manufacturing Technology', *International Journal of Production Research, 26*, (6): 1065–72.

Rajfer, J., Aronson, W. J., Bush, P. A., et al. (1992) 'Nitric Oxide as a Mediator of Relaxation of the Corpus Cavernosum in Respect to Nonadrenergic, Noncholinergic Neurotransmission', *New England Journal of Medicine, 326*, (2).

Ravenscraft, D. and Scherer, F. M. (1987) *Mergers, Sell Offs and Economic Efficiency*, Washington, DC: Brookings Institute.

Reader, W. J. (1970) *Imperial Chemical Industries – A History*, London: Oxford University Press.

Reich, L. S. (1977) 'Research, Patents, and the Struggle to Control Radio: A Study of Big Business and the Uses of Industrial Research', *Business History Review, 51*, (2): 208–35.

Reich, L. S. (1992) 'Lighting the Path to Profit – GE's Control of the Electric Lamp Industry 1892–1941', *Business History Review, 66*, (2): 305–28.

Reich, R. (1985) 'Bailout: A Comparative Study in Law and Industrial Structure', *Yale Journal of Regulation, 2*: 163.

Relman, A. S. and Angell, M. (2002) 'America's Other Drug Problem – How the Drug Industry Distorts Medicine and Politics', *The New Republic*, 16 December: 27–41.

Rickards, T. (2000) *Managing Creativity and Change*, Oxford: Blackwell.

Rip, A. Mise T. J. and Schot, J. (eds) (1995) *Managing Technology in Society: The Approach of Constructive Technology Assessment*, London: Pinter.

Robertson, M., Swan, J. and Newell, S. (1996) 'The Role of Networks in the Diffusion of Technological Innovation', *Journal of Management Studies, 33*, (3): 333–59.

Rose, M. B. (2000) *Firms, Networks and Business Values: The British and American Cotton Industries since 1750*, Cambridge: Cambridge University Press.

Rosenberg, N. (1976) *Perspectives on Technology*, Cambridge: Cambridge University Press.

Rosenberg, N. (1990) 'Why do firms do basic research (with their own money)?', *Research Policy, 19*: 165–74.

Rosenberg, P. D. (1975) *Patent Fundamentals*, New York: Clark Boardman.

Rosenbloom, R. S. and Cusumano, M. A. (1987) 'Technological Pioneering and Competitive Advantage: The Birth of the VCR Industry', *California Management Review, 29*, (4).

Rothwell, R. (1977) 'The Characteristics of Technically Progressive Firms', *R&D Management, 7*: 191–206.

Rothwell, R. and Zegveld, W. (1985) *Reindustrialisation and Technology*, London: Longman.

Rowthorn, R. (2001) 'Manufacturing Matters', *Guardian*, 5 November.

Sako, M. (2003) 'Modularity and Outsourcing: The Nature of Coevolution of Product Architecture and Organisation Architecture in the Global Automotive Industry', in Prencipe, A., Davies, A. and Hobday, M. (eds) *The Business of Systems Intergration*, Oxford: Oxford University Press.

Sampson, A. (1992) *The Essential Anatomy of Britain*, London: Hodder and Stoughton.

Sanderson, M. (1972) 'Research and the Firm in British Industry, 1919–1939', *Science Studies, 2*: 107–51.

Sanderson, M. (1988) 'The English Civic Universities and the "Industrial Spirit", 1870–1914', *Historical Research, 61*: 90–104.

Saxonhouse, G. and Wright, G. (1984) 'New Evidence on the Stubborn English Mule and the Cotton Industry, 1878–1920', *Economic History Review, 37*, (2nd series): 507–15.

Saxonhouse, G. and Wright, G. (1987) 'Stubborn Mules and Vertical Integration: the Disappearing Constraint', *Economic History Review, 40*, (2nd series, 1): 87–94.

Schiff, E. (1971) *Industrialisation without National Patents*, Princeton' NJ: Princeton University Press.

Schmidt-Tiedemann, K. J. (1982) 'A New Model of the Innovation Process', *Research Technology Management, 25*, (2): 18–22.

Schön, D. (1967) *Technology and Change – the New Heraclitus*, New York: Delacorte Press.

Schonberger, R. J. (1982) *Japanese Manufacturing Techniques*, New York: Free Press.

Schroeder, R. G. (1989) *Operations Management – Decision-making in the Operations Function*, London: McGraw-Hill.

Schumpeter, J. (1950) *Capitalism, Socialism and Democracy*, London: Allen & Unwin.

Sengenberger, W. (1992) 'Vocational Training, Job Structures and the Labour Market – An International Perspective', in Altmann, N., Kohler, C. and Meil, P. (eds), *Technology and Work in German Industry*, London: Routledge.

Sharp, L. (1952) 'Steel Axes for Stone Age Australians', in Spicer, H. (ed.), *Human Problems in Technological Change*, New York: Wiley.

Sharp, M. (1991) 'The Single Market and European Technology Policies', in Freeman, C., Sharp, M. and Walker, W. (eds), *Technology and the Future of Europe – Global Competition and the Environment in the 1990s*, London: Pinter.

Simon, H. (1996) *Hidden Champions – Lessons from 500 of the World's Best Unknown Companies*, Boston, MA: Harvard Business School Press.

Smith, A. (1986) *The Wealth of Nations*, Harmondsworth: Penguin.

Smith, M. R. and Marx, L. (eds). (1994) *Does Technology Drive History? The Dilemma of Technological Determinism*, London: MIT Press.

Souder, W. E. (1987) *Managing New Product Innovations*, Lexington, MA. Lexington Books.

Steinberg, J. (1996) *Why Switzerland?*, Cambridge: Cambridge University Press.

Swedberg, R. and Granovetter, M. (2001) 'Introduction to the Second Edition', in Granovetter, M. and Swedberg, R. (eds), *The Sociology of Economic Life*, Cambridge, MA: Westview Press.

Takeishi, A. and Fujimoto, T. (2003) 'Modularization in the Car Industry: Interlinked Multiple Hierarchies of Product, Production and Supplier Systems', in Prencipe, A., Davies, A. and Hobday, M. (eds), *The Business of Systems Integration*, Oxford: Oxford University Press.

Taylor, C. T. and Silberston, Z. A. (1973) *The Economic Impact of the Patent System; A Study of the British Experience*, Cambridge: Cambridge University Press.

Teece, D. J. (1988) 'Technological Change and the Nature of the Firm', in Dosi, G., Freeman, C., Silverberg, G. and Soete, L. (eds), *Technical Change and Economic Theory*, London: Pinter.

Thagard, P. (1999) *How Scientists Explain Disease*, Princeton, NJ: Princeton University Press.

Townes, C. (1999) *How the Laser Happened – Adventures of a Scientist*, Oxford: Oxford University Press.

Tranæs, F. (1997) 'Danish Wind Energy Cooperatives', Danish Wind Turbine Owners' Association, http://www.windpower.dk/articles/coop.htm.

Twiss, B. (1992) *Managing Technological Innovation*, 4th ed, London: Pitman.

Tyson, L. D. A. (1992) *Who's Bashing Whom? Trade Conflict in High-Technology Industries*, Washington, DC: Institute for International Economics.

United Alkali Company (1907) *The Struggle for Supremacy*, Liverpool: United Alkali Company.

Usher, A. F. (1972) 'Technical Change and Capital Formation', in Rosenberg, N. (ed.), *The Economics of Technological Change*, London: Johns Hopkins University Press.

Utterback, J. M. (1974) 'The Dynamics of Product and Process Innovation', in Hill, C. T. and Utterback, J. M. (eds), *Technological Innovation for a Dynamic Economy*, Oxford: Pergamon Press.

Utterback, J. M. (1996) *Mastering the Dynamics of Innovation*, Boston, MA: Harvard Business School Press.

Vaitsos, C. (1976) *Technology Policy of Economic Development*, Ottawa: IDRC.

den Belt Van, H. and Rip, A. (1987) 'The Nelson-Winter-Dosi Model and Synthetic Dye Chemistry', in Bijker, W. E., Hughes, T. P. and Pinch, T. (eds), *The Social Construction of Technological Systems*, Cambridge, MA: MIT Press.

von Hippel, E. (1988) *The Sources of Innovation*, Oxford: Oxford University Press.

Walker, W. (1999) *THORP and the Politics of Entrapment*, London: Institute for Public Policy Research.

Watson, T. J. J. and Petre, P. (1990) *Father, Son and Co. – My Life at IBM and Beyond*, New York: Bantam.

Weick, K. E. (1979) *The Social Psychology of Organising*, Reading, MA: Addison-Wesley.

Weimer, S. (1992) 'The Development and Structure of Small Scale Firms', in Altmann, N., Kohler, C. and Meil, P. (eds), *Technology and Work in German Industry*, London: Routledge.

Weston, G. E. (1984) 'New Trends in the US Antitrust Law: The Patent-Antitrust Interface as an Example', *International Review of Industrial Property and Copyright Law*, *15*, (3): 269–92.

Westphal, J. D., Gulati, R. and Shortell, S. M. (1997) 'Customisation or Conformity? An Institutional and Network Perspective on the Content and Consequences of TQM Adoption', *Administrative Science Quarterly*, *42*: 366–94.

Whitley, R. (1999) *Divergent Capitalisms*, Oxford: Oxford University Press.

Whittaker, D. H. (1990) *Managing Innovation – A Study of British and Japanese Factories*, Cambridge: Cambridge University Press.

Whittington, R. (1990) 'The Changing Structures of R&D: from Centralisation to Fragmentation', in Loveridge, R. and Pitt, M. (eds), *The Strategic Management of Technological Change*, Chichester: Wiley, 188–204.

Wickens, P. (1987) *Road to Nissan*, London: Macmillan.

Wiener, N. (1987) *English Culture and the Decline of the Industrial Spirit*, Harmondsworth: Penguin.

Wiener, N. (1993) *Invention*, Cambridge, MA: MIT Press.

Wilson, J. W. (1985) *The New Venturers*, Wokingham: Addison-Wesley.

Winner, L. (1993) 'Upon opening the black box and finding it empty', *Science Technology and Human Values*, *18,* (3): 362–78.

Wolf, A. (2002) *Does Education Matter? Myths about Education and Growth*, London: Penguin.

Wolffe, R. (1999) 'Window on Microsoft's World Closes', *Financial Times*, 21 June: 4.

Womack, J. P., Jones, D. T. and Roos, D. (1990) *The Machine that Changed the World*, New York: Rawson Associates.

Woronoff, J. (1993) *The Japanese Economic Crisis*, London: Macmillan.

Zbaracki, M. J. (1998) 'The Rhetoric and Reality of Total Quality Management', *Administration Science Quarterly*, *43*: 602–36.

Index